D0621125

The Economics and Politics of Oil Price Regulation

MIT Press Series on the Regulation of Economic Activity

General Editor
Richard Schmalensee, MIT Sloan School of Management

1 *Freight Transport Regulation*, Ann F. Friedlaender and Richard H. Spady, 1980

2 *The SEC and the Public Interest*, Susan M. Phillips and J. Richard Zecher, 1981

3 *The Economics and Politics of Oil Price Regulation*, Joseph P. Kalt, 1981

4 *Studies in Public Regulation*, Gary Fromm, ed., 1981

The Economics and Politics of Oil Price Regulation
Federal Policy in the Post-Embargo Era

Joseph P. Kalt

The MIT Press
Cambridge, Massachusetts
London, England

This book was set in VIP Times Roman by Achorn Graphic Services, Incorporated, and printed and bound by Halliday Litho in the United States of America.

Library of Congress Cataloging in Publication Data

Kalt, Joseph P.
The economics and politics of oil price regulation.

(MIT Press series on the regulation of economic activity; 3)
Bibliography: p.
Includes index.
1. Petroleum—Prices—United States. 2. Price regulation—United States. 3. Petroleum industry and trade—United States. 4. Petroleum law and legislation—United States. I. Title. II. Series.
HD9564.K34 338.2′3 81-4411
ISBN 0-262-11079-2 AACR2

To Curtis E. Green

Contents

List of Figures

List of Tables

Series Foreword

Government regulation of economic activity in the United States has grown dramatically in this century, radically transforming government–business relations. Economic regulation of prices and conditions of service was applied first to transportation and public utilities and recently has been extended to energy, health care, and other sectors. In the 1970s, explosive growth occurred in social regulation focusing on workplace safety, environmental preservation, consumer protection, and related goals. The expansion of regulation has not proceeded in silence. Critics have argued that many regulatory programs produce negative net benefits, while regulation's defenders have pointed to the sound rationales for, and potential gains from, many of the same programs.

The purpose of the MIT Press series, Regulation of Economic Activity, is to inform the ongoing debate on regulatory policy by making significant and relevant research available to both scholars and decision makers. The books in this series will present new insights into individual agencies, programs, and regulated sectors, as well as the important economic, political, and administrative aspects of the regulatory process that cut across these boundaries.

In this study, Joseph Kalt does a thorough and careful job of untangling the effects of federal regulations on both crude oil and refined products. He then computes estimates of both the efficiency and distributional impacts of these regulations. His finding of large wealth transfers goes a long way toward explaining not only the political importance of this regulatory program but also the structure of political concern. The study contains an innovative analysis of voting behavior on energy legislation. Kalt finds that ideology matters, along with the interests of constituents. This study makes important contributions to the literature on regulation and political behavior. It should have a significant impact on the ongoing policy debates about petroleum regulation in particular and energy policy in general.

Richard Schmalensee

Preface

In the years since the 1973 Arab oil embargo, significant changes and expansions have taken place in the governmental regulation of domestic energy markets. This book examines the causes and consequences of one case of this regulation—post-embargo federal regulation of petroleum prices. The consequences of petroleum price regulation may be generally classed as allocative and distributional; and this study offers estimates of the efficiency gains and losses and the wealth transfers corresponding to these categories. It turns out that the allocative and, especially, the distributional effects of post-embargo policy depend critically on the impact of regulation on the prices of refined petroleum products. Considerable attention is thus given to assessing this impact. In addition to measuring the consequences of petroleum price regulation, this study provides an analytical and empirical explanation of the political-economic factors that have shaped post-embargo oil policy. Within the context of general economic theories of regulation, the recent voting of US senators on issues of petroleum pricing is investigated econometrically.

Insofar as this study looks at only one component of US energy policy, it appears to be fairly narrow in scope. I have intended its underlying theme and ultimate conclusion, however, to be more broad: At least within the domestic public-policy arena, the Energy Crisis is primarily a battle over the appropriate distribution of income within society, rather than the manifestation of some massive failure of markets and institutions to allocate the nation's resources efficiently in this era of increasing energy scarcity. This observation is not meant to imply that allocative failures have not occurred, but only that the energy policies actually yielded by the political process have seldom been responses to the possibility or actuality of such failures.

The paramount importance of distributional issues in energy policy arises from the sheer magnitude of the income transfers promised by the energy price increases that have occurred since the fall of 1973. The magnitude of these transfers has not only called forth the self-interested political involvement of prospective winners and losers, but has also given a significant policymaking role to social and political ideologies. The continuing debates over appropriate energy policy turn, to a significant degree, on systems of belief about social and political justice. Not unlike the Civil Rights Movement and the Vietnam War, the Energy Crisis has become an object of public preoccupation because it forces us to consider the kind of society we have and would like to have. Hopefully, this book will contribute information on and perspective to these considerations.

The method and language of this work clearly put it in the camp of economics; and academic economists, particularly energy specialists, should find the subject matter (and, hopefully, its treatment) of interest. Even economists without strong appetites for the study of energy issues will, I think, find the book to be a useful example of applied policy analysis. Political scientists may also find it instructive (entertaining?) in its attempt to bridge the gap between a policy's effects and its political-economic causes. But whatever interest the book holds for academics should not deter others who are concerned with energy policy from reading it; I have tried to make the book accessible to business, regulatory, and legal practitioners in the field of energy. Mathematical developments have been paralleled, where possible, with graphical analyses. Noncrucial mathematical derivations and demonstrations have been relegated to footnotes and appendixes; the interested reader should be able to follow the major lines of argument without reference to these.

Acknowledgments

This book is based on my doctoral dissertation, *Federal Regulation of Petroleum Prices: A Case Study in the Theory of Regulation*, which was presented to the Department of Economics, University of California, Los Angeles, in May 1980. My work in the area of energy policy began, however, in 1974–1975, when I was on the staff of the President's Council of Economic Advisers. I owe a debt of thanks to former Council Member Paul W. MacAvoy, who gave me a free hand to participate in Administration policymaking on oil pricing, and to fellow Junior Staff Economist Robert S. Stillman, who provided collaboration and innumerable insights while working on energy issues at the Council.

At UCLA, my greatest debt is owed to Harold Demsetz, who supervised my dissertation. His constant intellectual stimulation and personal guidance since 1972 have not only raised the quality of this work substantially, but have also made the study of economics worthwhile. The other members of my thesis committee, Edward E. Leamer and Clement G. Krouse, have also been helpful with their criticisms and suggestions. I am further indebted to Harry P. Bowen, William W. Hogan, Arnold B. Moore, Richard Schmalensee, Michael P. Ward, Philip K. Verleger, J. H. Harry Watson, Arthur W. Wright, and an anonymous referee for useful comments.

The financial support of the Foundation for Research in Economics and Education, the Chancellor's Intern Fellowship Program at UCLA, the Hoover Institution on War, Revolution and Peace, the UCLA Foundation, and the Department of Economics at Harvard University has been indispensable and sincerely appreciated. Ms. Dorothy Yamamoto and Mrs. Jacqueline McManus deserve considerable gratitude for their careful typing of the many editions of the manuscript. I am also grateful to Anthony J. Abati and Mark A. Zupan, both fine economists in their own right, for many hours of outstanding research assistance.

I owe deep personal appreciation to my wife, Judy, for her unfailing support and encouragement during the long period consumed by this study. Finally, this book is dedicated to Curtis E. Green in appreciation for his profound and pervasive contributions to this and all my work.

Joseph P. Kalt
Cambridge, Massachusetts
October 1980

The Economics and Politics of Oil Price Regulation

Chapter 1

Introduction: An Examination of Causes and Consequences

1.1 Overview and Perspective

The last decade has witnessed the rise of a remarkable public preoccupation with energy. The impetus behind this preoccupation appears to have been the Arab oil embargo and the quadrupling of world crude oil prices by the Organization of Petroleum Exporting Countries (OPEC) in 1973–1974. Since that time, the general public has experienced occasional shortages of refined petroleum products, persistent and severe shortages of natural gas, a strongly upward trend in energy prices, sharp increases in the wealth and influence of foreign oil producers, and many other reverberations of a world adjusting its consumption and production behavior to an environment of increasingly expensive energy. These events have generated the publicity that has fostered widespread concern with energy developments. This concern has been accompanied, in the United States at least, by significant changes and expansions in the role played by governmental regulation in energy markets.

Doomsday hysteria, media sensationalism, and political rhetoric suggest that it is possible—indeed, not hard—to exaggerate the significance and misrepresent the implications of energy market developments and energy regulatory policy. Nevertheless, the increasing importance of these subjects should be of interest to academic researchers, well-meaning policymakers, and concerned citizens. The so-called Energy Crisis is more than a media event or pop-culture fad. The energy price increases that have occurred since 1973 are, in fact, causing dramatic social, political, and economic change. Even if we restrict ourselves to narrowly economic concerns, it is clear that the size of the adjustments required to ensure the efficient allocation of both energy and nonenergy resources in the presence of continuing change in world energy markets is not inconsequential. The impact of governmental regulatory policy on this process of adjustment has been and will continue to be of utmost consequence. The adverse or salutary allocative effects of actual and prospective policies, therefore, deserve careful study. The examination of these effects is one of the goals of this book.

Virtually every change in energy prices and energy policy results in a transfer of income and wealth among energy market participants. Energy-related transfers may, in some instances, be relatively small and diffuse. Very often, however, they are enormous. Indeed, one of the reasons energy price increases have stood out in a world of generally rising raw material prices is the sheer magnitude of the associated wealth transfers. These transfers are the source of the social and political change wrought by the Energy Crisis.

The transfers of income and wealth associated with change in energy markets put distributional issues at the heart of energy policy, in both positive and normative senses. For the positive scientist, interested in understanding the economics of politics and regulation, recognition of the central role played by distributional interests in policy formation is essential. For the policymaker and regulator, charged with the design and administration of energy policy, failure to account for distributional effects may lead to either unrealistic policy proposals or actual policies that overlook basic conceptions of social and political justice. The distributional effects of alternative policies not only serve as the primary incentive for interest groups to organize and exert their influence, but also bring social and political ideologies into the policymaking process. Ideologies enter the political arena because the question whether it is appropriate to offset, mitigate, or promote energy-related transfers cannot be answered without reference to an underlying system of ethical beliefs about the legitimate role of government and the proper distribution of wealth. Ethical questions are beyond the scope of this study. Nevertheless, those willing to undertake examination of such questions should not proceed without information about the distributional implications of the policies they consider. Consequently, a second broad goal of this book is to study the direction and magnitude, as well as the causal political role, of the distributional effects of alternative energy policies.

Notwithstanding the generality of the goals and themes of this work, no pretense of an in-depth study of all aspects of US energy policy will be made here. This book focuses on one component of post-embargo US energy policy—federal regulation of the petroleum industry. Certain aspects of US energy policy, such as the regulation of the nuclear power industry and the legal status of vertically and horizontally integrated energy companies, have received great attention from economists. Some of the more significant components of petroleum industry regulation, however, have been relatively overlooked. In particular, the regulations by which the prices of crude oil and refined petroleum products have been controlled since the early 1970s are wanting in-depth systematic assessment. In terms of their ability to attract the attention of policymakers, their prospective allocative and distributional effects, and the relative importance of petroleum as a fuel source, these regulations can be appropriately viewed as forming the core of recent US energy policy. Accordingly, the primary subject of interest here is post-embargo federal regulation of petroleum prices.

In the case of petroleum pricing, the most important and, as yet unresolved, questions are distributional and concern the differential impact of current federal price regulation on the well-being of crude oil producers, refiners of crude oil, and consumers of refined petroleum products. Although distributional issues in petroleum policy receive the most interest, particularly in the political arena, allocative questions also abound and are typically phrased in terms of the effects of regulation on petroleum imports, prices, and availability. The allocative issue (albeit with important distributional implications) that has received the most attention and the least consensus concerns the impact of post-embargo policy on the prices of refined petroleum products. Specifically, have federal price regulations raised or lowered domestic refined product prices? Questions of this sort are empirical; and much of the work of this study entails empirical testing of competing claims regarding allocative and distributional consequences as well as measurement of these consequences.

Research into the economics of petroleum price regulation should not be motivated solely by its policy relevance or its ability to yield a case study of economic behavior in a regulated environment. The question of the *causes* of recent policy is at least as interesting as the question of the *consequences*. Seldom has an entire industry been placed under governmental regulation or gone through such sharp reversals in the thrust of regulation as rapidly and with as much apparent ease as the US petroleum industry in the 1970s; and seldom have such policy changes taken place in the presence of such a relative abundance of data as that available in the petroleum sector. Investigations into the political-economic roots of post-embargo petroleum industry regulation have the potential for producing important insights into the nature of both the modern market-government nexus and the political behavior of economic agents. In the case at hand, such investigations can only proceed within the context of some general theory of regulation. Within such a context, this study undertakes an empirical examination of the causes of US energy policy and, specifically, attempts to explain the recent voting behavior of policymakers charged with designing federal petroleum price regulation. In the process, certain variants of economic theories of regulation and government are tested.

1.2 The Institutional Setting

Accurate assessment of the consequences of the current regulation of domestic petroleum markets, as well as any attempt to decipher political-economic causes, requires substantial knowledge of institutional details concerning the regulatory setting. Such details, after all, define many of the constraints and incentives that form the de facto context within which economic behavior manifests itself. Application of economic theory in the absence of institutional information has already led numerous analysts (including, with acknowledgment, a Nobel Prizewinner) to unwarranted conclusions concerning many of the effects of current federal petroleum policy.[1] Thus, an overview of the central components of this policy is essential.

1.2.1 The Historical Context of Regulation

Public policy has played a significant role in the development of the petroleum industry since the late nineteenth century. In an initial round of widespread concern with petroleum issues near the turn of the century, public policy focused on the competitiveness of petroleum markets and the appropriate scope for antitrust policy. The official emphasis of federal policy on issues of monopoly arose directly from the high visibility and much publicized behavior of the Standard Oil Company and had its most impressive moment in 1911, when the Supreme Court issued its dissolution decree in the Standard Oil case. In addition—and perhaps with more important long-term consequences—arguments in behalf of competition served in 1907 as the justification for expansion of the jurisdiction of the Interstate Commerce Commission to coverage of petroleum and gas pipelines. Less well-known, but significant, antimonopoly policies were also pursued by state governments in the late nineteenth and early twentieth centuries.[2]

Following an initial period in which public policy exhibited central concern for problems of monopoly, governmental involvement in petroleum markets took a decidedly anticompetitive, proproducer turn. This was accomplished primarily through three types of policies: limitations on domestic crude oil output (through so-called conservation or prorationing regulation), special tax treatment of crude oil extraction, and barriers to competition from foreign suppliers. The generally pro-producer tone set by these policies lasted, with occasional waxings and wanings, until the early 1970s.

Prompted by concern over excessively rapid depletion of oil reserves under a rule-of-capture system of property rights, Oklahoma enacted the first state prorationing regulation in 1915. This regulation was designed to limit crude oil production through output quotas. The Oklahoma regulation was followed by prorationing controls in Texas in 1930, and some form of prorationing eventually spread to almost all oil-producing states. The federal government provided direct and indirect support to state efforts to limit production through a number of actions and institutions. The Oil Division of the US Fuel Administration of 1918–1919, for example, promoted both production control through producer planning committees and antitrust immunity for committee members. The Federal Oil Conservation Board (1924–1929), the National Recovery Act's Petroleum Administration (1933–1935), and the Oil and Gas Division of the Department of the Interior (after World War II) backed roughly similar measures designed to promote predominantly self-regulated "cooperation and mutual forebearance," rather than competition, in the petroleum industry.[3] The success of such measures was often undermined by competitive pressures and weak enforcement capabilities, but their intent was clear. Truly effective federal involvement began in 1935 when the federal government assumed enforcement responsibility for state prorationing rules under the Connally Hot Oil Act and the Interstate Oil Compact was created to provide for federal coordination of state prorationing plans. While the resulting success of prorationing may have helped to overcome some of the problems associated with the use of a common pool resource, analysis by economists has highlighted the role of state-administered and federally enforced prorationing in achieving cartel-type output restrictions.[4]

Beneficial tax treatment for crude oil producers began with the Revenue Act in 1913, which provided a depletion allowance permitting tax deductions of 5% of the gross value of annual output. After several modifications, the depletion allowance was raised to 27.5% in 1926 under the pressure of heavy lobbying by the industry. This allowance served for over forty years as a subsidy to domestic oil producers; and its extension to the foreign activities of domestic firms subsidized their international operations by permitting the application of percentage depletion to production outside the United States.[5]

In the mid-1950s, pressure from domestic crude oil producers and concern, in the Cold War atmosphere, over possible negative national security implications of the increasing use of foreign oil prompted policymakers to consider restricting petroleum imports. After numerous reports and recommendations by special commissions and committees, President Eisenhower implemented a voluntary import control program in July 1957. This program attempted to reduce crude oil imports into most of the United States (except the West Coast) by approximately 20% below the level that had been planned for the second half of 1957. Imports of petroleum into the geographically isolated West Coast and imports of refined petroleum products were not subject to the voluntary constraints. Success of the voluntary import control program required the equivalent of a collusive agreement. The lack of a coincidence of interests between producers, integrated companies, and refiner-marketers made such an agreement impossible. By early 1958, any initial support that had existed for the voluntary program had eroded and open defiance of its guidelines was apparent.[6]

After some attempts to force compliance with the voluntary program through federal boycotts of noncomplying importers, the voluntary program was replaced by the Mandatory Oil Import Program (MOIP) in April 1959. Under the MOIP, imports of crude oil, unfinished oils, and finished products (except heavy industrial, or residual, fuel oil) into regions other than the West Coast were limited to a fixed percentage of domestic crude oil and natural gas liquids production. Import controls for petroleum moving into the West Coast were considerably more lenient and were restricted to the difference between demand and supply, with this difference determined by the Department of the Interior's Bureau of Mines. Similarly, the import quotas for residual fuel were based on the demand-supply gap determined by the Bureau of Mines. Rights to import petroleum, which were known as import tickets, came to be allocated to refiners and other importers roughly in proportion to their size. A sliding scale of allocations, however, granted small refiners a disproportionately large amount of tickets. Tickets were exchanged between market participants, and their value represented the difference between the price of domestic oil, propped up by import controls, and the (delivered) price of imported oil.

Prior to its termination in early 1973, the MOIP was a boon to domestic producers. At the same time, however, the MOIP was apparently extremely costly to the economy in terms of the deadweight losses incurred as a result of the induced overproduction of domestic oil and domestic underconsumption of petroleum. Taking 1969 as a representative year, for example, domestic prices of crude oil were raised above world prices (of about $2.00 per barrel) by approximately $1.05–1.50 per barrel.[7] It has been estimated by Bohi and Russell (1978) that in 1969 the MOIP created an increase in domestic producer rents of $2.3 billion and a wealth transfer to those refiners who were allocated import rights of $0.8 billion. The loss in consumer surplus for the users of petroleum was approximately $5.4 billion, and the difference between this loss and the wealth transfers noted was a waste—a deadweight loss—of approximately $2.3 billion.

The late 1960s and early 1970s brought the mandatory quota program under increasing attack. The criticisms made at this time emanated from two sources. First, many refiners and producers themselves regarded the quota program as badly administered. This opinion arose as a result of a growing number of special programs carried out within the MOIP that made the program's intraindustry impact highly discriminatory and diversified the individual interests of firms. Examples of these special programs included decontrol from 1966 onward of East Coast residual fuel imports; numerous bonus quota allocations and exemptions for petrochemical feedstocks, asphalt, No. 2 (middle distillate) fuel oil, and other products; the active encouragement of imports from the Caribbean islands; and certain bonuses and exemptions for Canadian and Mexican imports.[8] The effect of these and similar programs was to disunite the petroleum industry and weaken its collective political influence.

The second source of criticisms directed at oil import controls arose from the general macroeconomic environment and, specifically, the growing concern about domestic inflation in the late 1960s and early 1970s. In the case of petroleum, world prices of both crude oil and refined products began to rise around the turn of the decade after years of virtual constancy. The effects of the quota program became more recognizable and offensive to consumer groups in this setting. At the same time, economists contributed their criticisms and provided consumer groups with ammunition.[9] The consequent attacks on the quota program reached a culmination in 1970, when a Cabinet Task Force on

Oil Import Control (1970) issued a widely read report recommending that the quota program be abolished in favor of a less restrictive and less discriminatory system of import tariffs.

From 1970 onward, quantitative restrictions on imports were gradually relaxed to allow domestic petroleum prices to attain equality with world prices; but the Mandatory Oil Import Program was not officially suspended until April 1973. It was replaced with a system of import license fees. These fees, however, were quite small, and numerous exemptions and exceptions were provided. Consequently, by early 1973, the importation of petroleum had returned to its pre-1957 regulatory status.

The first half of the 1970s marked a watershed in US petroleum policy. Not only were oil import controls eliminated, but the depletion allowance for crude oil production by large producers was reduced and eventually eliminated, prorationing controls were abandoned, and, for the first time, peacetime price controls were imposed on all stages of the industry. The federal depletion allowance was reduced from 27.5 to 22% on 1 January 1970 and was eliminated completely for large producers on 1 January 1975. In addition, several states followed the federal lead and eliminated state tax depletion allowances in the early 1970s.[10] The states also allowed prorationing controls to become ineffective, although such controls were not officially repealed. The de facto abandonment of prorationing controls arose as it became obvious that the increasing competition of imported oil, the development of substantial reserves around the world, and the beginnings of OPEC's cartelization of international petroleum markets were eliminating any price-making ability the domestic crude oil industry might have previously enjoyed.[11]

While import controls, depletion allowances, and prorationing controls could serve profitably as the objects of further economic analysis, the focus of this study is on the federal price controls placed on the petroleum industry in the 1970s. These controls were first imposed as part of an economy-wide anti-inflation program begun in August 1971. This original program went through numerous stages and modifications, and price controls on most sectors of the economy were removed by late 1973. The petroleum industry, however, was singled out for special treatment. In mid-1973, while the rest of the economy was under a regime of voluntary controls, the twenty-three largest domestic petroleum companies were subjected to mandatory price constraints on their producing, refining, marketing, and retailing operations. In the late summer of 1973, after a short period of economy-wide freezes, the entire petroleum industry became subject to an elaborate system of controls on prices, supply allocations, and business practices. Then, within weeks after the Organization of Arab Petroleum Exporting Countries (OAPEC) instituted its embargo on shipments of crude oil to the United States and several other western nations in the fall of 1973, these controls were extended and the power to regulate the petroleum industry was vested in an act of Congress. The general form of these controls has remained in effect up to the present, but is scheduled for termination in the fall of 1981. Notwithstanding scheduled termination, however, many of the price incentives and the bulk of the distributional effects of current controls will be preserved at least until the end of the 1980s. This comes as a result of the adoption in April 1980 of an elaborate system of crude oil excise taxes designed to extract the increased income that producers would otherwise realize upon the removal of explicit price controls.

The noted change in the tenor of petroleum policy that occurred in the 1970s has been accompanied by a marked expansion in the regulatory apparatus needed to administer expanded federal intervention. In early December 1973, the groundwork for a permanent energy bureaucracy was laid with the creation of a Federal Energy Office (FEO) charged with administering petroleum price controls as well as related regulations previously handled by the Cost of Living Council (CLC). The FEO was replaced in May 1974 by a Federal Energy Administration (FEA), which, in turn, was succeeded by the Department of Energy (DOE) in 1977. Prior to the creation of the DOE, more than ten years had passed without the creation of a cabinet-level federal department. The existence of the Department of Energy signals the permanence, if not the direction, of pervasive direct federal regulation of the domestic petroleum industry.

1.2.2 Petroleum Price Regulations since the Early 1970s

A history of the various phases of federal petroleum price regulation not only provides requisite institutional details; it also affords insight into the behavior and interests of the various parties involved in policy formation. A brief summary of the content of the significant petroleum price regulations since the early 1970s is presented in appendix 1.A.

Petroleum price regulation began on 15 August 1971 with the ninety-day economy-wide Phase I price freeze imposed under the presidental powers granted by the Economic Stabilization Act of 1970.

Under Phase I, some disruption in petroleum markets occurred as a result of strict limitations on the ability of firms to pass on increases in the prices paid for imported goods. For the most part, however, impacts on the petroleum sector were minor, as the Phase I freeze coincided with a period of generally soft petroleum prices.

When Phase I ended in November 1971, it was replaced by a more liberal set of Phase II regulations. These regulations allowed firms to raise prices above Phase I ceilings to reflect input cost increases (subject to a limit on profit margins); and large, multiproduct firms were given considerable flexibility in the pricing of individual product lines so long as they satisfied constraints on the weighted average of firm-wide price increases. Significant for the precedents it set, Phase II did not provide gasoline, heating oil, residual fuel, and crude oil prices the same flexibility given to most of the rest of the economy. The prices of these goods were effectively frozen at Phase I levels.[12] This posed particular problems in heating oil markets, where shortages began to appear (notably in inland areas without direct access to imported products) in the winter of 1972–1973.[13] Phase II ceiling prices were able to clear most other petroleum markets.

Phase II was replaced by Phase III price controls in January 1973. Phase III was essentially a voluntary form of Phase II. In spite of the official emphasis on voluntary compliance, however, considerable public pressure, jaw-boning, and the threat of recontrol were brought to bear on larger firms selling politically sensitive products (such as petroleum). In fact, prompted by a jump in heating oil prices and resulting pressure from Congress in early 1973, mandatory controls were reimposed on the twenty-three largest oil companies (via Special Rule No. 1) in March 1973. Although the affected companies accounted for approximately 95% of industry gross sales, failure to control simultaneously the prices of smaller firms left marginal prices and, hence, market-clearing prices uncontrolled in most petroleum markets. Market-clearing prices, moreover, were rising substantially in the spring of 1973 as world prices increased and the dollar went through a period of devaluation. As a result, smaller producing and refining firms were given a competitive advantage over the firms subject to Special Rule No. 1. The latter faced reduced incentives to import oil (because of restrictions on the passing through of higher import prices) and were constrained in their ability to expand production in response to rising market prices. Amid claims that the major oil companies were conspiring to hold back supplies, some customers—independent marketers, fuel oil distributors, and other bulk consumers—of the controlled firms experienced considerable short-term hardship when unable to acquire contracts and completely satisfy their demands at controlled prices. This prompted political pressure for federally directed allocation (that is, rationing) of refined petroleum products among refiners and marketers—a predictable response to the prospect of capturing the economic rent associated with forcing major firms to sell at below market-clearing prices. Similarly, pressure mounted for regulatory allocation of domestic crude oil as the prices of crude oil produced by the major companies continued to be held below market levels.[14] The regulatory allocation of petroleum eventually became a central aspect of federal policy.

With most of the economy under a voluntary program of price restraint, the first half of 1973 turned out to be a period of rapid general inflation. In this environment, the Nixon Administration once again decided to control prices on an economy-wide basis and instituted a sixty-day freeze on 14 June 1973. For the petroleum industry, the Phase III freeze superseded Special Rule No. 1 and applied to both large and small firms. Petroleum markets could still be cleared by imported oil, if such oil was not further refined in the United States. When imported oil was combined in a stream with domestic oil, however, the entire stream could sell for no more than the weighted average price of its foreign and domestic portions. With foreign prices rising, consumers often found themselves facing different prices—one for domestic oil, one for imported oil, and still others for commingled streams—for the same product at the same time in the same market.[15] The implied competition for acquisition of domestic oil through nonprice means was eventually worked out within a regulatory context provided by a major expansion of the federal role in the petroleum industry.

When the Phase III freeze ended on 12 August 1973, the petroleum industry was not freed from regulation. Rather, after a transition staggered over several weeks, the petroleum industry was subjected to a special set of Phase IV price controls. The Phase IV regulations were soon embodied in the Emergency Petroleum Allocation Act of 1973 (EPAA), which was signed into law on 27 November 1973. This act established pervasive controls on prices, production, marketing, and business practices in the domestic petroleum industry and has served, with varying degrees of modification, as the foundation for federal petroleum policy from late 1973 up to the present. Since these controls are of primary interest to this work and are referred to repeatedly, they are described in detail here.

The central element of Phase IV-EPAA regulation was a two-tier system of price controls on domestically produced crude oil.[16] Under the two-tier system, output of crude oil from a given producing property in each month of 1972 was defined as that property's base period control level (BPCL) for that month. For a property that had once produced more than its BPCL, the amount by which production in any subsequent month fell short of the BPCL was added into a property's current cumulative deficiency (CCD). Output in any month less than or equal to the sum of the BPCL and the CCD was defined as old oil. Output in any month in excess of the sum of the BPCL and the CCD and output from properties not producing until after 1972 were defined as new oil. For each barrel of new oil produced from a property brought into production before the end of 1972 (that is, for each barrel produced above the old quantity), a producer was permitted to release a barrel from its old oil classification. Output from a property that averaged less than ten barrels per well per day for any twelve consecutive months after 1972 was defined to be stripper oil. The prices of stripper oil not in excess of its BPCL plus CCD and old oil were limited to the levels they were at on 15 May 1973 plus $0.35 per barrel. New, new stripper, released, and imported oil prices were not controlled. In November 1973 all stripper oil was exempted from controls. In December 1973 the ceiling price on old oil was raised $1.00 per barrel. This left average old oil prices at approximately $5.03 per barrel.

The freezing of domestic old oil prices at a time when world oil prices were rising created a rationing problem for federal regulators. In the initial stages of Phase IV, refiner competition for acquisition of rent-bearing old oil led to attempts to bypass old oil price controls. It was reported, for example, that early in the Phase IV-EPAA era refiners often paid prices for uncontrolled, new domestic crude oil that were considerably above imported prices in order to secure a contract for old domestic oil.[17] Opportunities of this type, however, were foreclosed on 15 January 1974. On that date, the FEO implemented regulations that froze buyer-supplier relationships (at all stages except retail) into their 1972 status (1 December 1972 for producer-refiner relationships) and provided for the allocation of crude oil and petroleum products. In general, suppliers were required to continue to provide supplies to a customer in accord with the percentage of the supplier's total output provided to that customer in the base period. Supply obligations were generally not altered by changes in ownership.

Under the allocation regulations, federal regulators became enmeshed in many aspects of day-to-day business operations in the petroleum industry. Any substantive changes or terminations in buyer-supplier relationships, as well as any changes in ownership, required federal approval since such relationships typically carried far-reaching allocation requirements. In the case of crude oil, allocation was controlled by a special buy/sell program designed to guarantee refiner access to crude petroleum by requiring refiners with ratios of total crude oil to refining capacity greater than the national average ratio to sell (at the purchase price) crude oil to refiners with ratios less than the national average. In May 1974, the buy/sell program was revised to limit the sellers to the fifteen largest integrated refiners and to limit the buyers to small refiners. The net result of all of these programs was to expand the federal presence in the petroleum sector beyond the scope implied by price controls alone.

Refined petroleum price controls were substantially altered by Phase IV-EPAA regulations. After a series of hearings, legal challenges, delays, and rule makings in the fall of 1973, all refined product sellers were brought under a system of base prices (as of 15 May 1973) plus dollar-for-dollar passthroughs of cost changes. Under the dollar-for-dollar cost passthrough provisions of the Phase IV-EPAA refined product price controls, product sellers were permitted to recover any increases in average petroleum input costs (so-called product costs) and any increases in average operating costs (so-called nonproduct costs). Increased capital or interest costs could not be passed through to ceiling prices. Any eligible costs that were not recovered (for example, because market-clearing prices were below allowed ceilings) could be banked (that is, recorded with federal regulators) and drawn upon in future periods to justify further price increases. As a result of the banked cost regulations, refined product price ceilings were considerably more flexible than crude oil ceilings.

In the absence of some program to equalize their costs, a multitiered pricing system of the Phase IV-EPAA type would result in differences in the average crude oil costs of refiners. Refiners granted access (through the regulatory freezing of buyer-supplier relationships) to relatively large amounts of that crude oil subject to price controls would have lower average costs of production than, for example, refiners whose only sources of supply are in the uncontrolled imported, stripper, or new oil markets. Consequently, a multitiered system of crude oil price controls could be expected to produce disparities in the profits of refiners; or, at the least, it would produce differences in refined product prices if such prices were subject to a regulatory system of average cost pricing.

Ostensibly to remove the interrefiner differences in reported profits and refined product price ceilings that were created by crude oil price controls, the Federal Energy Administration proposed an "old oil entitlements" program in August 1974.[18] This program was adopted in early December 1974 and became effective, retroactively, in November 1974. With some modifications, the entitlements program is still in effect. Under the entitlements program, monthly issues of entitlements to controlled crude oil are made to each domestic refiner. The number of entitlements granted to any given refiner is equal to the number of barrels of controlled crude oil that that refiner would have used in the previous month had it operated using the national average proportion of controlled to uncontrolled crude oil. If a refiner has used more controlled crude oil than the number of entitlements it is issued (that is, if it has used controlled crude oil in a higher proportion than the national average), that refiner's average crude oil costs will be less than the national average of all refiners. Such a refiner must purchase entitlements to its above-average use of controlled oil from entitlements-selling refiners. If a refiner has processed fewer barrels of controlled crude oil than the number of entitlements it has been issued, that refiner's average crude oil costs will be greater than the national average and it will be permitted to sell its extra entitlements to entitlements-purchasing refiners.

Under EPAA regulations, the purchase price of an entitlement was set at the difference between the refiner acquisition cost of uncontrolled crude oil and the refiner acquisition cost of old oil. Consequently, refiners using more than the national average proportion of old oil were required to pay a premium equal to the difference between the cost of uncontrolled crude oil and the cost of their extra old oil. Refiners using relatively small proportions of controlled domestic oil, on the other hand, received an offsetting payment of the same size as compensation for their use of extra uncontrolled oil. In this way, crude oil costs could be equalized across refiners. Indeed, under the EPAA entitlements program it was as if ownership of controlled crude oil was allocated proportionately among all refiners since being granted an entitlement was tantamount to owning a barrel of controlled oil—although physical possession never had to occur.

Numerous special programs were introduced into the entitlements program during the term of EPAA. These programs primarily took the form of allocations of extra entitlements above those needed to equalize refiner crude oil costs. Since such extra entitlements could be sold, they were of value to the recipients. The most important of the special programs instituted under EPAA was the Small Refiner Bias. Under this program, which is still in effect, refiners with daily crude oil processing capacity of less than 175,000 barrels are granted extra entitlements according to a sliding scale favoring the smallest refiners. Special issues of entitlements were also made to importers of middle distillate and residual fuel oils during the first several months of the entitlements program. These importers received 30% of an entitlement for each barrel of the imported product. Last, numerous special issues of entitlements were granted to refiners as "hardship relief" under the FEA's Exceptions and Appeals process.

The bulk of EPAA price regulations was originally scheduled to expire in February 1975. This prospect touched off considerable debate in policy circles. The result was a series of congressional extensions of EPAA, threatened presidential vetoes of such extensions, and eventually a new act of Congress signed into law on 22 December 1975. The Energy Policy and Conservation Act of 1975 (EPCA) amended EPAA and officially took effect in February 1976. With some modification since that time, EPCA is still in effect.

EPCA left the general structure of petroleum industry regulation intact. Its major change was the establishment of a *three*-tier pricing program for domestically marketed crude oil.[19] The passage of EPCA also provided procedures for the removal of price and allocation controls on most refined products and introduced some small changes to the entitlements program. The crude oil pricing provisions of EPCA, as well as the entitlements program, are scheduled to expire on 30 September 1981. The act, however, gives the President the authority to place petroleum price controls on standby status any time after May 1979.

The three-tier pricing system established under EPCA extended federal price controls to most of the domestic production that had gone uncontrolled under EPAA. The basic BPCL for a property in any month is defined as the lesser of average monthly output of old oil in 1975 and average monthly output of all oil in 1972. Since July 1976, any property consistently failing to produce above its BPCL has been permitted slight reductions in its BPCL based on actual annual decline rates between 1972 and 1975. Under EPCA, lower-tier oil has been defined as all output from a property not in excess of that property's BPCL plus CCD. Upper-tier oil has been defined as production from pre-1976 properties in excess of the associated lower-tier output and production from properties that began producing after 1975. There is no released oil program.

Lower-tier oil sells at its 15 May 1973 posted price plus \$1.35 plus inflation and incentive adjustment factors determined by the Department of Energy. Upper-tier oil sells at its 30 September 1975 price less \$1.32 plus inflation and incentive adjustment factors. Stripper oil originally was treated as upper-tier oil under EPCA, but was decontrolled on 1 September 1976. Alaskan North Slope crude oil is treated as upper tier, although, prior to the increases in uncontrolled crude prices in 1979, transportation costs kept its price below upper-tier ceilings. Production from Federal Naval Petroleum Reserves and, since September 1978, incremental production attributable to tertiary oil recovery projects is not controlled. Imported oil, of course, is not subject to price controls. The average price of domestic oil was originally limited to \$7.66 plus incentive and inflation adjustment factors. Incentive adjustments were initially limited to a maximum of 3% annually, and the inflation adjustment was limited to the rate of inflation; the combination of the two adjustments, however, could not exceed 10% annually. In September 1976, EPCA was amended to allow average domestic prices to rise by 10% per year without regard to the inflation rate or limits on incentive adjustments.

Under the provisions of EPCA, the Carter Administration began a gradual decontrol of domestic crude oil prices on 1 June 1979.[20] Production from onshore properties not producing in 1978 and from offshore properties leased after December 1978 has been decontrolled. In addition, beginning in January 1980, 4.6% of a property's upper-tier output is made "market-level new" oil and decontrolled each month; and small amounts of lower-tier oil are allowed to sell at uncontrolled prices as an offset to certain expenses associated with newly undertaken tertiary recovery projects. On 17 August 1979, "heavy" crude oil of 16° API specific gravity or less (20° or less since December 1979) was decontrolled. In June 1979, 80% of the production from marginal (that is, almost stripper) lower-tier properties was decontrolled, with the remaining 20% decontrolled upon enactment of the Crude Oil

Windfall Profit Tax Act of 1980. Also in June 1979, producers were allowed to redefine the BPCLs of lower-tier properties to the average output in the six months ending March 1979 and to recalibrate CCDs to zero. Thereafter, BPCLs were reduced 1.5% per month in 1979 and are being reduced by 3% per month from the beginning of 1980 through the end of controls in October 1981. Output in excess of lower-tier BPCLs still sells for the upper-tier price.

Table 1.1 exhibits the history of both EPAA and EPCA pricing of domestic petroleum. For reference, the price of stripper oil represents an uncontrolled price except for the period from February through August 1976. The degree to which price controls have been binding on domestic crude oil producers is suggested by the differences in the stripper price and the prices of the several controlled categories. The average domestic price was typically $3–5 below uncontrolled prices prior to 1979. The 1979 increases in world prices, however, greatly increased this gap. Table 1.2 shows the relative importance of the several categories of domestic crude oil under EPAA and EPCA. Over most of the period of EPAA/EPCA controls, roughly 60–70% of domestic output has been subject to some form of binding price constraint.

The introduction of controls on upper-tier prices under EPCA necessitated modifications in the entitlements program. In order to equalize refiner crude oil costs, access to upper-tier oil, as well as lower-tier oil, had to be equalized. To accomplish this, each barrel of upper-tier oil, which is closer in price to uncontrolled oil than is lower-tier oil, is granted a fraction of the entitlement given to lower-tier oil. The price of an entitlement under EPCA is set at the difference between the per barrel refiner acquisition cost of uncontrolled oil and the acquisition cost of lower-tier oil less $0.21. This $0.21 adjustment is designed to maintain a $0.21 per barrel preference for domestic crude oil created by the tariffs imposed following the suspension of import quotas.[21] The Small Refiner Bias in the entitlements program has been continued under EPCA. Since April 1976, residual fuel imports into the East Coast have been eligible for partial entitlements. In response to the severe winter of 1976–1977, imports of middle distillates were granted similar partial entitlements in February and March 1977. Salable entitlements were also granted to middle distillate imports in May–September 1979. Special allocations of entitlements to refiners are also made through an Exceptions and Appeals Program for the refining of low-quality California crude oil, certain uses of nonpetroleum fuels in

Table 1.1 Domestic Crude Oil Prices: 1974–1980
(Dollars per Barrel at the Wellhead)

	Old	New	Domestic average			
1974 average	5.03	10.13	6.87			
1975 average	5.03	12.03	7.67			
1976 January	5.02	12.99	8.63			
	Lower tier	Upper tier	Stripper[a]	Alaskan North Slope[b]	Naval Petroleum Reserves[c]	Domestic average
1976 Average (Feb.–Dec.)	5.14	11.57	12.17	—	—	8.14
1977 Average	5.19	11.22	13.59	6.35	12.34	8.57
1978 Average	5.46	12.15	13.95	5.22	12.96	9.00
1979						
January	5.75	12.66	14.55	5.79	13.10	9.46
February	5.76	12.68	14.88	5.87	13.94	9.69
March	5.82	12.84	14.88	6.66	13.97	9.83
April	5.85	12.94	16.71	7.45	14.56	10.33
May	5.91	13.02	17.53	8.47	15.85	10.71
June	6.07	13.14	20.24	8.97	16.02	11.70
July	6.00	12.79	24.76	13.35	20.13	13.39
August	6.09	13.33	25.71	14.14	20.77	14.00
September	6.09	13.53	27.09	13.09	20.85	14.57
October	6.12	13.56	29.42	13.12	21.01	15.11
November	6.09	13.68	30.64	13.48	26.48	15.52
December	6.21	13.76	34.99	13.60	29.04	17.03
Average	5.95	13.20	22.93	10.57	19.40	12.64
1980						
January	6.23	13.82	35.92	13.77	28.94	17.85
February	6.37	14.03	36.14	13.77	34.96	18.81
March	6.35	13.99	36.33	13.77	34.67	19.36
Average	6.32	13.96	36.16	13.77	32.79	18.68

Source: US Department of Energy, *Monthly Energy Review*.
a. Stripper oil was exempt from price controls beginning 1 September 1976. From February through August 1976 stripper oil was subject to upper-tier price ceilings.
b. Alaskan North Slope (ANS) crude oil prices are treated as upper tier for determining the applicable wellhead ceiling prices.
c. The Naval Petroleum Reserves (NPR) are exempt from pricing regulations but have been reported here as upper tier prior to July 1977.

industrial processes, and the importation of Puerto Rican petrochemicals. Last, federal purchases of crude for the Strategic Petroleum Reserve have been covered by entitlements since 1976.

A final component of EPCA policies that is of interest here concerns the treatment of refined products. As noted above, EPCA allowed the relaxation of refined product price controls. During 1976, controls were eliminated for all major products except gasoline, jet fuel, and propane. Jet fuel was exempted from controls in early 1979. Although proposals for termination of remaining regulations have been made, gasoline and propane prices continue to be subject to DOE ceilings and the dollar-for-dollar cost passthrough provisions.[22]

EPAA/EPCA regulation has been notoriously complicated. A new standard of complexity in petroleum regulation appears to have been established, however, with the adoption of the Crude Oil Windfall Profit Tax Act of 1980.[23] The publicly stated purpose of the Windfall Profits Tax (WPT) is to extract the largest part of the differences between ceiling prices and uncontrolled prices as EPCA price controls are phased out. This is accomplished through a percentage tax on the difference between the price at which a barrel of crude oil sells and a governmentally determined base price. Through its effect on after-tax producer realizations, this tax substantially undoes decontrol.

The WPT establishes several categories of crude oil for tax purposes. Tier One includes EPCA's lower- and upper-tier oil, as well as recently decontrolled upper-tier oil and output from marginal properties. Tier Two includes stripper and Naval Petroleum Reserve crude oil. Tier One has a base price approximately equal to the May 1979 upper-tier ceiling price, adjusted upward for inflation. The Tier Two base price is approximately the adjusted Tier One base plus $1.00. The difference between selling price and base price (less state severance taxes) is subject to a 70% tax in Tier One and a 60% tax in Tier Two. Independent producers [with sales less than $1.25 million per quarter or less than 50,000 barrels per day (b/d) refining capacity] only bear taxes of 50% and 30% on the first 1,000 b/d produced under Tier One and Tier Two, respectively. Tier Three includes output of newly producing post-1978 properties, heavy crude oil, and incremental oil from tertiary recovery. The Tier Three base is approximately the May 1979 upper tier ceiling plus $2.00; and Tier Three is taxed at a rate of 30% (independents bear no tax on their first 1,000 b/d). Finally, crude oil produced from properties owned by state and local government bodies, charities, and Indian tribes, as well as oil from geographically isolated

Table 1.2 Crude Oil Production by Control Category: 1974–1980[a]
(Percentage Sold at Wellhead)

	Old oil	New oil[b]	Released[b]	Stripper[b]	
1974 Average	64	15	9	12	
1975 Average	62	16	8	13	
1976 January	54	21	10	15	
	Lower tier	**Upper tier**	**Stripper[b]**	**Alaskan North Slope[c]**	**Naval Petroleum Reserves[c]**
1976 Average (Feb.–Dec.)	54.40	31.50	14.10	—	—
1977 Average	45.92	36.11	13.32	4.14	0.51
1978 Average	37.54	34.41	14.03	12.96	1.08

	Lower tier	Upper tier	Stripper[b]	Alaskan North Slope	Naval Petroleum Reserves	Incremental tertiary[b]	Newly discovered[b]	Marginal[b]	Heavy crude[b]	Market-level new[b]	Tertiary incentive[b]
1979											
January	35.51	34.25	14.14	14.88	1.20	—	—	—	—	—	—
February	35.20	34.97	15.08	13.71	1.01	—	—	—	—	—	—
March	34.59	34.56	14.95	14.58	1.29	—	—	—	—	—	—
April	33.98	34.93	15.27	14.52	1.28	—	—	—	—	—	—
May	33.53	34.78	15.62	14.71	1.32	—	—	—	—	—	—
June	29.32	38.22	15.97	13.64	1.34	0.05	0.61	0.81	—	—	—
July	26.96	37.49	16.01	15.86	1.38	0.02	1.12	1.13	—	—	—
August	26.03	36.72	16.93	15.82	1.33	0.15	1.66	1.33	—	—	—
September	23.52	33.89	16.55	16.08	1.57	0.06	2.38	3.08	2.82	—	—
October	23.46	32.58	16.20	16.27	1.57	−0.01	3.04	3.39	3.46	—	—
November	23.11	32.76	15.35	17.49	1.61	NA	3.24	3.11	3.28	—	—
December	22.21	32.52	16.34	16.51	1.60	−0.03	3.61	3.05	4.04	—	—
Average	28.91	34.79	15.71	15.36	1.38	0.03	2.24	2.27	3.40	—	—
1980											
January	21.18	31.18	15.67	17.03	1.54	0.01	3.86	3.16	4.24	2.15	0.00
February	20.52	29.45	15.82	15.73	1.44	0.01	4.33	2.71	5.13	4.79	0.01
March	19.82	28.24	15.18	15.26	1.54	0.01	4.76	2.52	5.19	7.38	0.04
Average	20.51	29.60	15.53	16.02	1.51	0.01	4.32	2.80	4.85	4.77	0.02

Source: US Department of Energy, *Monthly Energy Review*.
a. See notes to Table 1.1. NA, not available.
b. Uncontrolled except February–August 1976.
c. Percentages for 1977 for Alaskan North Slope and Naval Petroleum Reserves are for July–December.

Alaskan fields and tertiary incentive cost-offset oil are exempt from the WPT.

The Windfall Profits Tax went into effect on 1 March 1980 and will begin a scheduled 33-month phase-out at the later of (a) January 1988 or (b) the first month (but not later than January 1991) after the accumulation of a net addition to federal revenues of $227.3 billion. Together with existing taxes, the WPT will collect roughly $0.70 of every $1.00 of crude oil price increase over the next decade.[24] Although price controls will apparently be eliminated, the 1980s do not look much better than the 1970s for crude oil producers. There will, however, be beneficiaries from the Windfall Profits Tax. The huge revenues to be raised have been earmarked for a variety of programs, including tax credits for residential and business investment in energy conservation equipment; a $3.00 per barrel oil-equivalent tax credit for the production (typically by integrated refiners) of synthetic fuels and oil from alternative sources; removal of excise taxes on gasohol; assistance for heating and cooling by low-income families; general income tax reductions; and subsidies for expansion of mass transit systems. The political battles over WPT funds promise to be among the most controversial in the coming years.

1.2.3 Summary of the Regulations

Phase IV-EPAA and EPCA regulation of the petroleum industry has been pervasive, complicated, and repeatedly altered. The description here of this regulation has focused on only those components that have had significant economic effects. Nevertheless, this description is itself complicated and warrants summary. Phase IV-EPAA regulations controlled the wellhead price of domestic crude petroleum from older properties that were producing prior to 1973. Oil from post-1972 properties and from stripper properties, as well as oil produced from old properties in excess of 1972 base levels, was not controlled. In addition, production in excess of 1972 base levels from old properties allowed the producer to release an equivalent amount of old oil from controls. Refined product prices were subject to control under the Phase IV-EPAA regime, but these controls provided substantial flexibility in ceiling prices by allowing recoupment of current and past changes in input costs. Ownership of controlled crude oil was eventually determined by an entitlements program that required refiners with relatively large amounts of controlled oil to make cross payments to refiners with relatively small amounts of controlled oil. Except for some special policies within the program, the entitlements program tended to equalize refiners' average crude oil costs.

To summarize EPCA regulation, multitier crude oil price controls have applied to all domestic production except federal and stripper oil over most of the period since EPCA began in February 1976. Most output from older properties is priced at the lowest tier and output above base levels of old production and from newer properties is priced at an intermediate tier. Uncontrolled domestic oil and imported oil sell at market-determined prices. Refiner crude oil costs are equalized by an entitlements program. Most refined petroleum prices have been decontrolled under EPCA, except for gasoline and propane prices. A gradual decontrol of crude oil prices was begun in mid-1979. As crude oil price controls are phased out through September 1981, excise taxes on resulting price increases are being phased in. If decontrol survives promised congressional challenges, these excise taxes will form the backbone of petroleum policy in the near future.

1.3 Plan of the Study

Federal petroleum policy is in the midst of a transition from controls-based to tax-based regulation. This confounds economic analysis. Insofar as multitier price controls and the entitlements program still dominate policy and because future policy will continue to embody current multitier crude oil production incentives, crude oil price controls and entitlements are given primary attention in the investigations undertaken here. The Windfall Profits Tax is also examined, although in somewhat less detail.

Chapters 2 and 3 present analytical frameworks for investigating the consequences of the entitlements program and crude oil price regulation. Although refined petroleum products have been subject to price controls over most of the period of interest, an analysis of their structure and impact indicates that, with some notable short-term exceptions, refined product price controls have generally not been binding. Crude oil price controls, however, have been effective. By holding the prices of most domestic crude oil below market-clearing levels, these controls generate inframarginal rents for refiners favored by the nonprice rationing used to allocate domestic crude oil. The analysis of chapter 2 indicates, however, that the entitlements program (implicitly) taxes away such rents and uses the resulting revenue to finance a number of regulatory programs. The most important of these is a substantial implicit subsidy, at the margin, for the domestic refining of uncontrolled domestic and imported crude oil. After demonstrating the existence of this so-called entitlements subsidy, its allocative and distributional implications are examined at length. The implications of multitier crude oil price controls and excise taxes are also examined intensively—in chapter 3. Most significantly, these controls and taxes are seen as closely analogous to a monopsonistic price discrimination scheme; their consequences are investigated in both a static, textbook model of price regulation and a dynamic model of price regulation in the case of a nonrenewable resource.

By far the most hotly debated issue in the economics of current petroleum market regulation is the incidence of the entitlements subsidy. The importance of this issue has its roots in the political struggles that arise every time removal or alteration of current regulation is considered in Congress. These struggles, which economists have not been unwilling to enter, have consistently centered around the impact of post-embargo regulation on the prices of refined petroleum products and, particularly, the well-being of consumers of refined products. The mechanism by which such consumers may share, through lower prices, in the inframarginal rent associated with price-controlled domestic crude oil is the entitlements subsidy. Chapter 4 examines the available literature on the incidence of this subsidy and presents the results of a broad range of econometric and noneconometric evidence bearing on the issue. The goal and result is an estimate of the split of the entitlements subsidy between refiners and consumers of refined products.

The analytical and empirical results of chapters 2, 3, and 4 provide the necessary background for estimating the allocative and distributional consequences of controls, entitlements, and the Windfall Profits Tax. Chapter 5 undertakes these estimates. Aggregate estimates of the economy-wide efficiency gains and losses of regulation are generated and subjected to a sensitivity analysis. Aggregate measures of the wealth transfers among crude oil producers, refiners, and consumers of refined products are also developed and dissected. Particular attention is paid to the allocative and distributional implications of the many suggested sources of nonoptimal resource allocation in an unregulated domestic crude oil market.

On the basis of the insight into regulatory consequences produced in earlier chapters, chapter 6 explores the causes of current federal petroleum policy. The analysis is structured around an economic theory of regulation that (a) treats individuals as utility maximizers in both their market and nonmarket activities; (b) recognizes the value to individuals of both pecuniary and nonpecuniary returns; and (c) incorporates the general problem of shirking in agent-principal relationships. Within this theoretical context, recent petroleum-related voting in the US Senate is econometrically analyzed. The results are a description of the political-economic factors that have determined the direction of policy and an improved understanding of the economics of regulation and politics.

Chapter 7 summarizes this study and offers a discussion of unanswered questions and promising areas for related research. The temptation to offer some general observations on US energy policy is not resisted.

Appendix 1.A History of Petroleum Price Regulation

	Period	Coverage	Agency
Phase I	8/15/71 to 11/13/71	Economy-wide	Cost of Living Council (CLC) (policymaking) Office of Emergency Preparedness (administration and compliance)
Phase II	11/14/71 to 1/10/73	Economy-wide	CLC (policymaking) Price Commission (administration) Internal Revenue Service (compliance)
Phase III (voluntary)	1/11/73 to 3/5/73	Petroleum firms with annual revenues >$250 million	CLC
	1/11/73 to 6/13/73	Economy-wide	

Price regulations	Special provisions	Noted effects
1. Frozen at higher of: a. Average price in all transactions on 5/25/70. b. Highest price at which 10% of transactions occurred in 30 days prior to 8/15/71.	1. Import price increases only passed on if imported goods not further transformed and kept physically segregated.	1. Import price rules discriminated against firms lacking duplicate facilities. 2. Gasoline price ceilings near seasonal peak; middle distillate ceilings near seasonal low.
1. Base price at Phase I ceiling. Price increases permitted only if cost justified. 2. Unit cost increases passed on subject to profit limit. 3. Profit margin not to exceed average of 2 highest margins in 3 fiscal years prior to 8/15/71. 4. Firms with annual revenues: a. >$100 million must prenotify intended price increases; such increases subject to CLC rejection. b. $50–100 million must file quarterly report with CLC. c. <$50 million subject to monitoring and spot checks by IRS and Price Commission.	1. Firms with annual revenues >$100 million allowed Term Limit Pricing (TLP): a. Weighted average price increase across all products not to exceed 2% per year (1.8% after 2/72). b. Individual price rise must be based on cost increase and not to exceed 8% per year. c. TLPs not to include gasoline, distillate, residual fuel, or crude. 2. CLC publicly stated that at least large firms' gasoline, distillate, residual fuel, and crude prices effectively frozen. 3. Imports treated as in Phase I.	1. Seasonal pattern to gasoline and middle distillate ceilings as in Phase I. 2. Some middle distillate shortages, particularly in inland markets, in winter of 1972–1973.
1. Base price at authorized Phase II level on 1/10/73. 2. Unit cost increases passed on subject to Phase II profit margin limit. 3. Firm's weighted average annual price increase to reflect cost increases allowed up to 1.5% without regard to margin limit. 4. CLC reserves authority to reimpose mandatory controls on any firm or sector.	1. Voluntary compliance. 2. Firms with annual revenues: a. >$250 million required to file quarterly price, cost, and profit report. b. >$50 million required to keep records of same. c. <$50 million subject to spot checks. 3. Imports treated as in Phase II. 4. Phase II TLPs still in effect and binding.	1. CLC held hearings in 2/73 to investigate oil price increases. 2. CLC jawboning against oil price increases.

Appendix 1.A (continued)

	Period	Coverage	Agency
Special Rule No. 1	3/6/73 to 6/13/73	23 largest petroleum firms (with annual revenues >$250 million)	CLC
Phase III freeze	6/14/73 to 8/11/73	Economy-wide	CLC (policymaking) Internal Revenue Service (compliance)
	6/14/73 to 8/18/73	Petroleum industry except retailers	
	6/14/73 to 9/6/73	Petroleum retailers	
Phase IV-EPAA (petroleum)	8/19/73 to 1/30/76	Petroleum industry except retailers	CLC (to 12/3/73) Federal Energy Office (12/4/73–5/6/74) Federal Energy Administration (5/7/74–1/30/76)
	9/7/73 to 1/30/76	Retailers	
	EPAA adopted 11/27/73		

Price regulations	Special provisions	Noted effects
1. Base price at authorized Phase II level on 1/10/73. 2. Weighted average annual firm-wide price increase not to exceed 1% except: a. Up to 1.5% allowed for cost increases since 3/6/73. b. More than 1.5% allowed if Phase II profit limit not exceeded, reflects cost increases since 3/6/73, and CLC prenotified.	1. Products covered by TLP given base at highest price at which 10% of transactions occurred during 30 days prior to 1/10/73. 2. No limit on individual price increases within weighted average limit. 3. Phase III (voluntary) reporting and recordkeeping requirements still in effect. 4. Restriction on segregation of imports eliminated 5/11/73 retroactive to 3/6/73; non-transformation restriction retained.	1. Only largest petroleum firms subject to Special Rule No. 1. 2. Smaller firms acted to clear market and led oil price changes. 3. Bulk customers of controlled firms were unable to satisfy their demands for some products at controlled prices; resulted in pressure for allocation regulation.
1. Frozen at highest price at which 10% of transactions occurred during 6/1–8/73.	1. Import price increases only to be passed on if imported good not further transformed. 2. Special Rule No. 1 price rules govern import cost-justified price increases.	1. Imported petroleum prices rising; higher imported crude oil costs could not be passed on because crude oil is transformed in refining. 2. Combined domestic and imported product streams had ceilings above domestic streams; led to charges of price discrimination.
1. Crude oil: a. Old oil price at 5/15/73 level plus 35¢/barrel. b. Stripper (after 11/27/73), new, and released oil uncontrolled. 2. Refined products: a. Refiner base price at 5/15/73 level; dollar-for-dollar cost adjustment. b. Distributor price at cost plus 1/10/73 markup; beginning 9/28/73, price at 5/15/73 base plus dollar-for-dollar adjustment. c. Retailer price at cost plus higher of 1/10/73 markup or 7¢/gallon; beginning 11/1/73, base at 5/15/73 level plus dollar-for-dollar cost adjustment. d. Unrecouped costs banked beginning 11/73.	1. All product prices frozen 10/15–31/73. 2. Import price increases passed on without restriction. 3. Old oil ceiling raised $1 on 12/21/73. 4. Buyer-supplier freeze (1972 status) beginning 1/15/74. 5. Buy-sell program effective 1/15/74. 6. Entitlements program into effect 11/74. 7. Crude output above 1972 old oil base called "new" and allowed "release" of equivalent quantity from controls.	1. Entitlements program: a. Equalized refiners' crude oil costs. b. Created subsidy for use of uncontrolled crude. c. Small refiners heavily subsidized. 2. Unrecouped banked costs imparted flexibility to refined product price ceilings and helped avoid shortages. 3. Released oil program raised incremental revenue above uncontrolled price for highly elastic old oil supply sources.

Appendix 1.A (continued)

	Period	Coverage	Agency
EPCA	2/1/76 to 9/30/81	Petroleum industry	Federal Energy Administration (to 9/30/77) Department of Energy (10/1/77–present)
WPT	3/1/80 to present	Petroleum industry	Internal Revenue Service

Price regulations	Special provisions	Noted effects
1. Crude oil: a. Lower tier oil base price at 5/15/73 price plus $1.35/barrel. b. Upper tier oil base price at 9/30/75 level less $1.32/barrel. c. Stripper oil in upper tier, 2/1/76–8/31/76; decontrolled 9/1/76. d. Ceilings adjusted for incentives and inflation. 2. Refined products: a. EPAA base prices plus dollar-for-dollar cost adjustment. b. Unrecouped costs banked.	1. Some products decontrolled: residual fuel, 6/1/76; middle distillates, 7/1/76; jet fuel, 2/5/79. Gasoline and propane remain subject to price ceilings. 2. Gradual decontrol of crude begun 6/1/79; full decontrol scheduled for 10/1/81.	1. Entitlements program same as in Phase IV-EPAA. 2. Special entitlements subsidy for use of Calif. crude (starting 1/1/78) raised low quality, lower tier Calif. crude prices toward ceilings. 3. No released oil program. 4. EPAA in effect for refined products.
1. Tax on difference in base price and market price. 2. Tax rates: a. Tier One (lower tier, upper tier, marginal, market level new): 70%. b. Tier Two (stripper, NPR): 60%. c. Tier Three (newly discovered, heavy, incremental tertiary): 30%.	1. Independents' first 1000 b/d taxed at lower rates. 2. Base prices (approx.): a. Tier One: 5/79 upper tier. b. Tier Two: 12/31/79 uncontrolled. c. Tier Three: $16.55. 3. Base prices adjusted for inflation (+2%/yr. for Tier Three).	1. Continues multitier price incentives. 2. Special large incentives for tertiary enhanced recovery. 3. Revenues earmarked for special programs. 4. EPAA in effect for refined products.

Chapter 2

The Economics of the Entitlements Program

2.1 Introduction

The two most important components of post-embargo price regulation in the domestic petroleum industry—and, perhaps, the most important components of all of post-embargo energy policy—have been crude oil price controls and the entitlements programs. For better or worse, price controls on crude oil have moved the determinants of the exploration, development, and production decisions that govern the nation's most important energy source out of the marketplace and toward the political arena. They enhance the position of foreign crude oil producers and have a severely adverse impact on the wealth and welfare of domestic producers. The entitlements program determines, through a complex regulatory system, the distribution of the very substantial benefits attendant to access to low-cost, price-controlled crude oil. It also has significant effects on the use of petroleum by refiners and ultimate consumers. This chapter presents an analytic examination of these effects of the entitlements program. The next chapter provides a similar treatment of crude oil price controls.

Price controls without an entitlements program characterized Phase IV-EPAA regulation until November 1974. It is concluded that the primary effect of such a regulatory regime is to confer inframarginal windfall gains on refiners with access to controlled crude oil. An entitlements program, however, affects marginal costs in the refining industry and, in fact, heavily subsidizes domestic refiners. The implications of this for the importation of petroleum, the prices of refined products, and the welfare of market participants are discussed at length.

2.2 The Market Setting: Some Important Assumptions

Throughout the analysis that follows, the *domestic* producing and refining industries are taken to be competitive. Although this work is not directed toward investigation of this important assumption, the current significance of the competitiveness of the petroleum industry to policymakers considering such measures as horizontal divestiture, vertical divestiture, and limitations on the expansion of petroleum companies into nonpetroleum energy forms suggests the need for special comment here. While policymakers do not seem to have reached a consensus on whether or not the US petroleum industry is competitive, it is fair to say that academic opinion is less divided. The majority of recent investigations by economists have concluded that, with certain exceptions in isolated cases, there is little evidence of significant monopoly in the present-day domestic petroleum industry.[1] Where the implications are important, the consequences of a contradiction to this conclusion are noted.

A second assumption in the following analysis is that the United States is a price taker with respect to the world price of crude oil. This assumption is important because many aspects of EPAA/EPCA regulations affect the US demand for imported crude oil. Hence regulation-induced changes in world oil prices are conceivable; and any such changes would carry allocative and distributional consequences. Full investigation of the direction and magnitude of the effect of US regulation on world oil prices constitutes a formidable area for research in its own right. The basic problem is to discern the behavior of an international producing cartel attempting to maximize the flow of value from its nonrenewable resource base in a dynamic setting while faced with a competitive fringe of other producers, competition from more or less close substitutes, changing customer demands, and an uncertain future. Fortunately, this behavior has been subjected to fairly extensive study already. The most in-depth studies are those by Pindyck (1978, 1979), Pindyck and Hnyilicza (1976), Salant (1976), and ICF, Inc. (1979). The general conclusion that emerges from these works is that the price of crude oil (or, more properly, the intertemporal price path) set by OPEC is not significantly responsive to the kinds of changes in the net demand for OPEC oil that US crude oil price regulation and entitlements programs introduce. Pindyck (1979), for example, finds that an addition to world crude oil output of 7 million barrels per day (mmb/d) from Mexico by 1985 (and, hence, a reduction in the net world demand facing OPEC) reduces the optimal OPEC price by 8–10% at most. Salant (in ICF, Inc., 1979) also reports the results of a dynamic optimization model of world prices. He finds that, in 1978 (when the United States consumed 14.7 mmb/d), a 1-mmb/d reduction in US crude oil demand at all prices (that is, a shift in demand) most likely would have reduced world oil prices, other things being equal, by only $0.048 per barrel (0.1¢ per gallon). In short, the world oil price is not likely to vary to any substantial degree in response to the domestic regulatory actions studied here. Accordingly, the United States is treated as a price taker in the world market throughout this work. Some of the implications of violations of this assumption are discussed in Chapter 5.

2.3 The Economic Impact of Controls without Entitlements

The implementation of the entitlements program in November 1974 was justified primarily on the grounds that the disparities in average costs among refiners that arose under EPAA pricing of crude oil were threatening the survival of those segments of the domestic refining industry with relatively heavy dependence on uncontrolled crude oil. The Federal Energy Administration asserted that access to controlled crude oil conferred cost advantages upon fortunate refiners that allowed them to gain at the expense of their competitors.[2] In fact, however, access to controlled crude oil confers windfall gains on fortunate refiners, but *does not* confer competitive advantages. Indeed, in the presence of binding ceilings on refined product prices, access to controlled crude might place "fortunate" refiners at a competitive disadvantage. The effects of controls without entitlements can be explicitly investigated by examining the implications for refiners' marginal revenues and marginal costs.

2.3.1 Effects on Refiners' Marginal Costs

The conclusion that interrefiner differences in access to controlled crude oil does not create competitive disparities is not self-evident. It might appear that refiners with access to controlled crude oil and lower average costs could undersell their competitors while still maintaining positive profits. The pricing and supply decisions of any profit-seeking enterprise, however, are based on the marginal costs of additional output, rather than the average costs of particular output levels. While EPAA-type crude oil price controls do introduce differentials in average refiner costs, they do not cause interrefiner differences in marginal costs.

Figure 2.1 presents a simplified version of the effects of crude oil price controls on the refining sector. In this figure, the supply of domestic crude oil is shown by the supply function S_c. The supply of uncontrolled oil is given by S_w. Domestic crude oil is controlled at a price of P_c. With the United States a price taker in world crude oil markets (see section 2.2), foreign crude oil (or otherwise uncontrolled

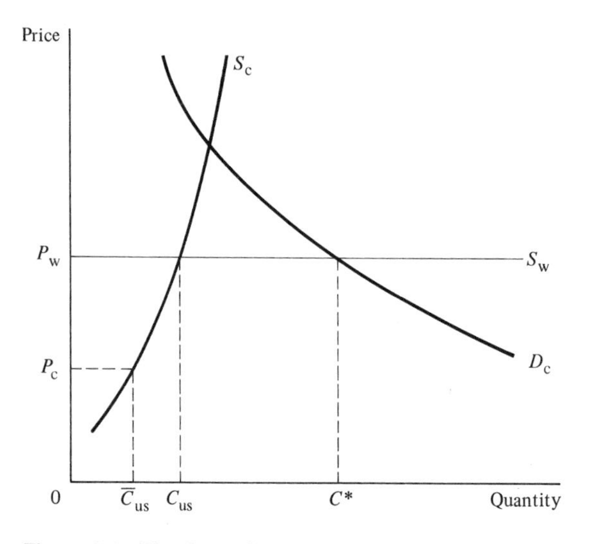

Figure 2.1 The domestic crude oil market: controls and marginal refining costs.

oil) may be purchased at P_w per barrel. In the absence of controls, domestic producers would produce C_{us}, refiners would purchase C^*, and $C^* - C_{us}$ would be imported. The imposition of domestic price controls causes domestic producers to supply $\bar{C}_{us}$. Refiners increase their importation of crude oil by $C_{us} - \bar{C}_{us}$ to $C^* - \bar{C}_{us}$, but do not alter their total purchases of crude oil because the marginal cost of crude oil remains the world price after the introduction of controls. That is, the crude oil cost that is relevant to refiners for the calculation of the marginal cost of their refined product output is the price of uncontrolled crude oil—just as it would be in the absence of any federal regulation of the crude oil market.

There is no doubt that the marginal barrel of crude oil input for the domestic refining industry was an uncontrolled barrel prior to the introduction of entitlements. In 1974, controlled old oil accounted for only about 64% of domestic production and 46% of domestic consumption of crude oil.[3] Consequently, the typical individual refiner could not buy all of the crude oil it desired at controlled prices. Hypothetically, an individual, small refiner could have found itself frozen into a relationship with a crude oil producer able to satisfy the refiner's entire demand with old oil in the short run (that is, given the refiner's capacity). Of course, if the hypothetical refiner's demand for crude could have been satisfied in the short run at an output of old oil short of the amount the producer would be willing to supply at old oil prices, an excess of other refiners would be willing to purchase the remaining old oil from the supplier. Further expansion on the part of the hypothetical refiner, interested in capitalizing on its apparent cost advantages, would require use of uncontrolled oil.

A vertically integrated refiner supplying itself with old oil also would have been unable to acquire the marginal barrel of crude oil at a cost less than the price of uncontrolled oil under Phase IV-EPAA controls. Such a refiner might have bypassed crude oil price controls (on those of its properties that did not carry second-party buyer-supplier or buy/sell obligations) by expanding production until marginal crude oil production costs equaled the associated marginal revenue from refining such crude.[4] This marginal revenue for any individual refiner would have been that given by the market and would have been just sufficient to cover the industry-wide marginal cost of crude oil (that is, the marginal revenue associated with incremental refined crude oil would have been made equal to the uncontrolled price of crude oil). Hence, the marginal cost of crude oil for the integrated refiner, even if the refiner were totally self-sufficient in controlled oil, would have been the price of uncontrolled oil under Phase IV-EPAA controls without entitlements.

Arguments that crude oil price controls have reduced domestic refined product prices by reducing domestic refiners' average crude oil costs are unfounded. Nevertheless, such arguments have been made repeatedly in the policy debates over the fate of controls. The flaw in these arguments arises from a failure both to understand the marginal character of input, output, and price determination and to distinguish between marginal and average costs. All refiners face essentially equal marginal costs under crude oil price controls without entitlements.

2.3.2 Effects on Refiners' Marginal Revenues

In an uncontrolled product market, the incremental revenue relevant to the decision to refine an additional barrel of crude oil is the sum of the prices actually received in the marketplace for those refined products that result from the additional crude oil. As discussed in chapter 1, however, federal ceilings have been placed on the prices of refined petroleum products. Under these price controls, legal ceilings have been allowed to adjust upward for increases in a federally calculated measure of average cost above a base price. Any such cost increases incurred in the previous month that are not immediately reflected in the current month's prices may be banked and drawn on in subsequent months in order to raise product prices above their base levels. Although most major products have been exempted from controls since mid-1976, all were subject to controls at the time the entitlements program was adopted in late 1974.

One of the sources of supposed competitive disparities noted in defense of the adoption of the entitlements program was the existence of interrefiner differences in refined product price ceilings. Such differences arose because of the average cost basis of the federal ceilings and the interrefiner differences in average input costs that resulted from uneven access to controlled domestic crude oil. Differences in refined product price ceilings, it was argued, introduced some heterogeneity to prices in the market and tended to prevent high-cost refiners from recouping their average costs. Low cost refiners, it was contended, were at a competitive advantage: A refiner unable to raise its ceiling to the level of the industry market-clearing price would sell at a lower price than its competitors, capture a greater market share, and determine the market price. This argument, in other words, held that the imposition of a binding price control on a single refiner in the market would force all other producers to lower their prices in an attempt to compete with the controlled refiner.[5] The opposite conclusion regarding the response of uncontrolled refiners, however, would be expected to hold. The imposition of a binding ceiling on an individual refiner in the presence of any nonzero supply elasticity would cause that refiner to reduce output, leave greater excess demand for the output of other refiners, and allow the market-clearing price faced by other refiners to rise rather than fall. Binding ceilings would be a disadvantage, not an advantage, to the affected refiner. It turns out, however, that refined product price ceilings have only been infrequently binding.

The refined product price controls under EPAA/EPCA regulation result in not only legal maximum prices that differ across refiners but also legal maximum prices that change over time. The availability of banked costs imparts substantial flexibility to refined product price ceilings. The question arises whether this flexibility has generally been sufficient to allow market forces, rather than federal regulation, to establish domestic refined product prices. This question has significant bearing on recent and current congressional debates concerning the decontrol or recontrol of selected refined products. Moreover, it is central to the attempt to analyze the effects of EPAA/EPCA regulation of the crude oil market on refiner price and output behavior. In particular, it is important to an assessment of the effects of refined product price controls on refiner marginal revenues.

Certain memorable instances in which shortages and queuing in refined product markets were widespread attest to the ability of EPAA/EPCA regulations to impose binding price controls on a broad basis. The period associated with the OAPEC embargo in the fall of 1973 and extending into the early months of 1974 and the period in early 1979 associated with the shutdown of Iranian production and relatively large increases in the price of oil on world markets were both accompanied by visible evidence of binding price controls.[6] Notwithstanding these two occurrences, however, the rarity of shortages, queuing, and seller closings suggests that refined product price controls have not been binding on a widespread basis over most of their tenure. Moreover, the presence of shortages at the retail level is not infallible evidence of binding controls on refiners. Systematic evidence on the impacts of refined product price controls on refiners can be acquired from data on refiner banked costs. The inferences that may be drawn from this data are investigated through examination of the relevant regulations.

Regulations governing the determination of refiners' price ceilings have been altered many times. The following representation captures all of the essentials of these regulations. In any month $t + 1$, the price of the ith refined product other than a "general refinery product" or an exempted product may be increased above the 15 May 1973 price P_i^0 by an amount D_i^{t+1}, where D_i^{t+1} is defined as

$$D_i^{t+1} = \frac{C^t}{V^{t+1}}(P_c^t - P_c^0) + \frac{S^t}{V_i^{t+1}}(P_s^t - P_s^0) + \frac{N^t}{V_i^{t+1}}(P_n^t - P_n^0) - \frac{G_i^t}{V_i^{t+1}} \pm \frac{R_i^t}{V_i^{t+1}} + \frac{B_i^t}{V_i^{t+1}}, \tag{2.1}$$

where C is refinery crude oil throughput; V, refinery output (or sales prior to March 1977) of all refined products; V_i, refinery output (sales) of the ith product; P_c, the weighted average price of crude oil inputs; S, refinery use of petroleum product inputs; P_s, the price of product inputs; N, refinery use of nonproduct inputs (for example, labor); P_n, the price of nonproduct inputs; G_i, any accumulated overcharge on the ith product; R_i, total banked costs reallocated to (+) or from (−) the ith product from or to other covered products; B_i, total available banked costs attributable to the ith product; and the superscript zero denotes the 15 May 1973 base period.[7] The R_i factor reflects the ability sellers have had to allocate the costs associated with some products to the banks of another product (notably gasoline).

The banked cost regulations do not appear to be designed to track changes in average costs exactly. There are several imperfections. First, D_i represents a sum of fractions of changes in the per unit costs of inputs used in the refining process, which is a classic joint-product process. Except in the unlikely case that there are no possibilities for substitution between inputs in response to input price changes or for varying the refinery slate of outputs, D_i will fail to reflect precisely changes in the true average cost of producing the ith product. Second, changes in the costs of certain inputs are excluded from D_i. The most notable exclusion is capital costs. Third, correspondence between D_i and changes in average production costs is confounded by the fact that changes in input prices that may be added to the base price of the product in the *current* month have occurred in the *previous* month. [Notice the pattern of time superscripts in (2.1).] Finally, when V could be defined as sales, the level of D_i varied with changes in input-sales ratios. If product inventories were rising (for example, due to seasonality), the ratios C/V, S/V_i, and N/V_i would be expected to increase. This effect would decrease D_i and tend to raise banks, although such an increase in banked costs would not reflect an increase in average production costs or an inability to cover such cost increases.

Notwithstanding any looseness in the connection between banked costs and economic costs, banks provide a cushion between a refiner's ceiling prices and actual selling prices. When market conditions dictate, selling prices may be raised until any accumulated banked costs are debanked (see appendix 2.A). Thus, to a first approximation, the presence of positive banked costs indicates that market forces, rather than federal legislation, are determining refined product prices at the refinery level. This conclusion must be qualified somewhat when refiners have the ability to let banks become negative. Negative banks can arise because of imperfect enforcement in any month and intentional or unexpected overcharges. As the G_i term in formula (2.1) indicates, any overcharges must be subtracted from selling prices in the subsequent month. Of course, there may be economically rational reasons for a refiner to engage in overcharging. The more swift and complete is the enforcement of the refined product price regulations, however, the less likely it is that refiners will accumulate negative banks and the more likely it is that controls will be binding.

In addition to the possibility of using negative banks, variability in ceiling prices is provided by the fact that changes that determine D_i are not beyond the control of individual refiners; that is, individual refiners are not price takers with respect to ceiling prices. For example, consider a refiner selling V_i units at a legal ceiling price $\bar{P}_i$ that is below the market-clearing price and is equal to a legally calculated average cost AC_i. This refiner can raise its ceiling by importing the ith product when imports can be had at a price of $P'_i > \bar{P}_i$. Upon the importation of one more unit of foreign product, the average cost of the refiner becomes

$$AC'_i = \frac{AC_i V_i + P'_i}{V_i + 1} = \frac{\bar{P}_i V_i + P'_i}{V_i + 1}. \tag{2.2}$$

The change in the price ceiling is $\Delta AC_i = AC'_i - AC_i$, or

$$\Delta AC_i = \frac{\bar{P}_i V_i + P'_i}{V_i + 1} - \bar{P}_i = \frac{P'_i - \bar{P}_i}{V_i + 1}. \tag{2.3}$$

Because such a change allows an equivalent increase in price ceilings on all units, the resulting marginal revenue MR_i is the sum of (a) the additional revenue on the original V_i units and (b) the $\bar{P}_i + \Delta AC_i$ received on the imported unit:

$$MR_i = V_i\, \Delta AC_i + (\bar{P}_i + \Delta AC_i) = (V_i + 1)\, \Delta AC_i + \bar{P}_i. \tag{2.4a}$$

Substituting from (2.3) for ΔAC_i indicates that

$$MR_i = P'_i, \tag{2.4b}$$

where P'_i is the marginal cost of the imported barrel. Hence the hypothetical refiner bound by controls would import the barrel of product.

This last conclusion holds so long as the calculated change in average cost is insufficient to cause the new ceiling price to rise above $\bar{P}_i$ by more than the difference between the domestic market-clearing price and $\bar{P}_i$; that is, (2.4a) assumes that the price realized by the refiner after importing rises by the full amount of ΔAC_i. This conclusion also holds regardless of whether P'_i exceeds or falls short of the domestic market-clearing price. This latter observation acquires relevance in chapter 4, where the effects of federal regulations on product imports are analyzed. The analysis here serves to indicate the type of responses a refiner might make upon finding itself without positive banked costs—refined product price controls may encourage refined product imports.

Table 2.1 presents the total banked costs of the thirty largest US refiners for 1974–1980. These data indicate that, on an industry-wide basis, banked costs have been substantially positive over most of the

Table 2.1 Total Banked Costs for the Thirty Largest Refiners: 1974–1980[a] (Millions of Dollars)

	1974	1975	1976	1977	1978	1979	1980
January	250	1,357	1,224	1,392	1,666	1,699	5,304
February	446	1,508	1,216	1,528	1,898	1,988	7,629
March	520	1,700	1,198	1,423	1,618	2,388	7,224
April	783	1,594	1,339	1,566	1,583	3,716	
May	963	1,433	1,503	1,542	1,535	4,093	
June	1,350	1,334	1,112	1,535	1,460	4,480	
July	1,327	1,075	1,100	1,470	1,298	5,079	
August	1,405	1,208	1,156	1,498	1,888	5,203	
September	1,490	1,342	1,093	1,470	1,200	5,527	
October	1,250	1,255	1,256	1,638	1,459	5,389	
November	1,315	1,497	1,332	1,660	1,351	5,794	
December	1,114	1,483	1,179	1,473	1,417	4,895	

Source: US Department of Energy, *Monthly Energy Review*.
a. Beginning with February 1977, data refer to only 29 largest refiners, reflecting the merger of Getty Oil Company and Skelly Oil Company.

period shown. To provide perspective, data on gasoline banked costs indicate that legal maximum gasoline prices have generally been more than 5¢ higher than market-clearing prices. The only noteworthy exception to this was in early 1974, when gasoline banks were less than 1.5¢ per gallon.[8] The rather small gasoline banks in this period correspond to the widespread appearance of gasoline lines, indicating that price controls were binding. Although industry gasoline banks were slightly positive in this period, individual refiners in many markets were undoubtedly compelled to price below market-clearing prices and apparently did not have positive banks. Although data on individual refiner banked costs are scarce, some information on middle distillate banks is available in the first five months of 1974; five of the ten largest refiners had negative banks in at least one month.[9]

One explanation why price controls were binding during late 1973 and early 1974 and generally not binding again until perhaps the first part of 1979 lies in the formula, (2.1), by which legal price adjustments are calculated. This formula allows input average cost changes to be added to the base price only after a one-month lag. In both periods of binding controls, crude oil input prices were rising very rapidly, and it is probably true that legal prices were not rising as fast as market-clearing prices. Indeed, there was a decidedly end-of-month pattern to the magnitudes of both the 1973–1974 and early 1979 shortages.[10]

Moreover, inspection of (2.1) indicates that only changes in the *average* price of crude may be included in the price adjustment factor. If the crude oil market were uncontrolled, all crude oil would be expected to sell at the same price. Hence, marginal and average crude prices would not differ for individual refiners. Under the crude oil price controls, however, domestically produced crude oil prices are effectively held below the prices of uncontrolled, imported crude oil. As discussed in section 2.3.1, in the absence of an entitlements program, imported crude oil prices are the relevant marginal crude input costs seen by domestic refiners; and the marginal crude price exceeds the average. When the marginal cost of crude oil is rising, it pulls up the average cost, although the average rises less rapidly than the marginal. Thus, because firms base their price and output decisions on marginal costs, average cost-based legal price restrictions are more likely to be binding in a period in which imported crude oil prices are rising as rapidly as they were in the winter of 1973–1974. A similar argument applies to the early 1979 period, although it is complicated by the entitlements program.[11]

The aggregate banked cost data for 1979 do not strongly suggest that price ceilings were binding on refiners in the first half of the year—despite memorable shortages at the retail level. As Verleger (1979) points out, numerous programs in addition to banked cost regulations contributed to the 1979 shortages. The buy/sell program, for example, required major refiners (which are more heavily concentrated on the East and West Coasts) to sell crude oil to independent refiners (which are more heavily concentrated in the nation's interior) at the major refiner's average purchase price, rather than the marginal opportunity cost as reflected in world transaction prices. As a result, the 1979 shortages were primarily confined to the East and West Coasts. Verleger concludes that the 1979 shortages might have been completely avoided by reallocating crude oil toward the coasts (and, by implication, aggregate banked costs were sufficient to allow market-clearing prices nationwide). This conclusion is reinforced by industry evidence from Jacobs (1980) and Erfle and Pound (1980) indicating that the "voluntary" price guidelines and jaw-boning of the Council on Wage and Price Stability were particularly successful in restraining major refiners' price increases in 1979.

There was not only a coastal pattern to the 1979 shortages. Federal allocation regulations biased refinery deliveries of products in favor of rural retailers; and shortages became primarily an urban phenomenon.

Urban, near-the-coast retailers ended up with regulation-induced supply constraints and, consequently, came up against price ceilings. The pattern of rising refiner banked costs in 1979 is consistent with a refining industry facing rapidly rising marginal input costs, but unable to raise output prices to downstream customers as a result of binding price controls at the retail level. Binding controls at the retail level make the retail sector's demand for refinery products more elastic (see appendix 2.B). By the results of standard incidence analysis, the incidence of any cost increase at the refining level may then be expected to fall more heavily on the refining sector.

With the winter of 1973–1974 and, possibly, early 1979 as exceptions, available evidence is consistent with the conclusion that refined product prices generally have not been controlled at the refinery level under EPAA and EPCA. With market competition determining refined product prices, all refiners face essentially equal marginal revenues for their output. As discussed above, all refiners also faced equal marginal crude oil costs under controls without entitlements. Thus no refiner was conferred a competitive advantage by multitier crude oil price controls without entitlements. The presence of positive banked costs and the absence of retail- or refinery-level shortages imply that this conclusion holds for the period immediately preceding the start of the entitlements program in November 1974. The official justification for this program, therefore, appears to have been invalid.

2.3.3 Effects on Refiners' Wealth

Crude oil price controls without entitlements did not create competitive disparities among domestic refiners. They did, however, create incentives for refiners to compete for access to lower-priced, controlled domestic crude oil. All refiners had essentially equal incremental costs and revenues; but fortunate refiners with access to controlled crude oil were able to refine barrels of low-priced controlled crude oil and sell the output at prices sufficiently high to cover the cost of their incremental products produced from uncontrolled crude oil. The difference between the buying and "reselling" prices of controlled crude oil was an economic rent conferred by controls on fortunate refiners.

From the refining industry's perspective, the imposition of EPAA crude oil price controls (without an entitlements program) can be portrayed as in figure 2.2A. The US refining industry is taken to be a price taker in the world market for uncontrolled crude oil. Moreover, in keeping with the discussion above, the domestic refined product mar-

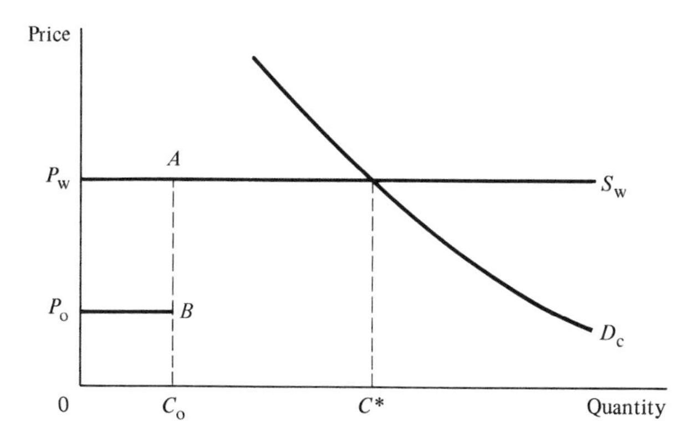

Figure 2.2A EPAA crude oil controls.

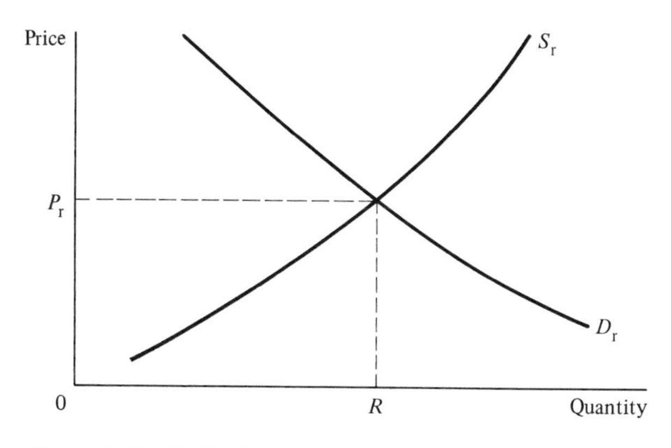

Figure 2.2B Refined product market.

ket is taken to be unconstrained by price controls. Finally, it is initially assumed that there are legal prohibitions on the importation and exportation of refined products, so that US refined product markets are isolated from world markets. This assumption is accurate in the case of exports, and the implications of the removal of the assumption of binding import controls are examined at length in chapter 4.

In figure 2.2A, the uncontrolled, world market price of crude is P_w. Prior to controls, the domestic refining industry, with a crude oil demand of D_c, faces a crude oil supply of S_w and purchases C^* units of crude. Corresponding to the input price of P_w is a supply schedule S_r of refined products and a refined product market equilibrium as shown in figure 2.2B. When old oil C_o is priced at P_o, refiners continue to purchase C^*. Rents of P_wABP_o accrue to those refiners who are given the rights (through the freezing of buyer-supplier relationships) to old oil.

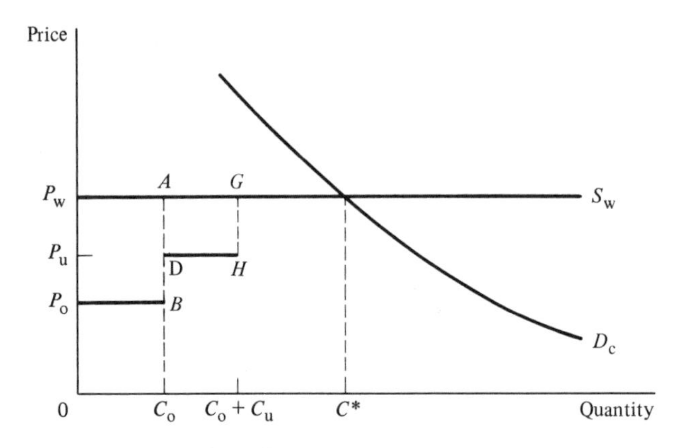

Figure 2.3 EPCA crude oil controls.

The cost of the marginal barrel of crude oil, however, remains unchanged. Consequently, the supply of refined products S_r remains unaltered in the range of the preregulation equilibrium; refined product prices and output are unchanged.

Although the entitlements program has been in effect throughout the period of EPCA regulation, it is instructive to isolate the EPCA crude oil price controls. This is done in figure 2.3. As above, the world price of crude oil faced by refiners is P_W. Prior to controls, the domestic refining industry purchases C^* units of crude oil. When lower-tier oil C_o is priced at P_o and upper-tier oil C_u is priced at P_u, refiners continue to purchase C^*. Rents of $P_wABP_o + AGHD$ accrue to refiners, but the marginal cost of crude oil is not altered—nor is the refined product market equilibrium.

The economic rents accruing to refiners, as shown in figures 2.2A and 2.3, are transfers of wealth away from domestic producers of controlled crude oil, since these producers would be able to sell their crude oil at uncontrolled prices in the absence of EPAA/EPCA regulation. An estimate of the size of these transfers of wealth can be found by calculating the difference between the expenditures refiners would have had to make in the absence of controls and their actual expenditures on controlled crude oil. On a per barrel basis, this difference is the refiner cost of uncontrolled crude oil less the average cost of crude oil under controls. Estimates of the total rents associated with controlled crude oil are shown in Table 2.2. These wealth transfers averaged approximately $21.1 billion (1980 dollars) over 1974–1979. Sharp increases in uncontrolled crude oil prices since early 1979 have

Table 2.2 Inframarginal Rent on Controlled Crude Oil: 1974–1980[a] (Billions of 1980 Dollars)

	Rent
1974	23.8
1975	23.1
1976	17.9
1977	17.5
1978	13.6
1979	30.7
1980 (Jan.–Mar.)[b]	45.0

Source: US Department of Energy, *Monthly Energy Review.*
a. Based on the difference between the refiner acquisition cost of uncontrolled crude oil and the average cost of crude oil under price controls. Deflation based on annual GNP Deflator as partially estimated by Data Resources, Inc., as of September 1980.
b. Annual rate.

been pushing rents upward; and as of the first quarter of 1980 they were running at an annual rate of over $45 billion. The annual figures amount to approximately 0.5–2% of GNP per year. They represent a measure of the prospective transfers from the users of crude oil to the producers of crude oil that would otherwise accompany increasing world oil prices, but that have been blocked by domestic crude oil price controls. It is easy to suspect that current federal regulations are the outcome of political competition between various interested parties seeking to capture these massive amounts of wealth. This proposition is examined in detail in chapter 6.

As table 2.2 indicates, the negative effects of recent dramatic increases in the world price of crude oil on the wealth of refiners can be offset by placing price controls on domestic producers. In the period during and immediately following the OAPEC embargo of late 1973, however, average cost-based refined product price controls were memorably binding; and at least some portion of the rents associated with access to controlled domestic crude oil was forced through to consumers. Indeed, the official justification of the refined product price controls rested heavily on their ability to prevent refiners from retaining all of these rents.[12] Notwithstanding stated objectives, the flexibility of the refined product price controls eventually allowed ceilings to rise above market-clearing levels. By the late spring of 1974, reports of product shortages had become infrequent and controls had ceased to be binding except in isolated cases. Consequently, the refining sector became the recipient of the rents shown in table 2.2—albeit with refiners sharing unequally in accord with unequal access to controlled crude oil. This boon to the refining sector, however, was short-lived.

2.4 Refiner Response to the Entitlements Program

Not surprisingly, the prospect of capturing billions of dollars of rents resulted in conflicts in the political arena over the rights to controlled oil. Both refiners without access to controlled crude oil and, by mid-1974, the consumers of refined products were effectively excluded from sharing in these rents under the original Phase IV-EPAA regulations that assigned rights in accord with precontrol buyer-supplier relationships. In November 1974, however, the entitlements program altered the property rights to controlled domestic crude oil. Through the entitlements program, the federal government has granted rights to controlled crude oil as an equal proportion of each refiner's total crude oil inputs. This redefinition of property rights has done more than grant domestic refiners equal access to controls-created rents. It has altered the marginal production decisions of refiners in ways that, at least potentially, have significant impact on refined product prices, outputs, producers, and consumers. In so doing, it has become a major element in US energy policy.

2.4.1 Rents, Taxes, and Subsidies

The economic effects of the entitlements program are conveniently illustrated by examination of a hypothetical program that might be adopted by federal regulators. Consider figure 2.4A. It is plausible to imagine that in addition to imposing a ceiling price of P_0 on domestic crude oil output up to C_0 under EPAA, the government decides that, because of equity or other considerations, refiners should not earn rents on inframarginal barrels of controlled crude oil. A policy is imposed that taxes away these rents and the Treasury captures P_WABP_0. Refiner output, crude input, and refined product prices are unchanged since marginal costs have not been altered in the refining industry. Policymakers, with political incentives to keep product prices low, however, might decide to spend their new revenues by providing subsidies to refiners who buy crude oil at uncontrolled prices. Under such a policy, refiners would face a marginal crude oil cost lower than P_W. In

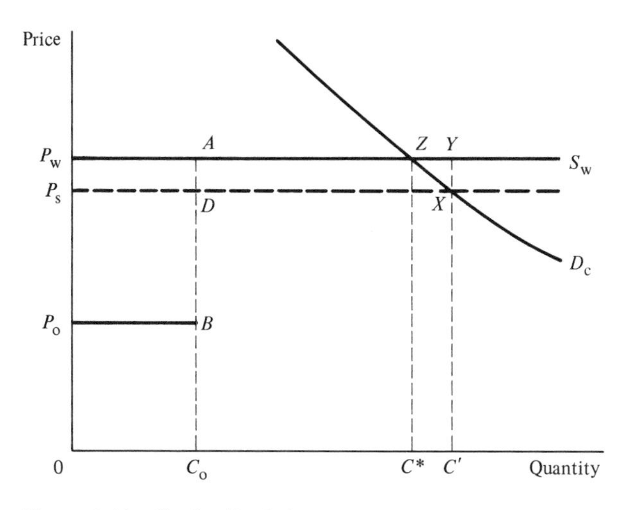

Figure 2.4A Crude oil subsidy.

the absence of collusion, competition among refiners would result in an expansion of their crude purchases along D_c in accord with the subsidy-inclusive marginal crude cost.

An equilibrium under the subsidy plan occurs at a subsidized price of P_s and crude oil consumption of C'. P_s is the marginal and average crude oil cost for domestic refiners. If the Treasury iteratively taxes away rents and distributes subsidies until it finds no rents remaining in the refining sector, the area of total rents P_wABP_o is equal to the sum of the subsidy across all C' units of crude oil refined. Thus $P_wABP_o = (P_w - P_s)C' = P_wYXP_s$. With a marginal value of crude oil in the United States of P_s, rents directly taxed away from refiners are P_sDBP_o. These rents are equal to $AYXD$ and refiners are able to cover the total cost of their crude oil inputs. In response to the marginal subsidy, refiners find marginal production costs reduced and expand their output. This expansion of the refining industry is represented by an outward shift of the supply of refined products to S_r' in figure 2.4B. Deregulation from this system of controls, taxes, and subsidies would tend to raise the price of refined products (back to P_r). A precisely analogous analysis would apply to the imposition of a rent tax and subsidy scheme under EPCA.

There may be some temptation to view the horizontal line at P_s in figure 2.4A as the supply curve of crude oil faced by the refining industry under the rent tax and subsidy scheme. This temptation should be avoided. The position of the demand for crude determines the size of

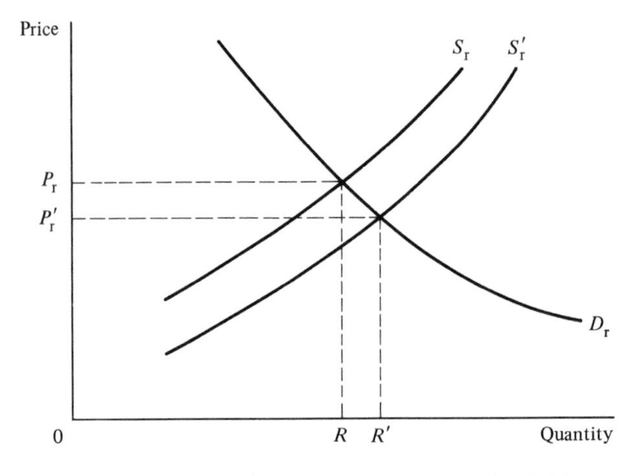

Figure 2.4B Refined product market with crude oil subsidy.

the crude oil subsidy. As demand for crude declines, for example, there are fewer units of uncontrolled crude over which to distribute tax-collected rents. Thus, the per unit subsidy increases. The converse is true for increases in demand. The price relevant to refiners' crude input decisions, P_s, is always less than the price P_w that would result under complete decontrol. P_s, however, would asymptotically approach P_w as D_c is shifted outward (that is, finite rents being distributed over ever increasing levels of uncontrolled crude use). The supply schedules of crude faced by the refining industry can be drawn as S_s in figures 2.5A and 2.5B, where S_s traces out the tax- and subsidy-exclusive industry weighted average price of crude at each level of crude use. Per unit subsidies, that is, the difference between S_w and S_s, decrease as the use of crude increases.

2.4.2 The Entitlements Program under EPAA

The entitlements program under EPAA regulation was essentially the same as this hypothetical rent tax with marginal subsidies program. Under the hypothetical program, the wealth transferred away from domestic crude oil producers is distributed among those who sell inputs to the refining sector at uncontrolled prices and domestic oil consumers. In the case of the actual system of price controls, the prospect of windfall wealth transfers induces political competition for its acquisition among producers, refiners, and consumers. The entitlements program is an outcome of this process of competition and is the

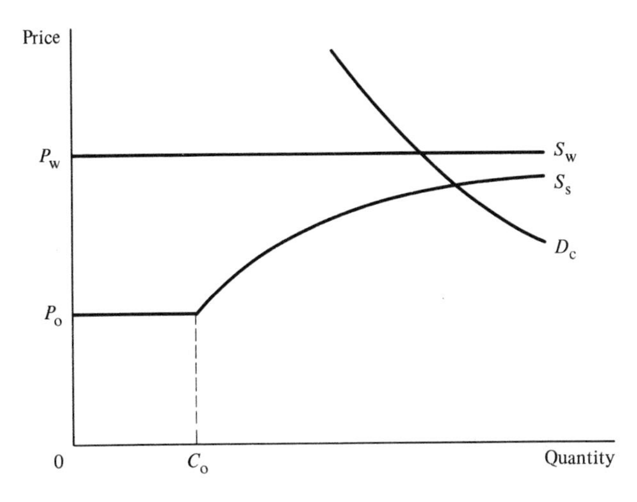

Figure 2.5A Crude oil subsidy under EPAA.

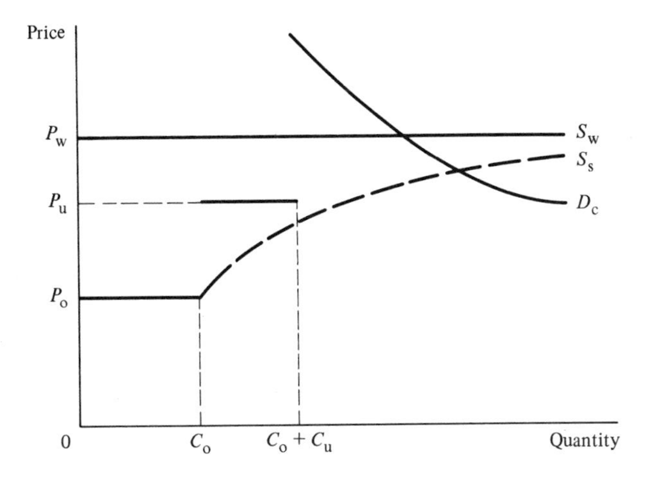

Figure 2.5B Crude oil subsidy under EPCA.

mechanism by which the conflict over eventual ownership of the rents that arise under crude oil price controls is resolved.

As discussed in chapter 1, the entitlements program is designed to equalize the average costs of crude oil across all refiners. The equalization of refiners' average crude oil costs is accomplished through cross payments from low-cost refiners, with access to relatively large amounts of controlled crude oil, to high-cost refiners, with greater dependence on uncontrolled crude. The weighted average cost of crude for the ith refiner is equal to the national average W when a cross payment or receipt E^i is transferred so that

$$\frac{P_o C_o^i + P_W C_W^i + E^i}{C_o^i + C_W^i} = \frac{P_o C_o + P_W C_W}{C_o + C_W} \equiv W, \tag{2.5}$$

where C_o^i and C_W^i are the ith refiner's use of old and uncontrolled crude, and $C_o = \Sigma_i C_o^i$ and $C_W = \Sigma_i C_W^i$. By solving for E^i, letting the national ratio of old to total crude consumption be A, and letting the ith firm's ratio be a^i, (2.5) can be rewritten as

$$E^i = (P_W - P_o)(a^i - A)(C_o^i + C_W^i) \gtreqless 0. \tag{2.6}$$

Expression (2.6) was, in all essentials, the formula used by FEA to calculate entitlements payments under the two-tier pricing regulations. The price of an entitlement was the difference between the market-clearing price and the controlled price, $P_W - P_o$. The number of entitlements that had to be purchased or sold was $(a^i - A)(C_o^i + C_W^i)$ according as $a^i - A$ was greater or less than zero, that is, according as the individual refiner used old oil more or less intensively than the industry as a whole.

The individual refiner that seeks to maximize its profits continues to buy crude oil until the addition to its revenues from the refining of the incremental barrel is just equal to the consequent addition to its costs. In an unregulated environment in which the individual refiner is a price taker, this condition occurs when the market price of crude P_W is equated to the value of the marginal physical product of crude (that is, the refined product price times the additional quantity of product yielded by the incremental barrel of crude oil). Under the EPAA controls and entitlements program, however, the ith refiner's profits π^i are

$$\pi^i = P_r R^i - P_o C_o^i - P_W C_W^i - (P_W - P_o)(a^i - A)(C_o^i + C_W^i), \tag{2.7}$$

where the last term represents the entitlements payment or receipt, P_r

is the price of refined product output, $R^i = f^i(C^i_o + C^i_w)$ is refined product output and is a function f^i of crude oil inputs, other symbols are as before, and nonpetroleum inputs are omitted for simplicity.

After substituting

$$a^i = \frac{C^i_o}{C^i_o + C^i_w} \quad \text{and} \quad R^i = f^i(C^i_o + C^i_w)$$

and expanding terms, the first-order conditions for a profit-maximizing input of crude can be found by differentiating (2.7) with respect to C^i_w. This yields as conditions

$$\frac{\partial \pi^i}{\partial C^i_w} = P_r\left(\frac{\partial f^i}{\partial C^i_w}\right) - P_w + A(P_w - P_o) = 0, \tag{2.8a}$$

or

$$P_r\left(\frac{\partial f^i}{\partial C^i_w}\right) = P_w - A(P_w - P_o). \tag{2.8b}$$

In the absence of the entitlements program, the counterpart to (2.8b) is simply

$$P_r\left(\frac{\partial f^i}{\partial C^i_w}\right) = P_w, \tag{2.9}$$

which is the usual condition for profit-maximizing use of an input. That is, in (2.9) the value of the marginal product is equated to the market price of the incremental unit of the input. Since $A > 0$ and $P_w > P_o$, however, (2.8b) indicates that the value of the marginal product would be equated to something less than the price of uncontrolled crude—less by an amount equal to an implicit per unit entitlements subsidy.[13]

EPAA crude oil price controls without entitlements created unsatisfied demands for old oil at the ceiling price. With the addition of the entitlements program, however, a penalty in the form of the implicit tax of entitlements purchase obligations became associated with the use of controlled oil. In fact, the entitlements program raised the marginal cost of controlled domestic oil, as perceived by refiners, to the weighted average cost of all domestically refined oil. Thus the marginal cost of old oil was the same as the subsidy-inclusive marginal cost of uncontrolled oil. This conclusion is derived by finding the conditions for a profit-maximizing use of old oil. The maximization of (2.7) with respect to C^i_o requires that

$$P_r\left(\frac{\partial f^i}{\partial C_o^i}\right) = P_o + (P_w - P_o) - A(P_w - P_o)$$

$$= P_w - A(P_w - P_o). \qquad (2.10)$$

In other words, the marginal cost of old oil was the old oil price plus an implicit tax that raised the effective cost to the world price less an entitlements subsidy of $A(P_w - P_o)$. The net effect was to leave the marginal cost of controlled crude oil equal to the subsidized price of uncontrolled oil [that is, the right-hand sides of (2.8b) and (2.10) are equivalent].

Under the rent tax with subsidy scheme invented above, refiners saw the national weighted average price of crude as their relevant marginal crude cost. Moreover, all rents associated with access to controlled crude oil were redistributed as subsidies to refiners' use of controlled crude oil. Because of the guarantee provided by (2.5) and (2.6) that all refiners would have the same crude oil costs, both of these results also applied to the EPAA controls and entitlements system (see appendix 2.C). This conclusion is instructive because the redistribution of rents as subsidies under the rent tax with subsidy program produced the schedule S_s in figure 2.5A as the locus of weighted average crude costs. As under the hypothetical program, S_s was the supply schedule of crude faced by the domestic refining industry under EPAA controls and entitlements.

In figure 2.4A, P_s can now be interpreted as the entitlements-equalized price (and the average cost) of crude oil perceived by refiners under EPAA. Refiners purchase C' units of crude oil in response to an effective subsidy of $P_w - P_s$ per unit. With the incremental source of uncontrolled crude oil being the international market, $C' - C^*$ consists entirely of imported crude oil. Total crude oil costs are $P_oBC_o0 + AYC'C_o$ and are equal to total revenues $P_sXC'0$ remaining after the compensation of all other factors used in refining. The lower perceived price of crude inputs P_s is translated into a shift in the supply schedule of refined products to S_r' in figure 2.4B and a refined products output of R' at a price of P_r'. In short, the EPAA entitlements program is predicted to have caused a decline in refined product prices, an expansion of the domestic refining industry, and an increase in crude oil imports.

2.4.3 The Entitlements Program under EPCA

The implementation of EPCA's three-tier system of crude oil pricing in February 1976 was accompanied by changes in the entitlements program. First, because EPCA holds upper-tier oil prices below market levels, entitlements had to be created for upper-tier oil in order to equalize average refiner crude oil costs. Second, ostensibly to maintain the \$0.21 per barrel preference for domestic crude oil embodied in the import fee system, refiners have been permitted to retain \$0.21 of the rent associated with each barrel of controlled domestic oil. Notwithstanding these changes, the effects of the entitlements program have remained the (approximate) equalization of refiner crude oil costs and the subsidization of refiner crude oil purchases.

Under EPCA, the weighted average cost of crude oil for the ith refiner is equal to the national average when a cross payment or receipt E^i is transferred so that

$$\frac{P_o C_o^i + P_u C_u^i + P_w C_w^i + E^i}{C_o^i + C_u^i + C_w^i} = \frac{P_o C_o + P_u C_u + P_w C_w}{C_o + C_u + C_w}, \tag{2.11}$$

where C_o^i, C_u^i, and C_w^i are the ith refiner's use of lower-tier, upper-tier, and uncontrolled crude oil, and $C_o = \Sigma_i C_o^i$, $C_u = \Sigma_i C_u^i$, and $C_w = \Sigma_i C_w^i$. The price of an entitlement to lower-tier oil is set equal to the difference between the price of uncontrolled (that is, imported) crude and the lower-tier price less \$0.21: $P_w - P_o - 0.21$. Since the upper-tier price is below the world price, entitlements to upper-tier oil must also be exchanged in order to equalize refiners' average crude costs. To accomplish this, each barrel of upper-tier oil is treated as a fraction of a lower-tier barrel. This fraction is $(P_w - P_u - 0.21)/(P_w - P_o - 0.21)$. By expressing this fraction as B, the national ratio A of lower-tier crude to total crude and the ith refiner's ratio a^i are

$$A = \frac{C_o + BC_u}{C_o + C_u + C_w} \quad \text{and} \quad a^i = \frac{C_o^i + BC_u^i}{C_o^i + C_u^i + C_w^i}. \tag{2.12}$$

Each month the ith refiner is issued $A(C_o^i + C_u^i + C_w^i)$ entitlements and its actual use of controlled crude is calculated to be $a^i(C_o^i + C_u^i + C_w^i)$. The number N^i of entitlements that an individual refiner must purchase or sell according as $a^i - A^i$ is greater or less than zero is then

$$N^i = (a^i - A)(C_o^i + C_u^i + C_w^i) \gtreqless 0. \tag{2.13}$$

The total monthly entitlements payment or receipt E^i is then

$$E^i = (P_w - P_o - 0.21)(a^i - A)(C_o^i + C_u^i + C_w^i) \gtreqless 0. \tag{2.14}$$

The effects of the entitlements program under EPCA can be investigated by examining the behavior of the profit-maximizing refiner.

Under the EPCA controls and entitlements program, the ith refiner's profits π^i are

$$\pi^i = P_r f^i(C_o^i + C_u^i + C_w^i) - P_o C_o^i - P_u C_u^i - P_w C_w^i$$
$$- (P_w - P_o - 0.21)(a^i - A)(C_o^i + C_u^i + C_w^i), \qquad (2.15)$$

where symbols are as before. After substituting for a^i as above and expanding terms in (2.15), the first-order conditions for a profit-maximizing input of uncontrolled crude can be expressed as

$$P_r\left(\frac{\partial f^i}{\partial C_w^i}\right) = P_w - A(P_w - P_o - 0.21). \qquad (2.16)$$

Since $A > 0$ and $P_w > (P_o + 0.21)$, (2.16) indicates that EPCA regulation causes the value of the marginal product of uncontrolled crude to be equated to an amount that is less than its price—less by the amount of the entitlements subsidy.

As was the case under EPAA, controlled domestic crude oil used by refiners under EPCA carries an implicit tax that raises the marginal cost of such oil toward the cost of uncontrolled oil. Unlike the EPAA case, however, this implicit tax under EPCA is not sufficient to raise the refiner cost of controlled oil to equality with the subsidy-inclusive cost of uncontrolled oil. Differentiating (2.15) with respect to lower- and upper-tier oil yields as first-order conditions

$$P_r\left(\frac{\partial f^i}{\partial C_o^i}\right) = P_o + (P_w - P_o - 0.21) - A(P_w - P_o - 0.21)$$
$$= P_w - A(P_w - P_o - 0.21) - 0.21 \qquad (2.17a)$$

and

$$P_r\left(\frac{\partial f^i}{\partial C_u^i}\right) = P_u + (P_w - P_u - 0.21) - A(P_w - P_o - 0.21)$$
$$= P_w - A(P_w - P_o - 0.21) - 0.21. \qquad (2.17b)$$

In both cases, the middle term indicates that an implicit tax is placed on controlled oil that causes its presubsidy cost to rise to within \$0.21 of the price of uncontrolled oil; that is, $P_u + (P_w - P_u - 0.21) = P_w - 0.21$. As a result, when the subsidy of $A(P_w - P_o - 0.21)$ is added, the cost of controlled oil remains \$0.21 below the subsidy-inclusive cost of uncontrolled crude oil. This \$0.21 is realized by refiners as rent on inframarginal controlled crude oil since price controls prevent refiners from bidding lower- and upper-tier prices up by \$0.21. The \$0.21 allowance does not provide a preference for domestic crude oil producers. It provides a preference for domestic refiners with access to controlled oil.

The \$0.21 per barrel allowance for refiners with access to controlled crude oil causes the EPCA entitlements program to fail to lower the cost of uncontrolled crude oil to the national weighted average cost. Reducing the right-hand side of (2.16) yields

$$P_W - A(P_W - P_o - 0.21) = W + \frac{0.21(C_o + C_u)}{(C_o + C_u + C_W)}. \tag{2.18}$$

The subsidy-inclusive cost of uncontrolled crude oil is greater than the national average cost of domestically refined crude oil by an amount equal to the \$0.21 allowance weighted by the ratio of controlled to total crude oil use.

The total rents associated with controlled crude oil are not spent on the entitlements program under EPCA. Writing the right-hand side of (2.18) as $W + \Delta$, total rents r and the total subsidy s are $r = (P_W - P_o)C_o + (P_W - P_u)C_u$ and $s = (P_W - W - \Delta)(C_o + C_u + C_W)$. Thus

$$\begin{aligned} r - s &= -P_oC_o - P_uC_u - P_WC_W + W(C_o + C_u + C_W) \\ &\quad + \Delta(C_o + C_u + C_W) \\ &= 0.21(C_o + C_u). \end{aligned} \tag{2.19}$$

Refiners are able to hold on to these rents despite the entitlements program. The seemingly insignificant \$0.21 refiner allowance was worth \$700 million (1980 dollars) in 1978.[14]

The spreading of the rents associated with controlled crude oil across input purchases by the domestic refining industry traces out a supply curve described by S_s in figure 2.5B, where the difference between the world price of crude oil and S_s is the EPCA entitlements subsidy of $A(P_W - P_o - 0.21)$. At a price of P_s in figure 2.4A, C' units of crude oil are refined and the supply of refined products expands as in figure 2.4B. EPCA regulations thus have allocative and distributional consequences that are in the same direction as the consequences of EPAA regulations.

In the absence of any special adjustments, such as the \$0.21 allowance, the entitlements subsidy under both EPAA and EPCA would equal the difference between the national weighted average cost of crude oil and the price of uncontrolled crude oil. The actual subsidies that result under the entitlements program, however, are subject to the effects of numerous special programs, exceptions, and exemptions that are "financed" out of the rents associated with controlled crude oil. These modifications to the basic entitlements program are examined below. Both the actual subsidy and the difference between the marginal cost of uncontrolled crude oil and the national weighted average cost of crude oil are shown in table 2.3. The latter is the maximum possible entitlements subsidy.[15] At the indicated levels of actual subsidy, the entitlements program has typically been paying for approximately one tenth to one fifth of each barrel of uncontrolled crude oil.

2.5 Special Programs within the Entitlements Program

The characterization of the entitlements program in section 2.4 is somewhat of a simplification insofar as it overlooks several special programs financed by the entitlements program. Smaller refiners, for example, receive a special subsidy in the form of extra allocations of salable entitlements. Some importers of certain refined products also have been the recipients of special allocations of salable entitlements, while numerous exceptions and appeals to entitlement purchase obligations are granted on a monthly basis to fortunate refiners. Since 1978, special entitlement subsidies have also been granted for the refining of low-quality California crude oil. Finally, since 1977, the federal government has allocated itself entitlements in order to lower the average and marginal costs of its Strategic Petroleum Reserve Plan, under which crude oil is stored as insurance against import supply disruptions.

Table 2.3 Maximum Possible and Actual Entitlements Subsidies: 1974–1980 (Dollars Per Barrel)

	Maximum possible subsidy[a]	Actual subsidy
1974 (Nov.–Dec.)	3.33	2.03
1975	3.55	2.85
1976	2.59	2.51
1977	2.57	2.27
1978	2.11	1.60
1979	5.58	3.08
1980 (Jan.–Mar.)	10.88	5.16

Source: US Department of Energy, *Monthly Energy Review*.

a. Based on difference between the price of uncontrolled crude oil and the national average price of crude input to refineries, adjusted for refiner acquisition costs.

2.5.1 The Small Refiner Bias

Under a program known as the Small Refiner Bias, certain domestic refiners receive allocations of entitlements in addition to the allotment determined by (2.6) or (2.14). Extra entitlements are based on a sliding scale that grants a decreasing number of entitlements per barrel of crude oil throughput as throughput increases up to 175 mb/d. As of 1977, the Small Refiner Bias applied to approximately 18% of US refining capacity and to 126 of the 148 domestic refining firms.[16] To a first approximation, allocations of entitlements to these affected firms under the Small Refiner Bias represent pure transfers. While these refiners' incremental crude oil use is subsidized by the general entitlements program, the Small Refiner Bias assigns each level of crude input extra entitlements that may be sold at the full entitlements price (that is, the difference, less $0.21, between the prices of imported and lower-tier crude oil). Table 2.4 shows the total subsidy made to various size refiners under the Small Refiner Bias in 1978. As of 1978, the subsidy increased from 0 to 30 mb/d crude oil throughput. It peaked at a throughput of 30 mb/d and an annual value of approximately $9.4 million ($11.2 million in 1980 dollars).[17]

Some aggregate estimates of the benefit of the Small Refiner Bias to the class of smaller refiners are derived in chapter 5; and examination of the role of this group in the political economy of current policy is undertaken in chapter 6. Here it is appropriate to note the incentive system embodied in the Small Refiner Bias. As shown in table 2.4, this bias has kept marginal crude oil costs of refiners with less than 30 mb/d crude oil throughput below the average costs of all larger refiners. Above 30 mb/d, however, small refiners have faced slightly higher mar-

Table 2.4 Benefits under Small Refiner Bias[a]

Refiner throughput (mb/d)	Marginal crude cost ($ per barrel)	Average crude cost ($ per barrel)	Total daily value of bias (thousands of dollars)	Total annual value of bias (millions of dollars)
0 → 10	11.08	10.61	0 → 18.9	0 → 6.9
10+ → 30	12.61	11.42	18.9 → 25.8	6.9 → 9.4
30+ → 50	13.39	12.01	25.8 → 17.2	9.4 → 6.3
50+ → 100	13.10	12.38	17.2 → 10.4	6.3 → 3.8
100+ → 175	13.10	12.48	10.4 → 0	3.8 → 0

Source: Calculated from data in US Department of Energy, *Monthly Energy Review*.
a. Using 1978 data, with an entitlements price of $8.265 and a marginal crude oil cost for large refiners of $12.96.

ginal costs for acquiring crude oil than those faced by large (that is, 175+ mb/d) refiners since incremental inputs of crude oil have carried the penalty of a reduction in the small refiner subsidy. Consequently, there has been an incentive, other things being equal, to move the scale of operations toward 30 mb/d. At the same time, the small refiner subsidy has been a lure to entry by refiners otherwise unable to cover their costs. In the first two years of the entitlements program, fourteen small refining firms entered the industry and three firms became large refiners. Of the fourteen new small refiners, eleven were in the 0–30-mb/d range. By comparison, over the five years preceding the entitlements program, three firms became large refiners, while the total number of small refiners declined by three. Within the 0–30-mb/d class, the number of firms declined by six. The 30–175-mb/d class showed an increase of three firms.[18]

Small refineries typically are most efficient at producing limited slates and specialty products. In general, the Small Refiner Bias encourages capacity expansion in units that are smaller than optimal, that is, that fail to take full advantage of the economies of scale available in refining. Evidence from Scherer et al. (1975) indicates that the minimum size refinery necessary to take full advantage of present production economies of scale is one designed to process 200 mb/d. Managerial and organizational economies of scale reinforce this tendency toward large-scale operations and are apparently not exhausted by the single-plant firm. Moreover, encouragement of expansion through operations by small firms that employ relatively unsophisticated and inflexible technologies and produce relatively large yields of heavy refined products (such as residual fuel) runs counter to the expansion implied by a continuing secular trend in the composition of demand favoring light products (such as gasoline). Thus the product mix, as well as the mix of firm and plant sizes, is distorted by the special entitlements subsidy to small refiners. Significantly, no new large grass-roots refinery (that is, a completely new plant as opposed to an expansion of an existing operation) has been started under the entitlements program.[19]

Among the objectives expressed as official justification for the Small Refiner Bias have been "economic efficiency" and the "minimization of economic distortion, unflexibility, and unnecessary interference with market mechanisms."[20] The foregoing discussion suggests that the Small Refiner Bias does not satisfy these goals. Other objectives have included "preservation of an economically sound and competitive petroleum industry; including the [preservation of] the competitive viability of . . . small refiners'' and an ''equitable distribution of crude oil . . . among all . . . sectors of the petroleum industry.''[21] At least some of these objectives appear to be served by the Small Refiner Bias. In light of the state of technology in refining and trends in the composition of demand, the small refiner entitlements subsidy may indeed preserve the ''competitive viability'' of some small refiners who might otherwise be unable to cover their total costs. Moreover, this preservation is accomplished through a distribution of crude oil among refiners that might be judged equitable under some standards of fairness (for example, a standard that attached value to smallness per se). Indeed, the Small Refiner Bias can be viewed as a mechanism for enriching smallness through special assignments of property rights to controlled crude oil. The extra allocations of entitlements allow small refiners to capture the rent associated with access to the controlled crude represented by such entitlements. For many small refiners, this rent amounts to millions of dollars per year.

2.5.2 Refined Product Entitlements

Another special entitlements program financed out of the rents associated with controlled crude oil has been the subsidization of certain refined product imports.[22] During the first three months of the entitlements program, residual fuel and middle distillate imports were granted 30% of the entitlements subsidy given to imported crude oil. This was accomplished by treating each barrel of imported residual fuel and middle distillate as equivalent to 30% of a barrel of uncontrolled crude oil when calculating entitlements allocations and purchase requirements. Consequently, imports of these products reduced required entitlements payments for entitlements-buying refiners and increased entitlements receipts for entitlements-selling refiners—with a resulting implicit subsidy of the same character as the crude oil subsidy.

From February 1975 through January 1976, product subsidies were dropped. The severe winter of 1976–1977, however, prompted an entitlements subsidy of $2.10 per barrel for heating oil imports in March and February 1977. Since February 1976, residual fuel imports into the East Coast have again been eligible for entitlements, receiving a subsidy equal to 30% of the crude oil subsidy until July 1978, at which time the subsidy fraction was raised to 50%. From May through September 1979, middle distillate imports were granted a subsidy of $5.00 per barrel in order to hold down diesel fuel price increases and to encourage the building up of heating oil inventories prior to the winter heating season. Finally, as an outgrowth of the Western Hemisphere preference that had existed under the oil import control program, Puerto Rican imports of naphtha for use as a petrochemical feedstock have been eligible for approximately the full entitlements subsidy to crude oil since May 1976.

Refined product entitlements directly subsidize product imports. The expected results are a decline in domestic product prices and an increase in imports. A price decline is felt as an increase in real income by domestic product consumers and as a decrease in the producer surplus generated in the domestic refining industry. In keeping with the results of standard incidence analysis, domestic product prices fall by the full amount of the subsidy when the subsidized imported product is supplied perfectly elastically. In the more general case in which the United States is not a price taker in international product markets, the incidence of product entitlements depends in a determinant way on elasticities of supply and demand. Examinations of this incidence and its allocative and distributional effects are undertaken in chapters 4 and 5.

2.5.3 Exceptions and Appeals

Most of the annual "budget" of the entitlements system (table 2.2) is dispensed through programs applicable to either the entire refining industry or major subgroups of refiners. The one program that provides the Department of Energy with the greatest degree of discretionary control over the rents associated with controlled oil is the program of Exceptions and Appeals Relief. This program allocates grants to individual refiners who satisfactorily demonstrate unusual hardship or a threat of bankruptcy. Grants are made in the form of additional allocations of salable entitlements or exemptions from entitlement purchase obligations. The value of an Exceptions and Appeals grant is the product of the entitlements price and the number of extra entitlements received or entitlement purchase obligations avoided.

Exceptions and Appeals Relief began as a relatively minor component of the entitlements system, but has become a substantial outlet for the funds under regulatory control. In the first twelve months of the entitlements program, for example, only about $85 million (1980 dollars) were distributed through Exceptions and Appeals. By the late 1970s, several hundred million dollars per year were being distributed through Exceptions and Appeals.[23]

The Exceptions and Appeals Relief system represents a unique case in which administrative regulation has created a system of ongoing discretionary subsidies for an industry of approximately 150 firms. Competition and the expenditure of resources (that is, lobbying) for acquisition of these subsidies can be expected to be substantial since the discretionary nature of the Exceptions and Appeals process is not subject to free rider problems. A successful applicant for Exceptions and Appeals grants is the sole beneficiary, whereas an individual refiner promoting, say, an expansion of the Small Refiner Bias must share any increase in the bias with an entire class of refiners. Consequently, despite their apparent lump-sum character, Exceptions and Appeals grants are likely to have allocative impact through, for example, the moral hazard effects of tying receipts of relief to "hardship" and through the consumption of legal and managerial services.

2.5.4 Entitlements for Refining California Crude Oil

As discussed in section 2.4, the entitlements program places an implicit tax on controlled domestic crude oil. This tax raises the effective refiner cost of such crude oil to approximately the weighted cost of all crude oil used in the United States. This implicit entitlements tax is based on the refiner acquisition cost of the *average* barrel of controlled oil. It does not vary with the deviations from this average that quality variations impart to the structure of crude oil prices. As a result, both low- and high-quality controlled crude oils bear the same tax.

For a relatively low-quality lower-tier crude oil, with a below-average price frozen by controls, the implicit entitlements tax raises the refiner cost by a larger percentage than it does for a barrel of relatively higher-quality and higher-priced lower-tier crude oil. Thus, the entitlements tax raises the refiner cost of low-quality lower-tier crude oil relative to the cost of high-quality lower-tier crude oil. The tax reduces the effective demand for low-quality lower-tier oil and places downward pressure on its price—despite the existence of a ceiling price that is not even half of the price of uncontrolled crude oil.

The tendency for the entitlements program to depress the prices of low-quality lower-tier (and previously old) oils below their ceiling prices has been of particular interest to California crude oil producers. California has an above-average proportion of its crude oil production classified as lower tier. In September 1976, for example, 69% of California crude oil was lower tier, compared to 53% nationally.[24] Moreover, California crude oil is considerably lower in gravity and higher in sulfur content than the national average crude oil.[25] Thus, California crude oil is both abnormally low in quality and abnormally confined to the lower tier.

By at least 1976, California crude oil producers were complaining of their inability to sell low-quality lower-tier oil at ceiling prices. In October 1976, the Federal Energy Administration increased the ceiling prices on low-quality lower-tier oil by $0.03 per degree of gravity below 34°. This change, however, left the nationally determined implicit entitlements tax unaffected, and many California prices did not rise all the way to the new ceilings. Finally, to remedy this perceived anomaly, DOE began issuing special entitlements in January 1978 for every barrel of low-quality California crude oil purchased by refiners. These entitlements are worth $1.74 per barrel. They effectively subsidize demand and have succeeded in moving prices toward ceilings. In 1978, approximately $60 million (1980 dollars) was devoted to special California entitlements.[26]

2.5.5 Strategic Petroleum Reserve

The Strategic Petroleum Reserve (SPR) Plan is a program under which the federal government stores crude oil in underground reservoirs. The SPR began building its stocks in late 1977 and reached approximately 66 million barrels by the end of 1978. To partially finance purchases of this crude oil, the federal government treats itself as if it were a domestic refining firm. In particular, SPR crude oil purchases are eligible for the basic entitlements subsidy shown in table 2.3. Thus part of the "tax" of price controls on crude oil producers is used to finance a portion of the federal stockpiling of crude oil. In 1978, the entitlements program paid for approximately $100 million (1980 dollars) of the crude oil placed in the Strategic Petroleum Reserve.[27]

2.6 Rents, Property Rights, and Subsidies: Conclusion

The entitlements program is a mechanism for establishing property rights to price-controlled crude oil. These rights are valuable because the refining of controlled crude oil yields economic rents. Prior to November 1974, property rights in controlled oil were established by a buyer-supplier freeze that essentially turned contractual agreements, existing as of a base date, into long-term contracts. The primary effect of this method of establishing ownership in controlled crude oil was to confer billions of dollars of inframarginal rents unevenly among domestic refiners, with the happenstance of historical buyer-supplier relationships determining the extent of each refiner's gain. Claims that controls without entitlements put unfortunate refiners with limited access to price-controlled crude at a competitive disadvantage were, for the most part, inaccurate. Claims, which continue into the present, that crude oil price controls have benefited domestic oil consumers by reducing refiners' average costs and, thereby, refined product prices were also generally invalid prior to the adoption of the entitlements program. Such claims overlook the dependence of refiners' price and output decisions on marginal crude oil costs. These costs were not affected by the system of controls without entitlements. Attempts to force reductions in consumer prices through price controls on refiners' products have seldom succeeded; refined product price controls have been only rarely and haphazardly binding.

The entitlements program is a property rights system in which each refiner is guaranteed proportionately equal access to controlled crude oil. From the point of view of the individual refiner, ownership in price-controlled crude oil and attendant rents is contingent upon the purchase of uncontrolled crude. An industry equilibrium is reached under this system of property rights when the costs of refiners' induced increases in the refining of uncontrolled crude oil have exhausted the rents created by controls.[28] Until the equilibrium is reached, however, individual refiners have incentives to expand the use of uncontrolled crude.

If a refiner is currently producing with a greater proportion of uncontrolled crude oil than the national average, that refiner can attempt to capture the windfall gains that might go to other refiners by expanding its use of uncontrolled crude oil. Such a refiner thereby becomes eligible for a greater number of entitlements and greater payments for those entitlements from entitlements-purchasing refiners. Consequently, the effective cost of incremental crude oil to such a refiner is the uncontrolled price *minus* the increase in receipts from the sale of entitlements. On the other hand, if it is currently producing with a smaller proportion of uncontrolled crude oil than the average refiner, a refiner can attempt to retain the windfall gains associated with the controlled oil to which it has access by expanding its use of uncontrolled crude oil, moving its proportion of controlled crude oil closer to the national average and thereby reducing the burden of required entitlements purchases. Thus the effective cost of the incremental barrel of crude oil to such a refiner is the uncontrolled price *minus* the incremental reduction in the burden of required entitlements purchases.

In short, under the entitlements program *all* refiners perceive the cost of using one more barrel of crude oil as being less than the market price of the incremental barrel. This makes expanded production more profitable. Consequently, refiners expand their output so long as effective incremental costs do not exceed the sum of the prices received for the incremental products in the refined petroleum product market. In the absence of entitlements, this competition for the controls-created windfall gains would be fruitless. When there are no entitlements payments to capture, the effective cost of uncontrolled crude oil is simply the uncontrolled market price. Profit-seeking refiners would not expand output as much as under the entitlements program. In other words, the entitlements program, coupled with crude oil price controls, has the same impact as a program of direct subsidies to refining. In fact, it is as if the federal government taxes away rents created by controls and redistributes them as subsidies for the expansion of refinery output through the use of uncontrolled crude oil.

The entitlements subsidy is strikingly ingenious. EPAA/EPCA crude oil price controls provide a method of financing the subsidy, while leaving a portion of crude supply uncontrolled. This uncontrolled supply is available to clear domestic crude oil markets and avoids the severe economic disruptions and bureaucratic headaches that would accompany crude oil shortages. In addition, the entitlements program induces refiners to base price and output decisions on average crude oil costs. This allows refiners, consumers, and policymakers to avoid the shortages that systematically binding, average cost-based price controls on refined petroleum products would produce.

Notwithstanding its ingenuity, the entitlements program does represent a significant move away from an unregulated approach to the pricing of oil. Both the method of financing the entitlements subsidy and the subsidy itself have significant, subtle, and perhaps unintended allocative and distributional effects. The next chapter examines the economic consequences of the method by which the entitlements program is financed—crude oil price controls. Chapter 4 investigates the incidence of the entitlements subsidy.

Appendix 2.A Refiner Cost Banking and Debanking

Changes in legal maximum prices can be divided into those changes $\hat{D}_i^{t+1}$ due to current input price changes and those changes D_i^{t+1} due to all factors by defining

$$\hat{D}_i^{t+1} = D_i^{t+1} - \frac{B_i^t}{V_i^{t+1}} \pm \frac{R_i^t}{V_i^{t+1}} + \frac{G_i^t}{V_i^{t+1}}. \tag{2.A1}$$

Accumulated net banks A_i^{t+1} can then be defined as

$$A_i^{t+1} = B_i^t \pm R_i^t - G_i^t. \tag{2.A2}$$

At any time, it is possible for the market for the ith product to clear at a price P_{mi} that differs from the legal maximum price. Banking occurs when $P_{mi} < P_i^0 + \hat{D}_i$. Total new banking or debanking b_i in any month is

$$b_i = (P_i^0 + \hat{D}_i - P_{mi})V_i. \tag{2.A3}$$

EPAA/EPCA refined product price controls are binding when

$$P_{mi} > P_i^0 + \hat{D}_i + \frac{A_i}{V_i}, \tag{2.A4}$$

which is to say that the right-hand side of (2.A4) is the ceiling price in any month. When product price controls are binding, the price actually received for the ith product is this ceiling price. Substitution of this into (2.A3) for P_{mi} indicates that

$$b_i = \left(P_i^0 + \hat{D}_i - P_i^0 - \hat{D}_i - \frac{A_i}{V_i}\right)V_i = -A_i. \tag{2.A5}$$

Thus all accumulated banked costs are debanked when ceiling prices are binding.

Appendix 2.B Binding Downstream Controls: Effects on Refiners

Assume, for the sake of simplicity, that the refining sector sells directly to retailers. Retailers' demand for refinery product has an elasticity given (after Hicks—see Bronfenbrenner, 1971) by

$$E = \frac{\sigma(\psi + \eta) + S\psi(\eta - \sigma)}{(\psi + \eta) - S(\eta - \sigma)}, \tag{2.B1}$$

where σ is the elasticity of substitution between refinery products and other retailing inputs (assumed to constitute a single homogeneous factor); ψ, the elasticity of supply of the non-refinery product input; η, the elasticity of consumers' demand for petroleum products; S, the factor share of refinery products; and retailing is assumed to be governed by a linear homogeneous production function. Binding retail ceilings make $\eta = \infty$ and increase the elasticity of demand E faced by refiners:

$$\frac{\partial E}{\partial \eta} = \frac{1}{M^2} S(\psi + \sigma)^2 > 0, \tag{2.B2}$$

where M is the denominator of E.

Appendix 2.C EPAA Entitlements Subsidy Lowers Marginal Cost to National Weighted Average Cost and Redistributes All Rents

The right-hand sides of both (2.8b) and (2.10) reduce to the national weighted average cost of crude oil W:

$$\begin{aligned} P_{\mathrm{w}} - A(P_{\mathrm{w}} - P_{\mathrm{o}}) &= P_{\mathrm{w}} - \frac{C_{\mathrm{o}}}{C_{\mathrm{o}} + C_{\mathrm{w}}}(P_{\mathrm{w}} - P_{\mathrm{o}}) \\ &= \frac{P_{\mathrm{w}}C_{\mathrm{o}} + P_{\mathrm{w}}C_{\mathrm{w}} - P_{\mathrm{w}}C_{\mathrm{o}} + P_{\mathrm{o}}C_{\mathrm{o}}}{C_{\mathrm{o}} + C_{\mathrm{w}}} \\ &= \frac{P_{\mathrm{w}}C_{\mathrm{w}} + P_{\mathrm{o}}C_{\mathrm{o}}}{C_{\mathrm{o}} + C_{\mathrm{w}}} = W. \end{aligned} \tag{2.C1}$$

Under EPAA regulations, the total implicit subsidy s and total rents r were

$$s = (P_{\mathrm{w}} - W)(C_{\mathrm{w}} + C_{\mathrm{o}}) \quad \text{and} \quad r = (P_{\mathrm{w}} - P_{\mathrm{o}})C_{\mathrm{o}}.$$

Thus

$$\begin{aligned} s - r &= (P_{\mathrm{w}} - W)(C_{\mathrm{w}} + C_{\mathrm{o}}) - (P_{\mathrm{w}} - P_{\mathrm{o}})C_{\mathrm{o}} \\ &= P_{\mathrm{w}}C_{\mathrm{w}} + P_{\mathrm{o}}C_{\mathrm{o}} - W(C_{\mathrm{w}} + C_{\mathrm{o}}) \\ &= P_{\mathrm{w}}C_{\mathrm{w}} + P_{\mathrm{o}}C_{\mathrm{o}} - \frac{(P_{\mathrm{w}}C_{\mathrm{w}} + P_{\mathrm{o}}C_{\mathrm{o}})(C_{\mathrm{w}} + C_{\mathrm{o}})}{(C_{\mathrm{w}} + C_{\mathrm{o}})} = 0. \end{aligned} \tag{2.C2}$$

That is, EPAA regulation was equivalent to the rent tax with subsidy scheme insofar as all rents were redistributed in the form of subsidies.

Chapter 3

The Economics of Crude Oil Price Regulation

3.1 Introduction

When the typical policymaker, political commentator, or voter thinks of the major issues in domestic petroleum policy, it is probably crude oil price controls, and certainly not the entitlements program, that come to mind. There is some justification for this. While chapter 2 has made it clear that it is a mistake to think that crude oil price controls by themselves would hold down the prices of the refined petroleum products delivered to consumers, it is true that crude oil price controls have given rise to the economic rents that have been tapped for the entitlements program's subsidies to refined product supply. If crude oil price controls had never been adopted and domestic producers had always been allowed to receive market-determined prices for their output, there would not have been any crude oil cost differences to be equalized across refiners—and there would not have been any financing for the entitlements program.

Notwithstanding these observations, it would be misleading to imply that the only important effects of EPAA/EPCA crude oil price controls arise from the support they have given to the entitlements program. These controls have had substantial independent impact. This chapter examines this impact. The approach adopted compares the multiple-tiered marginal revenue schedules faced by a producer under controls with the marginal revenue schedule that would be faced in an unregulated market. In a single-period static setting, it is concluded that crude oil price controls generally cause a reduction in domestic production. The exception to this was the released oil program of EPAA, which implicitly subsidized crude oil production from some properties. In a dynamic model that permits examination of the time pattern of production from crude oil reserves, the implications of price controls are somewhat ambiguous.

3.2 EPAA Crude Oil Price Controls in a Static Context

The political environment surrounding the adoption of Phase IV oil price regulations and the passage of EPAA grew directly out of a series of dramatic changes in the international oil market. These included increases in producing country off-take taxes, property expropriations by host governments, the Arab embargo, and, of course, memorable price increases. In the world wrought by these events, two conflicting goals dominated policy debates. On the one hand, domestic policymakers were under pressure to shelter the intermediate and final users of crude oil from the unwelcome income and wealth effects of insecure supply and sharply rising prices. On the other hand, turmoil in international oil markets, the increasing market share of imported oil, and the overt hostility of many producing countries toward the United States gave rise to a political rhetoric of energy independence. Crude oil price controls might serve the first of these goals by impeding the proferred transfers to domestic producers; but to appear to be holding back domestic oil output through the imposition of a draconian system of punitive price ceilings would be politically damaging. The multiple-tiered price incentives embodied in Phase IV-EPAA (and eventually EPCA) regulations were used by policymakers to minimize the conflict.

The crude oil price regulations in force from the second half of 1973 through January 1976 controlled only the price of old oil. Output from old properties in excess of 1972 base period control levels (BPCLs), from new properties brought into production after 1972, and from stripper properties was uncontrolled. Any output above an old property's BPCL released an equivalent amount of old oil from controls. These multiple-price categories were officially said to be designed to (a) prevent windfalls from accruing to older supply sources (that is, pre-1973 properties); (b) avoid discouraging the development of new domestic supply sources (that is, post-1972 properties); (c) expand production from properties that could increase output rates easily (that is, old properties with new and released production); and (d) avoid abandonment of marginal properties (that is, stripper properties). While these justifications were marketable in the political arena, crude oil price controls did not simply duplicate the output response that would have occurred had domestic producers been allowed to respond freely to the market conditions they faced.

Crude oil price controls affect production decisions by altering the marginal revenue schedules perceived by domestic producers. In the absence of controls, any individual producer sees the (landed world) market price of crude oil as the relevant marginal revenue. Maintaining the assumption from chapter 2 that the United States is a price taker in

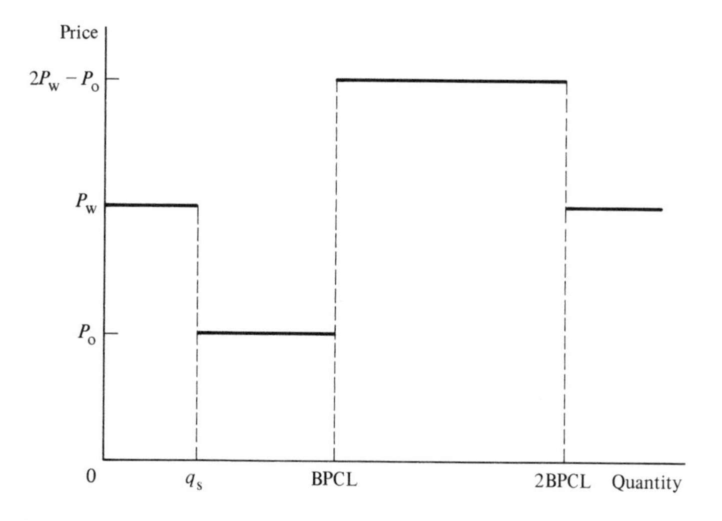

Figure 3.1 Producer marginal revenues under EPAA.

the international crude oil market, EPAA controls did not alter this marginal revenue for those producers bringing properties into production after 1972. A producer with an old property, however, faced the marginal revenue schedule shown in figure 3.1. Output from an old property up to a quantity q_s at which production averaged ten barrels per well per day was stripper oil and received an uncontrolled price and marginal revenue of P_w. From q_s to the BPCL, output received the controlled old oil price of P_o.[1] Above the BPCL, incremental output was uncontrolled and released a barrel of old oil from controls. This incremental output consequently had a marginal revenue of $2P_w - P_o$, that is, P_w, on the new oil *plus* the additional revenue $P_w - P_o$ received by releasing a barrel of old oil from controls. If a producer ever produced more than twice the BPCL, no old oil would have remained to be released and marginal revenue would simply be the uncontrolled price P_w.

3.2.1 The Old-Stripper Decision

The lack of controls on stripper oil undoubtedly prevented (and continues to prevent) some abandonments that would have occurred had such oil been treated as old oil. Uncontrolled stripper prices under EPAA permitted the application of enhanced recovery techniques to otherwise unprofitable properties. But the stripper provision also created some incentive for producers of old oil to reduce production in

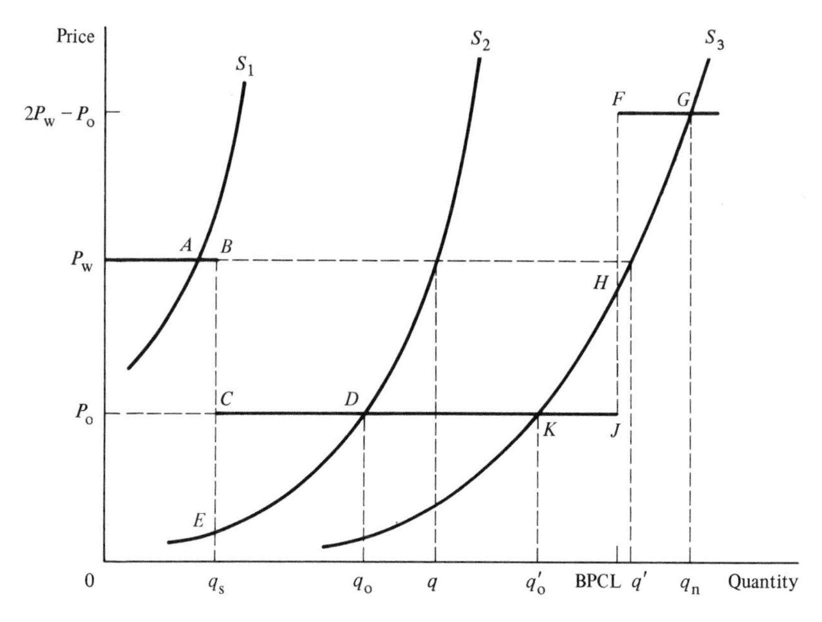

Figure 3.2 Alternative supply decisions under EPAA.

an effort to qualify for uncontrolled prices and avoid old oil price ceilings.

In figure 3.2, a producer with a supply schedule such as S_1 would find an equilibrium at A and would qualify for stripper prices. A producer with supply schedule S_2, however, would face a more complicated decision: Produce at q_o and receive a price of P_o or produce at q_s and receive P_w. With fixed costs on already producing properties sunk, the profit from the former would be $\pi_o = P_o q_o - \int_0^{q_o} S_2 \, dq$ and the profit from the latter would be $\pi_s = P_w q_s - \int_0^{q_s} S_2 \, dq$. The rational decision would be to produce q_o unless $\pi_s - \pi_o > 0$, where

$$\pi_s - \pi_o = (P_w - P_o)q_s - \left[P_o(q_o - q_s) - \int_{q_s}^{q_o} S_2 \, dq\right]. \tag{3.1}$$

Graphically, (3.1) says that the producer should only produce q_s if $P_w BCP_o > CDE$, since area CDE in figure 3.2 is the term in brackets in (3.1) and $P_w BCP_o = (P_w - P_o)q_s$. If output were stopped at q_s, producer surplus would equal the area above S_2 and below P_w. If output were carried on to q_o, producer surplus would equal the area above S_2 and below P_o. In either case, output up to q_s would yield producer surplus equal to the area above S_2 and below P_o. Hence the relevant comparison is between the additional surplus P_wBCP_o earned as a stripper property and the additional surplus earned CDE as an old oil property.

When the profits from being a stripper property exceeded those from being an old oil property, the stripper provision had the effect of magnifying the output reduction $q - q_o$ induced by price controls on old oil. The magnitude of such an effect, however, does not appear to have been large relative to the other output effects of controls. First, the likelihood that $\pi_s > \pi_o$ is inversely related to the elasticity of supply; that is, $\int_{q_s}^{q_o} S_2\,dq$ in (3.1) is a decreasing function of the elasticity of supply and $P_o(q_o - q_s)$ is an increasing function of the elasticity of supply. Thus, producers likely to select q_s rather than q_o would also have been likely to have inelastic supply sources and relatively small output changes in going from q_o to q_s. Second, the break even q_o at which $\pi_s = \pi_o$ was apparently quite small. Using 1975 prices, an old oil output of approximately 23 barrels per well per day produced revenue equivalent to 10 barrels per well per day of stripper oil. Third, there are regulatory restrictions designed to prevent otherwise more productive properties from being reclassified as stripper properties. To qualify as a stripper, "Each well on the property must have been maintained at the maximum feasible rate of production, in accordance with recognized conservation practices, and not significantly curtailed by reason of mechanical failure or other disruption in production."[2] Finally, any regulation-induced move toward stripper wells is not evident from the available data. Over 1960–1973, the average stripper well produced 3.59 b/d. Incentives to become a stripper property under EPAA (and EPCA) imply that, other things being equal, this average should rise toward 10 b/d; but, over 1974–1978, average stripper production was 2.94 b/d.[3]

3.2.2 The Old-New Decision

A producer with a supply schedule such as S_3 in figure 3.2, unable to take advantage of the stripper provision under EPAA, might have produced an output of q_o'. With a supply schedule of S_3, however, the producer could have elected to produce above the BPCL at q_n. Any output above the BPCL (that is, new oil from old properties) qualified for a price of P_w and resulted in additional revenue of $P_w - P_o$ on each barrel of associated released oil. This additional "released" revenue was an implicit subsidy for new oil production under EPAA. In 1975, this subsidy averaged \$7.00 per barrel and, when added to the price of uncontrolled oil, left producers owning old properties with a marginal revenue of slightly over \$19.00.

Consider the "new oil from an old property" production decision faced under EPAA-type price controls. Producing q_n in figure 3.2 results in an amount of new oil equal to q_n − BPCL and an equivalent amount of released oil. Old oil output is then the remainder: $q_n - 2(q_n - \text{BPCL}) = 2\text{BPCL} - q_n$. Profit from producing at q_n is

$$\pi_n = 2(q_n - \text{BPCL})P_w + P_o(2\text{BPCL} - q_n) - \int_0^{q_n} S_3\,dq.$$

This must be compared to the profit from producing only old oil at q_o': $\pi_o = P_o q_o' - \int_0^{q_o} S_3\,dq$. Output q_n is selected when $\pi_n - \pi_o > 0$, where

$$\pi_n - \pi_o = \left[(2P_w - P_o)(q_n - \text{BPCL}) - \int_{\text{BPCL}}^{q_n} S_3\,dq\right] - \left[\int_{q_o'}^{\text{BPCL}} S_3\,dq - P_o(\text{BPCL} - q_o')\right]. \tag{3.2}$$

Graphically, the first term in brackets in (3.2) is area *FGH* in figure 3.2; and the second term in brackets is area *HJK*. With marginal costs described by S_3, losses of *HJK* are incurred on output from q_o' to the BPCL; but profits of *FGH* are earned on the output from the BPCL to q_n. If the latter area exceeds the former, total profits are increased by producing q_n rather than q_o'. Significantly, $q_n > q'$, where q' is the output level that would result in the absence of regulation. Thus the released oil program under EPAA crude oil price controls tended to increase the output of domestic crude oil relative to the uncontrolled market.[4]

The likelihood that any producer would take advantage of the implicit subsidy of the released oil program and increase output varied directly with the difference between the world price and the old oil price as well as the elasticity of crude oil supply. Thus, during a period of price controls, which might be expected to reduce production, a subsidy was paid to relatively more elastic supply sources. Of course, for producers who did not find it profitable to take advantage of the released oil subsidy, the old oil price ceiling resulted in reductions in output relative to the uncontrolled market (that is, $q_o' < q'$). At least in the static context, then, the net effect of EPAA controls on domestic crude oil output was theoretically ambiguous. Estimates presented in chapter 5 indicate that the net output effect was most likely negative.

3.3 EPCA Crude Oil Price Controls in a Static Context

EPCA crude oil price controls replaced EPAA controls in February 1976. Although most media and policymaker attention in the second half of 1975 focused on the possibility of complete decontrol (supported by the Ford Administration and heavily opposed in Congress), the legislation that eventually emerged moved federal policy toward a more comprehensive price control regime. EPCA essentially redefined old oil as lower-tier oil and brought all previously uncontrolled oil into an upper tier. Upper-tier ceilings were initially set at approximately $1.50 below the price of uncontrolled crude oil; and both lower- and upper-tier prices were allowed small monthly inflation and incentive adjustments. The released oil program was abolished and stripper oil was, at least initially, brought under controls. Stripper oil was decontrolled under EPCA in September 1976. The gradual decontrol begun in June 1979 has not altered the basic system of lower-tier, upper-tier, and uncontrolled prices, but has operated by successively moving output toward higher tiers through reclassifications.

The marginal revenue schedules for crude oil production under EPCA are shown in figures 3.3A and 3.3B (with stripper oil shown as uncontrolled). Producers with properties that had begun operations by 1975 (that is, lower-tier properties) generally face the environment described in figure 3.3A. Output of q_s or less qualifies as stripper oil, and output above the lower-tier BPCL qualifies for the upper-tier price P_u. Between q_s and BPCL, the lower-tier price of P_0 is received. Producers with properties brought into production after 1975 face the environment described in figure 3.3B. Except for the availability of uncontrolled prices on stripper oil and selected types of recently decontrolled output (see table 1.2), these producers receive the upper-tier price.

3.3.1 The Stripper Decision

The decision under EPCA to produce as a stripper property has the same character as the decision under EPAA. For a producer with a lower-tier property, the decision to produce as a stripper is precisely as described by (3.1), with P_0 now denoting the lower-tier price. During the period from February through August 1976, when stripper oil was subject to upper-tier ceilings, (3.1) was applicable, with P_u replacing P_w. For a producer with a property that began production after 1975 (that is, an upper-tier property), there was no incentive to alter output

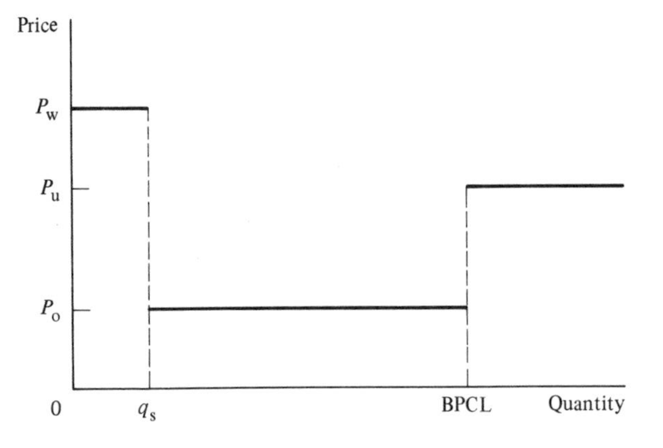

Figure 3.3A Producer marginal revenues under EPCA: old properties.

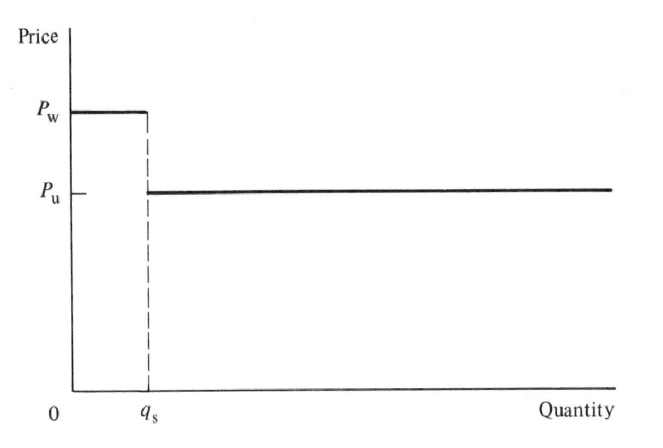

Figure 3.3B Producer marginal revenues under EPCA: new properties.

to qualify as a stripper property during the February–August 1976 period—stripper prices were upper-tier prices. With stripper prices uncontrolled, however, upper-tier producers have some incentive to reduce production to q_s in order to qualify for uncontrolled stripper prices. The decision is as described by (3.1), with P_u replacing P_o. For the reasons noted in section 3.2.1, distortions introduced by the EPCA stripper provisions appear to have been minimal.

3.3.2 The Lower-Upper Decision

A lower-tier producer with sufficiently elastic supply might find it profitable to produce in excess of the BPCL—qualifying output above

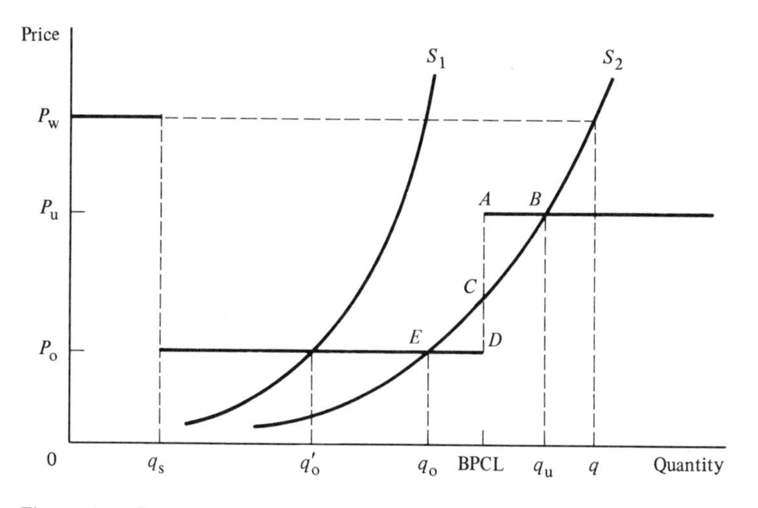

Figure 3.4 Lower-upper decision under EPCA.

the BPCL for upper-tier prices. With a supply schedule such as S_1 in figure 3.4, only lower-tier oil is produced. With schedule S_2, however, total output of q_u will be selected over q_o when the profits, π_u associated with q_u exceed the profits π_o associated with q_o. Analogous to (3.2) is the expression

$$\pi_u - \pi_o = \left[P_u(q_u - \text{BPCL}) - \int_{\text{BPCL}}^{q_u} S_2\,dq\right] - \left[\int_{q_o}^{\text{BPCL}} S_2\,dq - P_o(\text{BPCL} - q_o)\right]. \tag{3.3}$$

Graphically, (3.3) is the difference between area *ABC* and area *CDE* in figure 3.4.[5] When $\pi_u - \pi_o > 0$, output increases from q_o to q_u—but still remains below the output q that would result in the absence of crude oil price controls. Lower-tier supply sources such as S_1 produce q'_o under EPCA controls, which is also less than in the unregulated regime. In fact, even if S_1 and S_2 in figure 3.4 represent supply schedules for upper-tier properties, a ceiling price of P_u below the world price will leave output under controls less than output in the unregulated world. In short, EPCA controls do not contain any provisions like the EPAA released oil program that might offset the negative effect of ceiling prices on domestic crude oil production.

3.3.3 Phased Decontrol under EPCA

The phased decontrol of crude oil prices over June 1979–September 1981 operates by reclassifying various types of oil into the uncontrolled price tier. It does not alter the basic EPCA tiers. Thus for those properties still subject to controls, the marginal revenue patterns described by figures 3.3A and 3.3B, as well as the accompanying analyses, continue to apply. The outputs that are being freed from controls, however, face new and different production incentives. New categories of decontrolled crude oil include heavy oil, incremental tertiary recovery oil, newly discovered oil, marginal oil, tertiary incentive cost-offset oil, and the 4.6% per month of upper-tier oil that is reclassified as market-level new oil.

For the most part, the selected properties that have had incremental output freed from price ceilings operate as they would in an unregulated environment (disregarding, for the moment, the Windfall Profits Tax). The only current exceptions to this arise in the cases of tertiary incentive oil and market-level new oil.

Tertiary incentive oil is uncontrolled after December 1979 as an offset to a maximum of 75% of the expenses of tertiary enhanced recovery. A producer bound by the 75% limit would not have marginal revenue equal to the price of uncontrolled crude and would face some incentive to engage in excessively expensive tertiary recovery. (That is, greater total expenses would allow more crude to be decontrolled.) Regulations requiring certification of allowable expenses and the short duration of the tertiary incentive provisions (until 1 October 1981) work against any major impact from this incentive.

Under the provisions regulating the decontrol of market-level new oil, 4.6% of upper-tier oil is decontrolled each month after December 1979. This has the effect of slightly escalating the marginal revenue received by upper-tier properties every month. In any month after December 1979, incremental upper-tier output receives an effective price that is part upper tier and part uncontrolled. Specifically, marginal revenue in the nth month P_{u}^{n} is the sum of the upper-tier ceiling and the uncontrolled price, with each weighted by the percentages of controlled and decontrolled oil in the upper tier, respectively. In the first month after December 1979, for example, $P_{\mathrm{u}}^{1} = (1 - 0.046)P_{\mathrm{u}} + 0.046P_{\mathrm{w}} = P_{\mathrm{u}} + 0.046(P_{\mathrm{w}} - P_{\mathrm{u}})$.[6] The net effect of the market-level new oil provisions is thus gradually to raise the effective P_{u} in figure 3.3B toward P_{w}, covering the distance at the rate of 4.6% per month.

The market-level new oil provisions do not show producers fully uncontrolled prices at the margin.

3.3.4 Marginal Price Incentives under the Windfall Profits Tax

When EPCA was signed into law in late 1975, crude oil price decontrol was promised for the fall of 1981. The Windfall Profits Tax has falsified this promise by imposing excise taxes on domestic crude oil production. These taxes extract large portions of the differences between market prices and current ceiling prices. In so doing, they guarantee the continuation of a modified system of multitier price regulation for at least another decade.

Table 3.1 describes the marginal price incentives embodied in the Windfall Profits Tax. Effective tax rates are as described in chapter 1 for other than the first 1,000 b/d of an independent producer's output. The first 1,000 b/d of an independent's production is subject to slightly lower tax rates and slightly higher effective prices.[7] As of March 1980, lower-tier and upper-tier ceiling prices had not risen enough to trigger the imposition of the WPT, although market-level new crude was subject to a sizable tax. By the results of section 3.3.3, the Windfall Profits Tax on market-level new oil resulted in a rollback in effective upper-tier marginal revenue. The WPT also produced a slight rollback in Alaskan incentives and substantial reductions in the effective after-tax prices faced by the producers of marginal, stripper, NPR, newly discovered, heavy, and incremental tertiary crude oil. Indeed, the WPT has extended effective price penalties to most of the domestic output that would otherwise be uncontrolled. In addition, the preservation of multiple tiers in effective prices guarantees that, until full decontrol is reached, the general form of the lower-upper decision will remain as described in section 3.3.2 and figure 3.4. The special incentives to become a stripper property will continue even after complete decontrol.

The third row of table 3.1 reports the effective after-WPT prices that would have prevailed in March 1980 if all price ceilings had been removed at that time. Except for certain minor types of exempt oil (see chapter 1), all domestic crude oil will be subject to stiff excise taxes under the WPT, and producers will not face market-equivalent marginal revenue schedules. The Windfall Profits Tax can be expected to have a generally negative effect on incentives for crude oil production.

This last conclusion must be qualified by noting a possibly significant

Table 3.1 Crude Oil Price Incentives under the Windfall Profits Tax (WPT): March 1980
(Dollars per Barrel)

	Lower tier	Upper tier	Market-level new	Marginal	Stripper & NPR	Alaskan North Slope	Newly discovered, heavy, & incremental tertiary
Actual wellhead price[a]	6.35	13.99	36.33	36.33	36.33	13.77	36.33
Actual effective price (ex-WPT)	6.35	15.02[b] (spans Upper tier and Market-level new) (13.99)	(21.75)	21.75	24.64	13.41	30.89
Decontrolled effective price[c] (Estimated, ex-WPT)	21.75	21.75	21.75	21.75	24.64	19.40	30.89
Actual WPT	0	0	14.58	14.58	11.69	0.36	5.44
WPT upon decontrol[c]	14.58	14.58	14.58	14.58	11.69	8.93	5.44

a. Represents national average ceiling price for lower tier, upper tier, and Alaskan North Slope (ANS). Other categories are uncontrolled and are assumed to be priced at the uncontrolled stripper price as reported in US Department of Energy, *Monthly Energy Review*.
b. The effective price for upper-tier properties during phased decontrol is the weighted average of upper-tier and market-level new oil prices (ex-WPT). See section 3.3.3.
c. Assumes hypothetical full decontrol as of March 1980, with all wellhead prices (except ANS) equating to actual March 1980 price of uncontrolled stripper oil. State severance tax adjustment in calculation of WPT (except ANS WPT) is based on national average severance tax rate of 6.5% employed in US Congress, Congressional Budget Office (1979). ANS effective price is based on estimated ANS controlled price (ex-transportation) of $28.33, with WPT severance tax adjustment based on 11.5% tax rate and WPT Trans-Alaskan Pipeline System adjustment of $0.08.

exception to the price penalties imposed by the WPT. Verleger (1980) has pointed out that the special treatment of incremental tertiary oil may result in a surge of tertiary recovery projects. Output from tertiary recovery above the level that would be realized in the absence of the enhanced recovery project qualifies for a Tier Three tax rate of 30%—compared to the 70% and 60% rates in Tiers One and Two. The effects noted by Verleger arise because the base output levels that would be forthcoming in the absence of tertiary recovery are understated by the WPT provisions. (That is, physical output decline rates in the absence of tertiary recovery are overstated.) As a result, tertiary recovery projects not only qualify truly additional production for a relatively low 30% tax rate, but also qualify substantial amounts of Tier One and Tier Two oil for the lower rate. The amount of such reclassified oil is independent of the level of truly incremental tertiary production, so long as such production is positive. This independence means that, once a tertiary project begins to yield output, marginal revenue is the effective price as reported in table 3.1. The first barrel of tertiary recovery, however, has an extremely high payoff—the present value of the tax savings on the reclassified Tier One and Tier Two oil. This can amount to millions of dollars for a single property.

The large inframarginal return to tertiary recovery may result in the development of projects that would be unable to cover their total costs in an unregulated market. For the producer able to cover variable tertiary recovery costs at the effective after-tax price for tertiary recovery oil, the investment decision will involve a comparison of any uncovered fixed costs and the present value of the resulting tax savings. Even the producer unable to cover the variable costs of tertiary recovery at the effective after-tax price might undertake tertiary recovery—if the present value of the associated tax savings exceeds the present value of the sum of fixed costs and unrecovered variable costs. Of course, producers in such cases would produce the minimum amounts necessary to qualify their operations as incremental tertiary projects. In short, existing domestic oil reserves may be worked even more intensively (and expensively) than they would in the absence of any federal price regulation.

Contrary to Verleger (1980), the inframarginal character of the tax incentives makes it inappropriate to estimate the net supply effects of the WPT's incremental tertiary recovery provisions by treating average returns to such recovery as marginal incentives and comparing these returns to marginal cost-based supply schedules. Still, some marginal effects can be noted. As suggested, otherwise uneconomic tertiary recovery projects will be biased toward small sizes. Moreover, all projects—even those that would be adopted in the unregulated market setting—will produce at lower output rates than in the unregulated case because the WPT will depress *marginal* effective prices below market levels (see table 3.1). In short, the Windfall Profits Tax will most likely result in more, but smaller, enhanced recovery projects. Assessment of the net effect on production is confounded by the nonmarginal character of relevant supply decisions and the project-specific nature of the fixed costs of enhanced recovery. This assessment will have to wait for the passage of time to generate empirical evidence on the responses of producers.

3.4 Price Controls and Intertemporal Production Decisions

Most of the important allocative effects of crude oil price regulation that emerge from the preceding analyses are clear. The analysis of price controls in a static setting concludes that EPCA regulations reduce domestic crude oil output. Although qualified by the released oil program, the conclusions regarding EPAA controls are similarly straightforward. Except for the tertiary recovery provisions, the output effects of the Windfall Profits Tax also flow directly from the impact of regulation on producers' marginal revenues. Before pressing the conclusions of the static analysis too far, however, it is appropriate to consider the intertemporal effects of price regulation on the extraction of a nonrenewable resource stock. Although phrased in terms of price controls, the analysis below is equally applicable to excise taxes of the WPT type.

A static analysis of price regulation fails to account for the negative effects that controls can have on the present value of future production. As Sweeney (1977), Lee (1978), Frederiksen (1979), and Henderson (1978) have noted, the imposition of a price ceiling may actually *increase* production in the near term by reducing the reward producers receive for conserving their stocks for use in future periods. A price-taking producer facing a future of legally frozen nominal and declining real prices, for example, might find it consistent with the profit maximization objective to increase current output rates in order to "get while the gettin's good." Of course, with a given stock of crude oil such behavior implies reduced future production and alteration of the entire time path of resource use. Indeed, evidence from MacAvoy (1971) indicates that this pattern of behavior has been exhibited in response to price controls on natural gas, where current period shortages did not appear for more than a decade after the imposition of ceilings.

In the case at hand, it is important to distinguish between the effects of price controls on the extraction of a nonrenewable resource from an existing stock and the effects of price controls on the exploration and development of new supply sources. Both types of effects arise from the impact of price regulation on the "Hotelling rents" (after the seminal work of Hotelling, 1931) associated, albeit not uniquely, with non-renewable resources. These quasi-rents derive from the difference, in any period, between selling price and production cost. They represent the economic opportunity cost of foregoing future revenues when resources are sold today. As Hotelling has pointed out, Hotelling rents serve to allocate production of a nonrenewable resource over time; if these rents rise in the future because, for example, future prices rise above previously expected levels, extraction paths are shifted toward the future by profit-maximizing producers. Moreover, the discounted present value of the (expected) stream of Hotelling rents serves to allocate exploration and development efforts; an increase in this present value encourages exploration and development.

The following conclusions emerge from these considerations: (a) because price controls can either raise or lower the value of current rents *relative* to the value of future rents, their impact on the time path of extraction from existing reserves is *a priori* ambiguous; and (b) because, other things being equal, price controls reduce the *absolute* present value of the streams of rents accruing to producers, they discourage exploration and development of new supply sources. In short, the supply of crude oil from new sources responds (in direction, at least) to price controls as predicted by the static model. Ambiguity in the response of extraction from existing reserves, however, leaves the net effect on total domestic output uncertain.

3.4.1 The Determinants of Extraction Decisions

The time path of Hotelling rents, and hence extraction from an existing stock of any nonrenewable resource, depends on such factors as current prices, future prices, and the discount rate. Development of a full-blown model that simultaneously takes account of these factors, as well as factors specifically pertinent to the crude oil case (for example, the joint-good character imparted by associated natural gas, the dependence of extraction costs on remaining reserves, the effect of current extraction rates on ultimate aggregate recovery, and the impact of policy uncertainty on producer price, cost, and property right expectations), is beyond the present frontiers of economic science. Nevertheless, significant work on certain aspects of the problem relevant to current policy has been done. Sweeney (1977) and Sheerin (1977), for example, have addressed the issue of a single-tier price ceiling in the context of a dynamic model of extraction from a known stock. These studies have concluded that the time pattern of the difference between (a) the unregulated market-clearing price and (b) the controlled price of the resource plays the central role in determining whether near-term extraction is increased or decreased by price ceilings.

Montgomery (1977) has reached similar conclusions and applied variants of them to multitier oil price regulation. He has argued that, as in the static case, the ability (under EPCA) to qualify marginal output from lower-tier properties for upper-tier prices provides some producers with enough incentive to increase current extraction above BPCLs. This incentive, however, has been offset by the ability of producers, after March 1976, to increase the next year's marginal revenues by holding this year's lower-tier production below this year's BPCL, thereby qualifying for a smaller BPCL the next year. Montgomery has also concluded that the prospect of deregulation increases the likelihood that producers will hold back current production. The extreme paucity of relevant data on crude oil by tier and by property apparently has prevented Montgomery from quantifying the incentives he has identified. He has cautiously concluded that, as of 1977, EPCA controls were most likely increasing current extraction.

In order to address the question of what effect crude oil price controls have had on the time path of crude oil extraction, the analysis here proceeds along the lines suggested by the work of Burness (1976) and Sweeney (1977). Let $C(q(t),t)$ and $\mathrm{MC}(q(t),t)$ denote the total and marginal extraction costs, respectively, associated with a representative producer's output rate of q at time t. If domestic producers are price takers with respect to a world price of $P_{\mathrm{w}}(t)$ at any time t and an interest rate r, then the representative producer exhibits a supply price schedule $P^{\mathrm{s}}(t)$ given by

$$P^{\mathrm{s}}(t) = \mathrm{MC}(q(t),t) + Ue^{rt}, \qquad (3.4)$$

where MC is a nondecreasing function of the output rate $q(t)$ and U is the present value of the opportunity cost (Hotelling rent or so-called user cost) of using crude oil at time zero rather than at an alternative time and is independent of $q(t)$ (see appendix 3.A).

$P^s(t)$ in (3.4) is the economic marginal cost schedule at time t. This marginal cost schedule consists of two components: (a) the resources used up in extracting crude oil; and (b) the value of the extracted crude oil in alternative uses (that is, in later time periods). At any period t, $P^s(t)$ is a familiar rising marginal cost schedule when marginal extraction costs are a strictly increasing function of output. For any given user cost and interest rate, higher current prices induce greater current output. Moreover, at any given price and interest rate, a higher user cost leads to reductions in current extraction; that is, changes in user cost due to, for example, a change in future prices can be represented graphically by shifts in the supply schedule described by $P^s(t)$.

This last observation is significant because it is through the user cost component of marginal cost that the effects of price controls on intertemporal production plans operate. Specification of the direction and magnitude of any controls-induced supply shifts, however, is not straightforward. On the one hand, the prospect of price controls that are expected to become increasingly stringent in the future could reduce the perceived gain from future production and, hence, reduce the user cost of current production. The result would be an outward shift in the current period supply schedule $P^s(0)$. At any given price, current extraction would increase. On the other hand, if future price controls are expected to become less stringent than current ceilings, the relative values of current and future production may be tipped toward the latter and user cost may increase. The result would then be an inward shift of current period supply; and at any given price, current extraction would decline.

Results reported by Burness (1976) and Sweeney (1977) can be used to give meaning to the notions of more and less stringent price controls. The direction of controls-induced changes in the current output of a nonrenewable resource depends centrally on the relationship between the interest rate and the proportionate rate of change in the difference between the market-clearing price $P_w(t)$ and the control price $P_c(t)$. Specifically, at the current time $t = 0$

$$\hat{q}(0) \lesseqgtr q(0) \quad \text{as} \quad \frac{\mathrm{GAP}(t)}{\mathrm{GAP}(t)} \lesseqgtr r, \tag{3.5}$$

where $\mathrm{GAP}(t) = P_w(t) - P_c(t)$, $\mathrm{GAP}(t) = \partial \mathrm{GAP}(t)/\partial t$, $\hat{q}(0)$ denotes extraction under controls, $q(0)$ represents extraction in the absence of price regulation, and r is the rate of interest (see appendix 3.A). Thus, whether or not current extraction is increased by controls depends on the rate at which the implicit tax of controls (that is, GAP) is changing over time.

The results of (3.5) can be viewed intuitively as arising from a relative price effect. A tax that rises more rapidly than the rate of interest implies that the present values of later taxes exceed the present values of earlier taxes—or, equivalently, the ratio of early to late taxes (in present value terms) is such that earlier looks relatively better than later. Later taxes can be avoided by increasing near-term production. Conversely, if the implicit tax of controls rises at a rate less than the rate of interest, the present value of later taxes is less than the present value of earlier taxes, and later becomes relatively more attractive than earlier. Early production is therefore reduced and output is shifted toward the future. If the implicit tax of controls is constant in a present value sense (that is, $\dot{\text{GAP}}/\text{GAP} = r$), a wealth loss is imposed on producers and/or royalty recipients, but there is no incentive to alter the time path of production—unless the tax is so large that extraction costs cannot be covered.

3.4.2 The General Time Path of EPAA/EPCA Controls

Expression (3.5) provides a framework for addressing the empirical issue at hand: Have price controls on domestic crude oil increased or decreased production from existing supply sources? The use of (3.5), however, requires knowledge of the discount rate. Selection of "the" discount rate is confounded by the large number of possible candidates. With capital markets that are efficient in the sense that unanticipated increases or decreases in the value of a firm are quickly capitalized through the market, one measure of the relevant interest rate for investments with nondiversifiable risk equivalent to the average nondiversifiable risk in the firm is the mean rate of return to equity in the firm. Mitchell (1974) reports that stockholders' average annual rate of return (stock appreciation and dividends) in the domestic crude oil production industry over 1953–1972 was 9.0%. Inflation over this period averaged 2.7% per year, indicating an annual real rate of return in crude oil production of roughly 6.3%. If, as is often asserted regarding economic regulation, the expansion of federal energy regulation in the 1970s has increased the riskiness of the petroleum sector, this estimate of the discount rate that is relevant to (3.5) would be an underestimate.

Expression (3.5) indicates that the time path of crude oil production depends on the rate of change of GAP, as well as the rate of interest. Figures 3.5A and 3.5B show the year-to-year rates of change in GAP_u and GAP_o, where GAP_u and GAP_o refer to the deflated uncontrolled upper-tier and uncontrolled lower-tier (and old oil) price GAPs, respectively.[8] These figures indicate that the GAP for old (and subsequently lower-tier) oil tended to increase at a higher rate than the interest rate of 6.3% (dotted line) until mid-1976. After mid-1976, GAP_o increased less rapidly than 6.3% and, in fact, tended to decrease for much of the time. An upper-tier producer in February 1976 could have looked forward (with consternation) to approximately two years of GAP_u increasing more rapidly than the interest rate. In 1978, however, this trend was reversed, with GAP_u declining consistently. The huge increases in uncontrolled international oil prices in 1979 then sent both GAP_u and GAP_o shooting upward—literally off the graph. The December 1979 GAP_u, for example, was 1,141% higher than the GAP_u in December 1978. GAP_o rose 199% over the same period. Stabilization of uncontrolled prices in 1980 is apparently halting the rise in GAPs.

The time pattern of GAPs is not well behaved. The application of (3.5) consequently becomes problematical. How should a producer respond when $(\dot{GAP}/GAP) - r$ changes signs every year or two? Two considerations shed light on this question. First, as work by Sweeney (1977) would suggest, some average rate of change in GAP is likely to be more relevant to production decisions than observation of the current GAP when $(\dot{GAP}/GAP) - r$ is changing signs over time. For example, it makes little sense to exhaust rapidly the resource stock in the current period because at present $\dot{GAP}/GAP > r$ if all or most of the future promises $\dot{GAP}/GAP < r$. The behavior of a price-controlled producer faced with a period of $\dot{GAP}/GAP > r$ followed by a period of $\dot{GAP}/GAP < r$ can be described qualitatively: Early in the initial period, alter production plans (relative to original precontrol plans) so as to shift extraction toward the present; but, as the period of $\dot{GAP}/GAP < r$ approaches, reduce extraction rates so as to save output for the future (in anticipation of less stringent controls). An extreme example of this kind of behavior can arise in the hypothetical case in which controls are imposed and become increasingly stringent (that is, $\dot{GAP}/GAP > r$) until the moment of preannounced decontrol. If the life

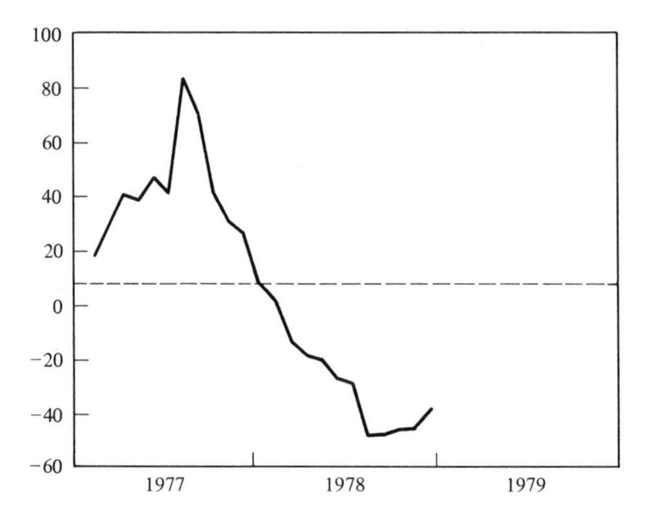

Figure 3.5A Annual rate of change: uncontrolled upper-tier price difference (year-to-year change). Price data are deflated by the Wholesale Price Index. (*Source:* US Department of Energy, *Monthly Energy Review.*)

of controls is relatively short, extraction falls during controls [notwithstanding (3.5)] and then surges upward upon decontrol to a level exceeding the extraction that occurs when there have never been any price controls.

The second point to be made concerning producer behavior in a regime of controls that vary between increasing stringency and increasing leniency is that application of (3.5) takes place in a setting of uncertainty. Production decisions must be based on *expectations* of controlled prices, uncontrolled prices, and price GAPs. This observation is particularly relevant to the events of 1979. If the very sharp increases in uncontrolled oil prices in 1979 had been anticipated by unregulated producers, the large positive values of the price GAPs in 1979 could have been expected to induce regulated producers to reduce 1979 extraction relative to the rates that would have been observed in the absence of controls. Indeed, the incentive to shift extraction toward the future that was provided by increasingly lenient controls for the upper tier in 1978 and the lower tier in 1976–1978, as well as the promise of decontrol, might have been ignored. Prescient domestic producers originally (that is, absent controls) planning to save output until 1979 (because they foresaw the jump in world prices) would be induced, upon the imposition of controls, to shift more output than planned to the pre-1979 period. Then the negative output effects of

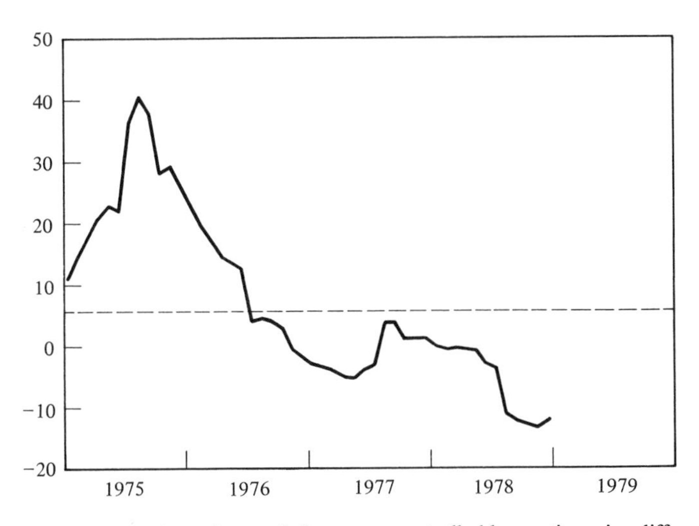

Figure 3.5B Annual rate of change: uncontrolled lower-tier price difference (year-to-year change). Price data are deflated by the Wholesale Price Index. (*Source:* US Department of Energy, *Monthly Energy Review*.)

controls that are predicted by a static analysis would not have materialized until after 1978.

Although it is difficult to know the expectations that producers hold at any time, the foregoing scenario strains the bounds of plausibility. Available evidence strongly suggests that the international political and economic turmoil that gave rise to the oil price increases of 1979 was universally unanticipated. Throughout late 1978, for example, the petroleum trade press focused on official OPEC price discussions between Saudi Arabia and the so-called OPEC price hawks and saw prospective 1979 price increases in the 10–20% range. Market-clearing prices actually tripled. The trade press also showed little *ex ante* recognition of the severity of the political disruptions in Iran that led to the cutoff of Iranian oil exports in late December 1979.[9] Domestic and international governmental officials in the major oil-consuming countries were similarly late in recognizing the realities of the Iranian situation.[10] In fact, the US State Department and Central Intelligence Agency (which has substantial responsibility for monitoring and forecasting developments in world oil markets) tacitly accepted widespread media and congressional criticism for not foreseeing the Iranian revolution and its aftermath.[11]

Oil markets themselves poorly anticipated the 1979 price shocks. The international spot market led the move to higher oil prices early in

1979. Domestic uncontrolled posted prices followed only gradually, apparently reflecting some disbelief that the steep international price hikes would really stick. Indeed, the major producer in OPEC, Saudi Arabia, persisted in the vain belief throughout 1979 that it could bring prices down from the mid-$30s range toward the low-$20s range. Finally, stock market behavior suggests little anticipation of the events of 1979. A randomly selected sample of ten nonintegrated crude oil-producing firms shows an unweighted average annual rate of stock price appreciation of 122% over the first six months of 1979, compared to a rate of −9% in the six previous months.[12] The abruptness of this change belies accurate expectations.

In short, virtually every important source of information and expectation formation that might have been expected to foresee the increases in uncontrolled oil prices and, concomitantly, GAPs in 1979 failed to do so. Thus, it is unlikely that (a) domestic producers in an uncontrolled regime would have planned for the actual post-1978 environment by shifting extraction rates so as to create an appropriately timed bulge in output; and that (b) the imposition of price controls that were particularly stringent after 1978, therefore, caused domestic crude oil extraction from existing reserves to be shifted back toward the pre-1979 period by reducing the relative value of later (post-1978) periods.

In fact, the prospect of gradual decontrol and the precedent of OPEC's inability to keep world oil prices abreast of inflation in the post-embargo period (real OPEC prices declined by approximately 16% over 1974–1978) with good reason could have led domestic producers looking forward to 1979 and beyond to expect GAPs to decline and controls to become more lenient.[13] Even a windfall profits tax was not going to tax away 100% of the gains from decontrol. Such expectations would lead to controls-induced reductions in early-period extraction rates (relative to the rates that would have been planned in the absence of federal price regulation). Moreover, such expectations would be consistent with the general pattern of GAPs under EPAA and EPCA.

Drawing on the data behind figures 3.5A and 3.5B, the annual rate of increase in the difference between uncontrolled and old, lower-tier oil prices (that is, GAP_0) averaged 6.1% from 1975 through 1978, compared to a discount rate of roughly 6.3%. The annual rate of increase in the difference between uncontrolled and upper-tier prices averaged 4.7% over 1976–1978. Thus, on average, it appears that $\dot{GAP}/GAP < r$, although it would be hard to dispute $\dot{GAP}/GAP = r$ in the case of lower-tier oil. In the early period of controls on old oil and upper-tier oil, sufficiently rapidly rising GAPs may have created some incentive to increase current extraction. But if domestic producers have generally expected that on average $\dot{GAP}/GAP < r$ would hold, current rates of crude oil extraction have tended to be held down relative to the rates that would prevail in the unregulated market. The expectation of producers that lies behind this conclusion is supported by the balance of the empirical evidence on the time paths of the differences between ceiling and uncontrolled prices. It has also been repeatedly reinforced by policymakers' willingness to show the carrot of decontrol to domestic producers. Hence, over most of their lives, EPAA/EPCA controls have most likely restricted crude oil output from existing domestic reserves. Estimates of the magnitudes of this reduction are provided in chapter 5.

3.4.3 Further Qualifications and Complications

While the available empirical evidence on the general pattern of the stringency of controls provides an assessment of the overall impact of EPAA/EPCA price regulation on domestic crude oil extraction, a number of the special provisions of the control system deserve comment. Among these are the released oil program, the provision permitting lower-tier properties to qualify output in excess of BPCLs as upper-tier oil, and the stripper provision. In addition, the impact of major alterations in regulation, such as the switch from EPAA to EPCA, warrant investigation.

Consider a major alteration in crude oil price regulation. Producers anticipating the imposition of controls (or the removal of the released oil provision, or the inclusion of formerly uncontrolled new oil in EPCA's lower and upper tiers) could be expected to increase production from existing properties immediately preceding such a policy change; that is, there would be an anticipation effect.[14] This kind of effect appears to have preceded the implementation of EPCA in February 1976. Daily production of uncontrolled new oil from existing new properties in January 1976, for example, was more than 28% above 1975 daily average production and was almost 22% above December 1975 daily production. Similarly, production of new and released oil from old properties as a percentage of total output from old properties reached an all-time high of 27.0% in January 1976, after showing a downward trend during most of 1975. Comparable percentages for January 1975 and January 1974 were 25.6% and 25.0%, respectively.

These surges in output of uncontrolled crude oil immediately before the implementation of EPCA are more likely to have been in response to the impending imposition of ceilings than to an increase in world oil prices in late 1975 or January 1976 since real world prices actually declined over this period.[15]

The foregoing evidence suggests that released oil production increased as the end of the released oil provision approached. Independently of this anticipation effect, however, the released oil provision provided an implicit subsidy to production from old properties while it was in effect.[16] The intertemporal impact of the subsidy depended in a manner analogous to (3.5) on the rate of change in the subsidy relative to the interest rate. Specifically, $\hat{q}(0) \lesseqgtr q(0)$ according as $\dot{\mathrm{GAP}}/\mathrm{GAP} \gtreqless r$, where GAP is now defined as the difference between the marginal revenue on released oil and the marginal revenue in the absence of controls. If the subsidy's rate of increase was (expected to be) larger than the interest rate, producers had an incentive to wait for future subsidies and reduce near-term production. If the subsidy's rate of growth was (expected to be) less than the interest rate, the opposite conclusion holds. The magnitude of the released oil subsidy was the difference between the marginal revenue on production above the BPCL, $2P_{\mathrm{w}}(t) - P_{\mathrm{o}}(t)$, and the marginal revenue which would be realized in an unregulated market, $P_{\mathrm{w}}(t)$. The subsidy was thus $P_{\mathrm{w}}(t) - P_{\mathrm{o}}(t)$ and was as shown in figure 3.5B.

During the period (1974–1975) when the released oil program was in effect, the implicit subsidy increased at a rate greater than the rate of interest. This suggests that producers with supply schedules that allowed them to reach an output rate extending into the released range were induced to hold back current production [that is, $\hat{q}(0) < q(0)$] and wait for future subsidies. Thus, the released oil program may have reduced output under controls, contrary to the conclusion reached in the static analysis. But this conclusion must be tempered by the recognition that producers may have anticipated the end of the released oil program in February 1976, or at least expected $P_{\mathrm{w}}(t) - P_{\mathrm{o}}(t)$ to follow its actual course after January 1976. It may have paid to postpone the release of some 1974 output from old properties until 1975, but it did not pay *ex post* to shift any 1975 output to 1976.

Under EPCA, production above lower-tier BPCLs qualifies for upper-tier prices, but does not release any lower-tier oil. The ability to qualify marginal production from lower-tier properties as upper-tier oil does provide incentive to increase production above the BPCL in the intertemporal context. But because $P_u < P_w$, there is still an implicit tax of GAP_u on production relative to the unregulated case; and the relationship between production under the controlled and uncontrolled cases is as described by (3.5) for the producer with upper-tier oil from a lower-tier property. The general time pattern of price incentives for such a producer is as described in section 3.4.1.

Finally, because stripper oil has generally been uncontrolled, the decision to produce stripper oil and the time path of such production from properties with supply price schedules in each period as described by S_1 in figure 3.2 has not been substantially affected by EPAA/EPCA regulation. During the brief period in 1976 when stripper oil was controlled at upper-tier prices, a reasonable anticipation of quick decontrol implied $\hat{q}(0) < q(0)$ during controls. For properties of the type described by S_2 in figure 3.2 or S_1 in figure 3.4, the character of the decision at any time to reduce production and qualify for stripper prices is as described in the static analysis, with supply schedules including a user cost component as in (3.4). Following the analysis behind figures 3.2 and 3.4, the willingness of an old, lower-tier or upper-tier producer to reduce production in order to qualify as a stripper property [that is, the likelihood that (3.1) will be satisfied at any point in time] increases if $\dot{GAP}_0/GAP_0$ or $\dot{GAP}_u/GAP_u < r$ [that is, if (3.4) is shifted inward by controls]. From the evidence on average GAPs, it appears that the incentive to reduce production and qualify as a stripper has generally been magnified in the dynamic context. If the implicit tax of controls tends to rise less rapidly than the interest rate, it pays to reduce current production; and it pays even more if, by reducing current output, current production can be sold at uncontrolled prices.

3.4.4 The Windfall Profits Tax

As table 3.1 indicates, the Windfall Profits Tax only gradually takes over from EPCA regulation. Lower- and upper-tier oil, which constituted roughly half of US output as of early 1980, are the last to be subject to the WPT. When the takeover is complete, however, the current provisions of the WPT will make the analysis of intertemporal price incentives in domestic crude oil markets relatively straightforward for both producers and researchers. Of course, recent experience with federal energy regulation suggests that the provisions of the WPT are not carved in stone.

The excise tax on crude oil production from newly developed

properties will reduce exploration and development of new reserves. Extraction from existing properties, however, will depend on the course of GAPs, where GAPs now denote explicit taxes. During phased decontrol, the effective prices of lower-tier, upper-tier, and Alaskan North Slope crude oil have been held well below the effective after-tax prices expected upon full decontrol (see table 3.1). Consequently, some incentive has been created for such properties to hold back output until the expiration of EPCA controls in October 1981. After full decontrol, the provisions of the Windfall Profits Tax will leave GAPs, and hence a significant component of intertemporal production effects, depending only on the course of world oil prices. For tax purposes, the WPT calls for the domestic base prices of all but Tier Three (heavy, incremental tertiary, and newly discovered) oil to be held constant in real terms. The real base price of Tier Three oil is to be increased at a rate of slightly over 2% per year. If domestic producers expect real prices of uncontrolled crude oil, as determined in the international market, to remain constant on average over the life of the WPT, the anticipated GAPs between uncontrolled and effective Tier One and Tier Two prices will remain constant in real terms. The expected real Tier Three GAP will decrease. With any positive discount rate, these patterns imply $\dot{\text{GAP}}/\text{GAP} < r$. Thus (3.5) indicates that, if the WPT were repealed at some arbitrary date after it takes effect, extraction from existing reserves would be shifted toward earlier periods and current period output from existing developed reserves would rise.

The foregoing is based on the assumption that producers expect real international oil prices to remain constant. With the resulting GAPs, the WPT makes the expected present value of later taxes less than the expected present value of earlier taxes. Of course, the opposite pattern may arise if world oil prices rise in real terms. From the tax rates and regulations governing inflation adjustments to base prices, it is possible to calculate the minimum rate of increase in real uncontrolled crude oil prices that would have to be expected by domestic producers in order to make $\dot{\text{GAP}}/\text{GAP} > r$ for each tax category. These minimum rates of price increase are approximately Tier One, 3.7%; Tier Two, 3.5%; Tier Three, 4.25%; and Alaskan North Slope, 2.9%.[17] The projection of world oil prices is far beyond the scope of this study and, indeed, presents a continuing challenge to the best economic forecasters. Suffice it to say that the minimum average rates of increase in real international oil prices needed to make $\dot{\text{GAP}}/\text{GAP} > r$ are well within the bounds of reasonableness. If these rates are met or exceeded in producers' expectations, repeal of the Windfall Profits Tax at some date in the future would be met by a drop in current period crude oil extraction rates and a shifting of extraction efforts to later periods. The drop in current output would be reinforced by a termination of the special incentives noted in section 3.3.4 that encourage the development of tertiary recovery projects.

3.4.5 Summary: Intertemporal Extraction Decisions

A shortage of data on such items as changes in base period control levels, current cumulative deficiencies, new oil from old and new properties, upper-tier oil from lower- and upper-tier properties, and stripper oil from lower- and upper-tier properties confounds the empirical assessment of the intertemporal effects of EPAA/EPCA crude oil price regulation. Nevertheless, it has been possible to examine the likely directions of any inaccuracies generated in a static analysis. Crude oil price controls unambiguously discourage the exploration and development of new supply sources—just as predicted by a static model. The effects of controls on extraction from existing reserves with capital already in place, however, are *a priori* indeterminant and depend on the discount rate and the (expected) time path of the stringency of controls (where stringency is measured by the GAP between uncontrolled and controlled prices).

The empirical evidence presented indicates that, on average, EPAA/EPCA crude oil price regulation has most likely reduced extraction from developed reserves. In the early stages of old, lower-tier and upper-tier controls, a tendency toward increasing (in present value) stringency may have created some incentive to get oil out of the ground before things got worse. This was reinforced by the released oil program. Overall, however, oil price control policy has embodied a general trend toward declining stringency, special benefits for stripper properties, and promises of either complete deregulation or an eventual move to higher effective prices for the bulk of domestic output under the Windfall Profits Tax. As a result, the decontrol of domestic crude oil prices at some arbitrary date would most likely have been met with an increase in output from existing reserves. This would have been complemented by an increase in the exploration and development of new reserves. Finally, the impact of the Windfall Profits Tax on intertemporal extraction decisions will depend on the course of world oil prices. Unless world prices rise in real terms by an average of 3–5% per year, the WPT will have the effect of holding down current-period extraction efforts. With one potentially significant exception—tertiary enhanced recovery—the WPT will discourage the exploration and development of new supply sources by reducing the net returns to these activities.

3.5 Multitier Controls, Price Discrimination, and Rent Extraction

It was argued in chapter 2 that the entitlements program can be viewed as a mechanism for distributing the rents associated with access to price-controlled domestic crude oil among the various direct and indirect users of crude oil. This section argues that, for all the apparent arbitrariness of categories and price levels, multitier crude oil price controls constitute an effective form of price discrimination. This argument has been made previously, with varying degrees of explicitness, by both policymakers and policy analysts.[18] For the most part, however, previous analyses have asserted that the price discrimination embodied in multitier controls has been designed to overcome the deleterious effects of price regulation on output and efficiency. It is argued here that, while multiple tiers do tend to dampen the undesirable allocative consequences of price controls, the price discrimination they produce is a method of maximizing the economic rents available to crude oil users. Chapter 6 explicitly tests the hypothesis that a coalition of crude oil users has more or less controlled the design of the controls-and-entitlements policy package.

3.5.1 Multiple Tiers and Monopsony

Perhaps the most important and consistently maintained tenet of postembargo crude oil price policy has been the relatively more favorable treatment of newer producing properties. At least following EPAA, the regulatory distinction made between older and newer properties has been similar to the pricing scheme that a price-discriminating monopsonist, interested only in profits, might adopt. In fact, the interests of crude oil users (likewise concerned with their own welfare and able to exercise monopsony power through the regulatory apparatus) can be well served by imposition of the same policies that the price-discriminating monopsonist would follow.

The objective of self-interested crude oil users is to maximize the total rents captured on the crude oil they purchase at controlled prices. These rents are the cost savings realized by holding the price of domestic oil below the price of incremental supplies purchased on the world market at the given price P_w. With two domestic sources (for example, upper and lower) that may be controlled, these cost savings are $(P_w - P_o)C_o + (P_w - P_u)C_u$. The hypothetical monopsonist, facing a marginal cost of P_w for incremental purchases of crude oil, would seek to maximize the same cost savings by depressing domestic prices.[19] To accomplish this, the monopsonist must discriminate across supply sources so as to equate the marginal contribution of each supply source to total crude oil costs. With supply elasticities $\epsilon_o < \infty$ and $\epsilon_u < \infty$, domestic lower and upper supply sources have marginal costs of

$$P_o\left(\frac{1}{\epsilon_o} + 1\right) \text{ and } P_u\left(\frac{1}{\epsilon_u} + 1\right),$$

respectively. Equating these to the marginal cost of foreign oil (and rearranging) tells both the monopsonist and domestic crude oil users as a whole how to set domestic prices so as to maximize rents:

$$P_o = P_w\left(\frac{\epsilon_o}{1 + \epsilon_o}\right) \quad \text{and} \quad P_u = P_w\left(\frac{\epsilon_u}{1 + \epsilon_u}\right). \tag{3.6}$$

From (3.6), it follows that there will be multiple tiers under monopsony or monopsonistic regulation only if $\epsilon_o \neq \epsilon_u$. Moreover, $P_u > P_o$ is rent maximizing only if $\epsilon_u > \epsilon_o$. Of course, if the price discrimination involved in setting $P_u > P_o$ is rent maximizing, it has supplemental desirable allocative effects (relative to, say, a single intermediate price ceiling) because it depresses price least where supply is most responsive.

Whether, in fact, the elasticity of newer, upper-tier supply sources is greater than the elasticity of older, lower-tier supply sources is an empirical issue. The numerous econometric studies of the supply of crude oil from new sources have typically yielded estimates of the price elasticity of additions to reserves and, hence, long-run supply that are near unity or slightly less than unity.[20] FEA-DOE estimates, as implied in supply projections under alternative price assumptions, appear to be in the range of 0.5.[21] Thus, a value in the range 0.5–1.0 for the elasticity of supply from new sources (that is, ϵ_u) appears to be reasonable. The price-output response of existing wells with capital in place is evidently considerably less elastic than the supply from newly developed sources. Cost studies (such as Steele, 1969) indicate supply schedules that are very steep over wide ranges of relevant prices; and Kennedy (1974) finds an elasticity of 0.1 to be reasonable. Arrow and Kalt (1979) consider an elasticity of 0.2 for existing wells to be in the high range. At the very least, it can be said that there is no available evidence or objections to contradict the contention that the supply of oil from existing supply sources is less elastic than the supply from newly developed sources, that is, the contention that $\epsilon_o < \epsilon_u$. Consequently, the price distinction between the lower tier and the upper tier is consistent with a monopsony model of multitier price controls.

The behavior described by (3.6) is that of a monopsonist able to practice only third-degree price discrimination, that is, able to distinguish between separate supply sources but unable to engage in any intrasource discrimination. A first-degree discriminating monopsonist could present suppliers with a pricing schedule that traces their supply schedules up to the world price and thereby extracts all producer surplus in the domestic crude oil extraction industry. Although first-degree discrimination is not embodied in multitier controls, some intermediate second-degree discrimination is evident in the provisions that permit output above lower-tier BPCLs to sell at upper-tier prices. Since the willingness of producers to take advantage of these provisions is positively related to the elasticity of supply (see section 3.3.2), the ability to qualify output in excess of BPCLs for upper-tier prices has encouraged production from lower-tier properties that happen to have relatively elastic supply characteristics. From the monopsonist's point of view, this is preferable to simple third-degree discrimination; and has the advantage of encouraging relatively elastic existing supply sources to identify themselves.

In a model characterizing multitier price controls as monopsonistic price discrimination, the stripper oil provisions constitute an anomaly. Except for the period February–September 1976, stripper oil has been uncontrolled. On the one hand, stripper properties are often old properties near exhaustion and subject to extremely inelastic supply conditions. Thus the monopsonistic model of controls would predict a low ceiling price. On the other hand, evidence presented in support of stripper decontrol has been used to emphasize an all-or-nothing attribute of at least some stripper operations: Without uncontrolled pricing, wells might be completely abandoned.[22] This characterization suggests that monopsonistic users of crude oil would prefer a relatively high price for stripper oil, but it does not imply that they would prefer uncontrolled stripper prices. Uncontrolled stripper prices do not benefit monopsonistic users. The monopsonistic model of controls, therefore, does not explain the treatment of stripper oil.

Montgomery (1977) argues that stripper oil represents a special case in the regulatory scheme; and, indeed, it has been singled out for special treatment for decades. An explanation of this treatment is beyond the scope of this study. By way of observation, though, it can be pointed out that the treatment of strippers has been consistent with (a) a long-standing favoritism toward smallness in petroleum policy and (b) an unwillingness on the part of policymakers to appear to be forcing the abandonment of domestic oil wells.[23]

The model of crude oil price controls as monopsonistic price discrimination is more obviously applicable to EPCA regulation than earlier EPAA regulation. EPAA placed price controls on old oil only and hence went only part of the way toward the discrimination described by (3.6). It is probably inappropriate to try to force every aspect of post-embargo crude oil policy into the monopsonistic discrimination model, since such an attempt obscures dynamic aspects of petroleum policy formation and the complicated details of refiner, consumer, and producer political competition. Nevertheless, on the basis of a survivor test, it might be inferred that EPAA regulation was only a step along the road to more full-blown price discrimination in crude oil policy. Not only has EPCA been in effect three times longer than EPAA, but the Windfall Profits Tax that will replace EPCA will continue the policy of price discrimination until at least 1988. Indeed, the WPT will retain the general pattern of discriminating most heavily against already developed properties, while providing the highest price incentives to those sources with relatively more elastic supply (that is, new discoveries and self-selected properties able to undertake tertiary enhanced recovery).

The characterization of crude oil price regulation as monopsonistic rent extraction may not explain every detail of post-embargo policy. But it would appear to have considerable power as a useful model for describing and interpreting the essential attributes of a body of complicated regulations consisting of continually changing individual components. At the very least, it can be said that post-embargo crude oil policy has extracted enough rents to keep any monopsonist satisfied—over $150 billion (1980 dollars) since 1975 (see table 2.2).

3.6 Conclusion

In its barest essentials, the thrust of post-embargo regulation of the domestic petroleum industry can be summarized as follows: Crude oil price controls have extracted inframarginal rents from crude oil producers, and the entitlements program has distributed these rents among refiners and consumers. The crude oil price controls used to extract domestic producer surplus amount, at least in the case of EPCA, to monopsonistic price discrimination carried out by federal regulatory institutions. While this discrimination has conferred benefits on the users of crude oil, domestic crude oil producers have been substantially harmed. The Windfall Profits Tax will not alter this conclusion.

The distributive results of crude oil price controls are not accomplished without alteration of domestic crude oil supply decisions. In the static setting, the effect of controls has been to reduce domestic production. The one qualification to this conclusion arises out of the released oil program under EPAA, which presented old oil producers with incentives to expand output beyond the level they would have chosen in the absence of price regulation. When the intertemporal aspects of crude oil production decisions are taken into account, the analysis of the effects of price controls becomes considerably more complicated than in the static setting. Consideration of the resulting complications indicates that, on average, over the life of controls, domestic production has most likely been discouraged by federal policy (again qualified by the released oil program under EPAA). The Windfall Profits Tax will discourage the exploration and development of new domestic reserves, although enhanced recovery from existing reserves may be stimulated. The impact of the WPT on rates of extraction from already developed reserves will depend on the path of world oil prices.

Measurement of the magnitudes of both the supply-side and demand-side allocative consequences of post-embargo oil price regulation is taken up in chapter 5. Before proceeding to this, however, one major question remains concerning the effects of post-embargo policy. While it is clear that crude oil price regulation harms domestic producers and the entitlements program subsidizes crude oil consumption, it is not clear how the rents extracted from producers are divided among the various users of crude oil. Indeed, of all the issues raised by federal petroleum policy, the question of this division has been the most hotly debated. Accordingly, the next chapter enters into this debate and measures the incidence of the entitlements subsidy.

Appendix 3.A The Time Path of Resource Extraction

Expression (3.4) describes the supply price schedule of a profit-maximizing producer of crude oil. It is derived by maximizing the present value of a reserve at time $t = 0$ subject to the constraints that (a) total extraction $\int_0^\infty q(t)\,dt$ not exceed the total available reserve R_0; and (b) the price received $P(t)$ be invariant with respect to $q(t)$. This last constraint implies that the producer is operating as a perfect competitor, as maintained throughout this study (see chapter 2). The relevant lagrangian form of the producer's problem is

$$L = \int_0^\infty [P(t)q(t) - C(q(t),t)]e^{-rt}\,dt - U\left(R_0 - \int_0^\infty q(t)\,dt\right), \qquad (3.A1)$$

where U is the lagrangian multiplier. Differentiation of L yields (3.4) as a necessary condition for present value maximization (Sweeney, 1977).

Expression (3.5) describes the impact of a price control on crude oil. It is derived by differentiating the hamiltonian form H of (3.A1) with respect to extraction. In the case where there are no price controls, so that $P(t) = P_w(t)$,

$$H = [P_w(t)q(t) - C(q(t),t) - Uq(t)]e^{-rt} \qquad (3.A2)$$

and

$$\frac{\partial H}{\partial q} = [P_w(t) - \mathrm{MC}(q(t),t) - U]e^{-rt} = 0. \qquad (3.A3)$$

The time path of extraction $\dot{q}$ is described by differentiating this last expression with respect to time:

$$\dot{q} = \frac{\dot{P}_w - r(P_w(t) - \mathrm{MC}(q(t),t))}{(\partial \mathrm{MC}/\partial q)}. \qquad (3.A4)$$

Under controls, the analogous expression is

$$\dot{\hat{q}} = \frac{\dot{P}_c - r(P_c(t) - \mathrm{MC}(q(t),t))}{(\partial \mathrm{MC}/\partial q)}. \qquad (3.A5)$$

The deviation of $\dot{\hat{q}}$ from $\dot{q}$ is thus

$$\dot{q} - \dot{\hat{q}} = \dot{P}_w - \dot{P}_c - r(P_w - P_c), \qquad (3.A6)$$

from which (3.5) follows. This characterization of the effects of controls, like that of Montgomery (1977), is based on Burness (1976).

Sweeney (1977) reaches the same result, but requires that the resource be ultimately depleted. Sweeney (1977) also points out that for (3.5) to hold for any arbitrary $t = 0$, the sign of $(\dot{\text{GAP}}/\text{GAP}) - r$ must remain unchanged for all $t > 0$. If this condition is not met, the producer must weigh the present values of offsetting periods.

Chapter 4

The Effects of Crude Oil Price Controls and Entitlements on Domestic Refined Petroleum Product Prices

4.1 Introduction

The most politically sensitive issue in post-embargo petroleum policy has been the effects of federal regulation on refined product prices. The focus on refined product prices is due primarily to their importance to the constituents of federal policymakers, and thus to policymakers themselves. The most widely accepted, albeit incorrect, view among policymakers is that post-embargo petroleum industry regulations reduce refined product prices by holding down average crude oil costs. From this perspective, deregulation is viewed as having politically significant effects on the level of energy prices, the distribution of wealth among interest groups, and even the macroeconomic course of the economy.

While there is broad agreement among policymakers that current regulation of the petroleum industry is substantially reducing refined product prices, no such consensus exists among economists who have examined post-embargo federal energy policy. The major unresolved questions concerning the effects of post-embargo petroleum policy on refined product prices are empirical. Specifically, while it is clear that the entitlements program provides a subsidy at the margin to domestic refiners, the quantitative significance of the implied impact on refined product prices is not self-evident. This chapter presents an empirical analysis of the effects of the entitlements program on refined product prices. Evidence from numerous sources is developed; and the task at hand is to create, in the manner of a detective, a consistent and coherent description of the consequences of the entitlements subsidy to crude oil use.

4.2 Modeling Entitlements: International Considerations

As the analysis of chapter 2 indicates, crude oil price controls, by themselves, will not alter domestic refined product prices so long as the US refining industry is a price taker in the world crude oil market. Any tendency for current regulations to lower product prices will arise from the effect of the entitlements program on marginal crude oil costs in the

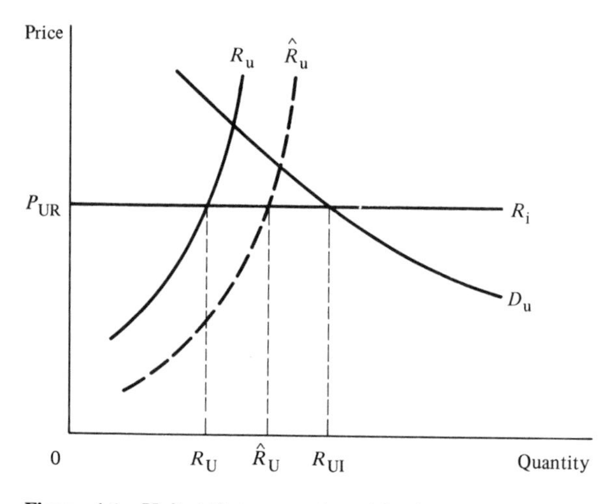

Figure 4.1 United States as price-taking importer.

domestic refining industry. In the characterization of chapter 2 (see figure 2.4B), the US refining industry is represented as a price taker in the world crude oil market and is implicitly isolated from foreign refined product markets. Consequently, controls and entitlements unambiguously reduce domestic product prices. It is plausible, however, that the United States is not insulated from foreign product markets. In fact, over the eighteen months preceding the introduction of the entitlements program in November 1974, the United States imported approximately 16% of its total refined product consumption.[1] To the extent the United States is not isolated from world product markets, domestic product prices are tied to foreign prices and the behavior of unsubsidized foreign refiners.

Figures 4.1–4.4 depict the domestic refined product market in the presence of international trade. In figure 4.1, the United States is assumed to face a perfectly elastic supply R_i of refined products from foreign sources at a price of P_{UR}, which is the world price plus transportation costs to the United States. Domestic demand is given by D_u. Prior to the introduction of controls and entitlements, the supply of domestically refined products is R_u; the equilibrium domestic price is P_{UR}; domestic production is R_U; and product imports are $R_{UI} - R_U$. Following the introduction of controls and entitlements, $\hat{R}_u$ is the (subsidized) domestic supply of products; domestic production is $\hat{R}_U$; domestic consumption is unchanged at R_{UI}; and product imports decline to $R_{UI} - \hat{R}_U$. In figure 4.1, the entitlements program has left

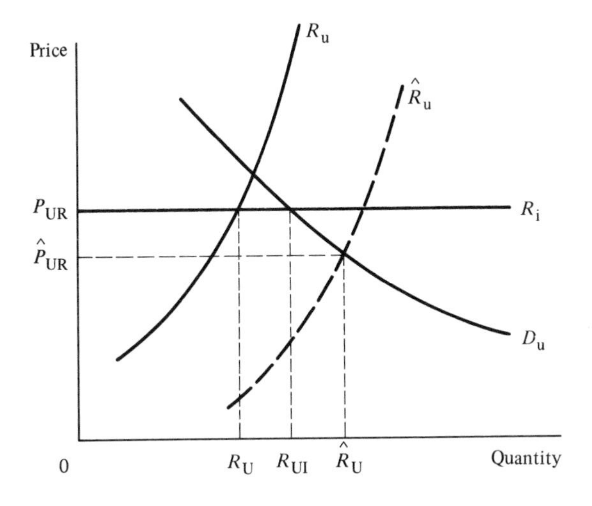

Figure 4.2 United States as price-making importer.

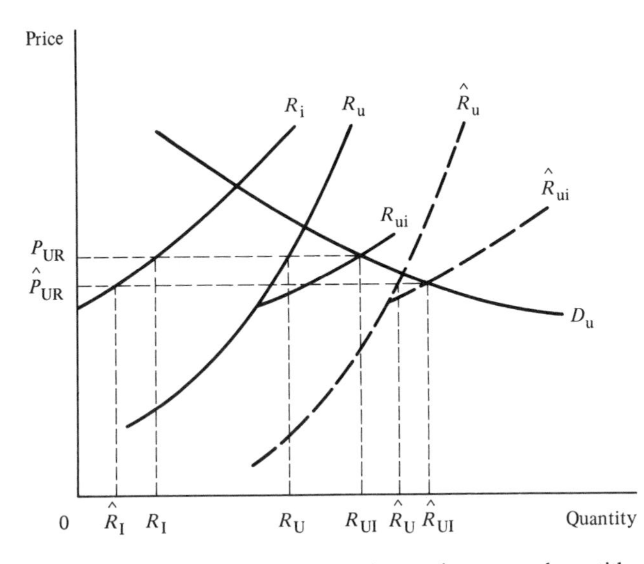

Figure 4.3 United States as price taker: no imports under entitlements.

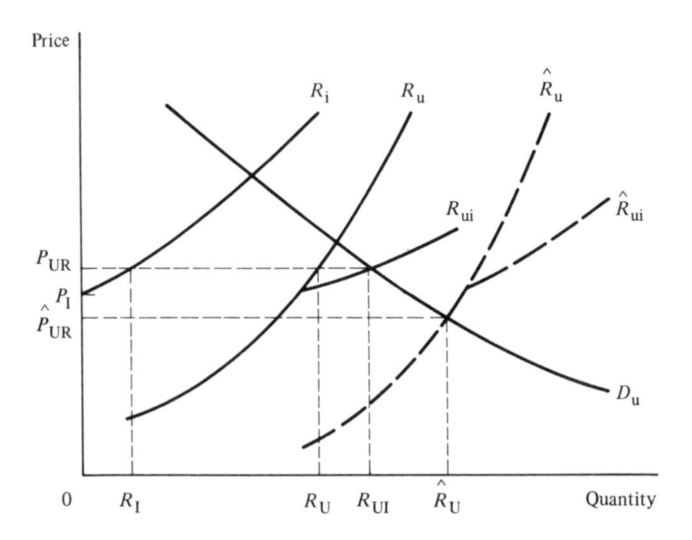

Figure 4.4 United States as price maker: no imports under entitlements.

domestic refined product prices unaltered relative to their levels under deregulation. The primary effect of regulation is to expand domestic production and to reduce product imports.

In figure 4.2, the United States is a price taker in world product markets, but the entitlements subsidy increases the supply of domestic products to $\hat{R}_u$ and reduces the domestic price of products to $\hat{P}_{UR}$. Domestic consumption and production increase to $\hat{R}_U$ and imports fall to zero. If the domestic price of products falls below the world price (that is, P_{UR} less transportation costs) by more than the cost of transportation to foreign markets from the United States, the domestic refining industry would export refined products. Export controls on petroleum products, however, have been in effect continuously since December 1973.[2] Thus $\hat{P}_{UR}$ persists as the equilibrium price. Foreign prices do not place a lower bound on the domestic price of products, and the entitlements subsidy reduces domestic refined product prices.

In figure 4.3, the United States is no longer assumed to be a price taker in world product markets. Accordingly, the quantity of foreign products supplied to the United States increases with price, as shown by R_i. The supply of products in the United States, R_{ui}, is the horizontal sum of R_u and R_i. The intercept of R_i is the foreign autarkic price (plus transportation costs). Prior to controls and entitlements, the domestic price of products is P_{UR}; domestic consumption is R_{UI}; domestic production is R_U; and imports are $R_{UI} - R_U = R_I$. Upon introduction of the entitlements subsidy, the supply of domestic products increases to $\hat{R}_u$ and the total supply of products increases to $\hat{R}_{ui}$. The domestic and world prices of refined products fall to $\hat{P}_{UR}$ and $\hat{P}_{UR}$ less transportation costs to the United States, respectively. The level of imports falls to $\hat{R}_I$.

With a sufficiently large shift in domestic supply in response to the entitlements subsidy (that is, an intersection of $\hat{R}_u$ and D_u at a price below the intercept of R_i), it would be possible for imports to fall to zero and for domestic product prices to fall below world prices, despite a decline in world prices. This case is shown in figure 4.4. Domestic consumption and production increase to $\hat{R}_U$; there are no imports; the domestic price of products falls to $\hat{P}_{UR}$; and the foreign price of products falls to $\hat{P}_I$ (less transportation costs). Export controls leave the United States isolated from world markets.

In short, when the United States is a price taker in world refined product markets, the entitlements subsidy encourages domestic production of refined products and reduces refined product imports. Domestic product prices are unaffected by the entitlements subsidy, unless the subsidy is sufficient to squeeze out all imports and move the United States to an autarkic equilibrium. When the United States is not a price taker in world refined product markets, the entitlements program unambiguously reduces domestic and world product prices. Domestic prices may or may not be pushed below world prices according as the United States is or is not moved to an autarkic equilibrium.

4.3 Evidence from Previous Research

Empirical investigations by economists of the effects of the entitlements subsidy on product prices are relatively scarce. What evidence has been generated reveals a substantial range of opinion and disagreement. Moreover, existing studies vary greatly in the attention they give to the methods of empirical research. In fact, only one study, produced by Phelps and Smith for the RAND Corporation (1977), has undertaken truly rigorous empirical analysis. The evidence from this and other studies is briefly summarized and reviewed in this section.

4.3.1 Selected Studies

Friedman (1975b) and Hall and Pindyck (1977) have argued, without attempting to provide empirical support, that the entitlements subsidy has reduced domestic refined product prices below the levels that would prevail in the absence of regulation. The market characterization that apparently lies behind their analyses is one in which the US refining industry is isolated from foreign markets and operates along a perfectly elastic supply schedule for products. Such a characterization implies a complete shifting of the entitlements subsidy to refined product consumers. Some support for this conclusion is provided by Montgomery (1977), who notes two pieces of data: (a) the domestic gross refining margin (that is, the difference between the price of a barrel of refined products and incremental crude oil costs) has declined since the introduction of entitlements; and (b) little new domestic refining capacity has been added, and rates of capacity utilization have not been unusually high under the entitlements program. Although numerous factors, such as refined product price controls and OPEC price increases, could affect margins, Montgomery finds the first datum supportive of the "complete passthrough" conclusion, since less than a 100% shifting of the subsidy to consumers would imply (other things being equal!) an increase in gross margins. The second assertion is meant to imply that the US refining industry has been operating under conditions of perfectly elastic supply. Evidence presented below in sections 4.4–4.6 indicates that this characterization, with its full shifting of the entitlements subsidy, may be roughly applicable to some products.

Writing in the *Wall Street Journal*, Lichtblau (1977) has argued that the entitlements program has reduced at least domestic gasoline and middle distillate prices over most of the period since November 1974. Figure 4.2 (no imports under entitlements) characterizes his view of gasoline and middle distillate markets. When "imports" from refineries located in US territories and possessions (which receive the entitlements subsidy for crude oil that yields products sold in the United States) are removed from official import statistics, Lichtblau finds that gasoline imports from unsubsidized sources in 1976 were only about 24 mb/d compared to domestic consumption of approximately 7,000 mb/d. These remaining imports Lichtblau attributes to "special factors," such as occasional disequilibria (for example, "distress" sales), local border demand for Canadian gasoline, and an occasional incentive for a refiner to raise ceiling prices by averaging higher-priced foreign products into domestic sales. Although proponents of the view that the entitlements subsidy has had no effect on domestic product prices have expressed skepticism about the usefulness of direct foreign-domestic price comparisons, Lichtblau reports that US gasoline prices (adjusted for tariffs and transportation costs) were 4.5–5.0¢ per gallon below European prices in 1976.[3] Adjusted US distillate prices are reported to have been below European prices by approximately 6.0¢ in the same period.

The general thrust of Lichtblau's conclusions is in agreement with the conclusions of Cox and Wright (1978). These conclusions are based on Chapel's (1976) tests of some of the implications of the model of chapter 2. Although results are marred by a failure to account for the subsidization of refineries located in US territories and possessions, Chapel reports that imports of most products declined significantly following adoption of the entitlements subsidy to US refiners. The notable exceptions were gasoline and middle distillate imports. From this, Cox and Wright argue that the imported supplies of these two products are relatively price inelastic. Inelasticity in import supply implies that the entitlements subsidy lowers both foreign and domestic product prices (see figures 4.3 and 4.4). Cox and Wright also point out that, while residual fuel import supply may be highly elastic, special entitlements for residual fuel imports under EPCA lower the effective price of such imports.

4.3.2 The RAND Corporation Study

The RAND Corporation (1977) study is the leading proponent of the view that EPAA/EPCA regulation has not reduced domestic refined product prices. The RAND study argues that the United States is a price taker in world product markets. As in figure 4.1, "U.S. petroleum prices are set, not by the FEA, but by world market conditions, and one could conceptually show this equality, once proper adjustments had been made for transportation costs, tariffs, true exchange rates, and other factors."[4] The study concludes that "the U.S. has negligible power in world refined product markets . . . [and the entitlements] program has altered trade flows, rather than product prices."[5]

The evidence that the RAND study relies upon most heavily is its observations that (a) domestic gasoline and heating oil prices did not decline in the early months of the entitlements program and (b) refined product imports did decline after the introduction of the entitlements program. As the analysis of figures 4.1–4.4 indicates, the second of these observations allows no discrimination between the competing characterizations of the domestic product market—in either case, the entitlements subsidy reduces product imports. Thus the crucial element in the RAND study's conclusion that entitlements have not reduced product prices is the observed pattern of product prices.

Although it observes no decline in gasoline and heating oil (that is, middle distillate) prices upon introduction of the entitlements program, the RAND study correctly points out that any direct examination of product prices that neglects other economic factors is highly suspect. It argues that changes in demand conditions should have reinforced any downward pressure from the entitlements program on product prices. The rate of inflation during late 1974 and all of 1975 was below 1% per month and real income declined during the fourth quarter of 1974 and the first quarter of 1975. Moreover, even after the imposition of increased crude oil tariffs in February 1975 and ongoing increases in world crude prices, a net subsidy of roughly 2.2¢ per gallon was created by the entitlements program.

The conclusions drawn by the RAND study from its data on product price behavior and associated market conditions are undermined by a number of observations. First, the data on gasoline and heating oil prices reported by the RAND study are not deflated. When adjusted for inflation, the data reveal that gasoline and heating oil prices did decline upon introduction of entitlements. Real gasoline prices were slightly below their October 1974 (that is, the last month without entitlements) level from November 1974 through February 1975. Real heating oil prices were below their October 1974 level from November 1974 through March 1975.[6]

Second, general market conditions apparently did not reinforce a product price reduction at the time the entitlements program began. Although real disposable personal income fell slightly from the fourth quarter of 1974 to the first quarter of 1975, real personal consumption expenditures rose by approximately 0.9% over this period. Real personal consumption expenditures on nondurable goods rose by approximately 0.3%.[7] Moreover, at least in the case of heating oil, the tendency (emanating from macroeconomic conditions) for prices to increase should have been reinforced by the seasonal pattern of demand.

Third, even in the context of its view of domestic product markets (figure 4.1), the RAND study should have expected to observe a decline in middle distillate prices in the first few months of the entitlements program. From November 1974 through January 1975, distillate fuel imports received a special entitlements subsidy equal to approximately 30% of the crude oil entitlements subsidy. This direct subsidy lowered the marginal cost of distillates to importers (that is, lowered R_i in figure 4.1). Thus, within the RAND study's preferred characterization of domestic product markets, a direct subsidy to distillate imports should have resulted unambiguously in lower prices.

Fourth, it may be that the particular price data reported by the RAND study are inappropriate for the purpose of identifying the effects of the entitlements program. The RAND study has selected an FEA series on national average dealer purchase prices. These prices occur in transactions that are typically removed from the first sale at refineries by at least one transaction (between refiners and distributors). If the refined product markets are similar to most other markets, prices are more volatile in the short run at the upper ends of product streams.[8] Moreover, the FEA data do not clearly distinguish between posted prices and transaction prices. Research by Stigler and Kindahl (1970) makes it clear that inferences drawn from posted prices over short periods of time are likely to be unreliable. To further complicate matters, the FEA switched its data source on gasoline prices from *Platt's Oilgram News Service* to the Lundberg Survey, Inc., in January 1975.[9]

Finally, the most basic shortcoming of the RAND study's use of the FEA price data is that it fails to address directly the issue at hand. If the goal is to assess the effect of the entitlements program on the level of product prices, the most appropriate comparison is not a comparison of product prices across time periods, but rather a comparison, within a given time period, of actual prices and the prices that would prevail in the absence of current regulations. When the United States is a price taker in world product markets, the latter comparison can be accomplished by examining the prices of products from unsubsidized (that is, foreign) refiners and the prices of products from subsidized (that is, domestic) refiners. The RAND study's conclusion that the introduction of the entitlements program did not change domestic product prices could be ruled out if foreign prices rose significantly in the same period. Evidence on foreign and domestic prices is presented in sections 4.5 and 4.6.

The RAND study argues that its observations on domestic product prices are supported by data showing that the United States imports significant quantities of all major refined products under the entitlements program and by Chapel's finding that product imports have been reduced by entitlements. The first contention is severely weakened by a failure to exclude from officially reported imports those shipments from US territorial refineries that qualify for the entitlements subsidy on crude oil refined for sale in the United States; and, as Cox and Wright have noted, Chapel's evidence is not particularly supportive of figure 4.1 as an accurate representation of either domestic or middle distillate markets.

In addition to direct observations of domestic product price changes and import levels, the RAND study reports evidence of a low degree or complete absence of US power in world product markets—as in figure 4.1. In support of this characterization of the US role in world product markets, the RAND study reports estimates of a highly elastic import supply. The elasticity of refined product import supply can be expressed as

$$\epsilon_{\mathrm{ir}} = \epsilon_{\mathrm{fr}}\frac{R_{\mathrm{F}}}{R_{\mathrm{I}}} - \eta_{\mathrm{fr}}\frac{R_{\mathrm{F}} - R_{\mathrm{I}}}{R_{\mathrm{I}}}, \tag{4.1a}$$

where ϵ_{fr} is the elasticity of non-US product supply; η_{fr}, the elasticity of non-US product demand; R_{F}, non-US product production; and R_{I}, the level of officially reported US product imports.[10]

Assuming values of -0.5 for η_{fr} and 0.2 for ϵ_{fr} and using data for 1974, the RAND study reports a product import supply elasticity of 12.4. The study considers this to be sufficiently large to justify the conclusions that the United States has no significant market power and that hence none of the entitlements subsidy is passed on in the form of lower product prices. From standard incidence analysis, however, it is known that the incidence of a subsidy depends on the elasticities of both supply and demand when neither is of infinite magnitude. Section 4.4 makes it clear that an import supply elasticity of 12.4 cannot clearly guarantee that the entitlements subsidy is not shifted to a significant extent toward the consumers of refined products.

As a final piece of evidence in support of the conclusion that domestic product prices have not been affected by EPAA/EPCA regulation, the RAND study reports the results of a regression analysis of the impact of the entitlements subsidy on the monthly flow of banked costs. The study contends that the entitlements subsidy has no effect on refined product ceiling prices, so that a subsidy-induced reduction in market-clearing prices should increase the flow of banked costs by widening the gap between ceiling and market prices. In support of its view that the entitlements program does not reduce the market prices of refined products, the study finds that the entitlements subsidy has had no effect on the flow of banked costs.

Given the premise that the entitlements subsidy has no effect on ceiling prices, the RAND study's reasoning about the subsidy's effect on banked cost flows is correct. The premise, however, is false. The entitlements subsidy is inversely related to refined product price ceilings. In fact, ceiling prices in any month are based on average crude oil costs *net* of entitlements payments and receipts.[11] As demonstrated in chapter 2, the entitlements program approximately equalizes average crude oil costs across refiners; and the subsidy is roughly the difference between the price of uncontrolled crude oil and the national weighted average price of crude oil. Thus, an increase in the entitlements subsidy, other things being equal, signifies a reduction in the average cost of crude oil for all refiners. Under the federal refined product price regulations, such a reduction reduces ceiling prices.

The foregoing implies that, in the RAND study's characterization of the US refined product market (figure 4.1), an increase in the entitlements subsidy, other things being equal, should reduce ceiling prices and leave market-clearing prices unaffected. Banked cost flows should then *decrease* and the study should expect a *negative* coefficient on the subsidy in its regressions—not an insignificant coefficient as asserted. On the other hand, if the entitlements program tends to reduce domestic market prices of refined products (figures 4.2–4.4), an increase in the entitlements subsidy should reduce both ceiling and market prices. Banked cost flows then depend on the relative sizes of the changes in ceiling and market prices (that is, increasing/decreasing if the subsidy-induced ceiling price change is smaller/larger than the induced change in market prices). If ceiling and market prices change by the same amount, banked cost flows per gallon should remain unchanged. Such a case would arise when market prices and ceiling prices both fall by the full amount of any subsidy increase; and such a case would produce a coefficient of zero in the RAND study's regression analysis. Of course, a characterization of the US product market as one in which 100% of the entitlements subsidy is shifted to consumers through lower product prices is the extreme statement of the market characterization against which the RAND study argues. Nevertheless, the study's regression analysis of banked cost flows yields an insignificant coefficient on the entitlements subsidy and does not discredit this extreme characterization.[12]

In summary, the RAND study has made the only concerted effort at deciphering the effects of the entitlements subsidy on refined product prices. Although none of the empirical evidence examined by the study clearly supports its claim that EPAA/EPCA regulations have failed to reduce domestic product prices, the study does suggest promising directions, as well as the need, for further research. Toward this end, the following sections of this chapter present additional evidence on the effects that EPAA/EPCA policies have had on refined product prices. It is found that the data consistently support a characterization of the US product market that shows the entitlements program as reducing domestic refined product prices. It appears that a significant portion of the entitlements subsidy is shifted to refined product consumers.

4.4 Imports and Exports of Refined Products under the Entitlements Program

The analysis of alternative characterizations of the US refined product market in section 4.2 indicates that imports of refined products should be reduced under the entitlements program, regardless of the effects of the program on domestic product prices. Moreover, figures 4.2 and 4.4 indicate that domestic product prices are unambiguously reduced if imports are driven to zero by the entitlements program. On the other hand, if imports of refined products are positive in the absence of entitlements and remain so under the entitlements program, any tendency for product prices to be reduced would have to arise from US price-making power in world product markets (figure 4.3). Of course, if imports are zero in the absence of the entitlements program, the effect of the program on product prices is negative (figure 2.4B). These implications suggest that a good place to begin to decide which characterizations of domestic product markets are most realistic is with an examination of US participation in international product markets.

4.4.1 The Policy and Institutional Setting of Petroleum Trade

In order to interpret the behavior of refined product imports over time, note must be made of the numerous governmental policies that have affected imports in recent years. During the late 1950s and all of the 1960s, crude and refined petroleum imports were substantially determined by the Mandatory Oil Import Program's (MOIP) quota system. While this program was officially abandoned by May 1973, the stringency of the quotas had been reduced gradually over the preceding several years and selected individual products had received special and often unequal treatment. Controls on residual fuel imports into the East Coast, for example, were eliminated entirely in April 1966. Middle distillates also received special treatment under the MOIP, with a gradual expansion of special import quotas beginning in late 1967, a more rapid relaxation of controls in the early 1970s, and a complete decontrol in January 1973. Similarly, restrictions on imports of liquefied petroleum gases (LPGs) and asphalt were relaxed in the early 1970s, with asphalt imports decontrolled completely in January 1973. In addition, several programs provided for special levels of crude and product imports from Canada, Mexico, Venezuela, and the Caribbean Islands.[13]

Even in the relatively short period since the end of import controls, numerous regulatory regimes and exogenous shocks have affected petroleum imports. In May 1973, MOIP was replaced by a system of import license fees. These fees were originally $0.52 per barrel in the case of gasoline and $0.15 per barrel for all other products except asphalt and LPG, which were exempt. Numerous initial exemptions from these fees were substantially eliminated and fees were gradually escalated to $0.63 per barrel.[14] Imports of residual fuel were effectively exempted from import fees beginning in May 1977.[15] In April 1979, product import fees were officially suspended.[16] Several products have been subject to special customs duties since February 1975. These total $0.525 per barrel for gasoline, $0.105 per barrel for middle distillates, kerosene, special naphthas, and jet fuels, $0.84 per barrel for lubricating oils, and $0.0525 per barrel for residual fuel. From June to August 1975, special supplemental import fees of $0.60 per barrel were applied to all products.

In addition to product import quotas and pecuniary measures such as product import fees, policy and exogenous changes in world crude oil markets have affected product imports and exports. The most notable exogenous shock of recent years was the Arab embargo on crude oil shipments to the United States, which began in October 1973 and was felt into the spring of 1974. Product import fees and tariffs can be expected to reduce product imports by raising delivered prices of foreign products. An embargo on crude oil sales to the United States, on the other hand, would tend to increase the importation of refined products, since the alternative to importing crude oil is to import crude oil in some processed or semiprocessed form. Similarly, crude oil tariffs (such as those imposed during February 1975–January 1976) could be expected to encourage product imports, unless countered by product tariffs (as in June–August 1975). In the same vein, the move away from the importation of products that is induced by the entitlements subsidy to crude oil imports is offset when refined products importers receive special entitlements as described in section 2.5.2.

Although it appears that the United States tends to be an importer of at least the major refined products in the absence of regulation (see below), export controls have been imposed on most domestically produced petroleum products. These controls were begun in December 1973, during the Arab embargo. Quota restrictions have been placed on gasoline, kerosene, middle distillates, jet fuels, aviation gasoline, residual fuels, liquefied petroleum gases, and most naphthas. Export licenses (which have little impact on export levels), but not quotas, apply to certain naphthas, carbon black feedstocks, petroleum coke, and asphalt. Exports of petrochemical feedstocks, lubricating oils, and waxes are uncontrolled.[17] In addition to export controls, substantial discouragement to exports of refined products is also provided by the entitlements program. Domestically refined crude oil to which exported products (other than lubricating oils) can be attributed is ineligible for the entitlements subsidy.[18] This discouragement to exports tends to isolate the United States from those international markets in which it would otherwise sell either as a result of underlying (that is, exregulation) supply and demand conditions or as a result of the entitlements subsidy to domestic refining.

4.4.2 The US Position in International Refined Product Markets: An Overview

Analysis of product imports under the entitlements program requires an assessment of the import levels that would prevail in the absence of the entitlements program. It appears to be the case that, in the absence of trade restrictions and the entitlements program, the United States would be an importer of at least the major refined petroleum products. Evidence of this is provided by the fact that imports of those products that had been subject to import controls under the MOIP generally rose upon termination of the quota program—despite the introduction of the import fee system. Imports of gasoline, for example, were 146% higher in the twelve months following April 1973 than in the last twelve months of the MOIP. Comparing the same periods, the market share of imported gasoline (calculated as the ratio of imports to domestic consumption) rose from 1.1 to 2.7%. Similarly, imports of middle distillates increased by 115% and the market share of middle distillates rose from 6.2 to 12.7% in the first year after import controls. Imports of residual fuel, which were effectively decontrolled in 1966, showed an increase of 9% in 1966 over 1965 and had a market share of 60.1% in 1966 as compared to 58.8% in 1965.[19] In short, it appears that domestic markets for gasoline, middle distillates, and residual fuel are not isolated from international markets in an unregulated environment.

The international position of the United States in markets for products other than gasoline, middle distillates, and residual fuel varies from product to product. In 1973–1974, the United States was a net importer of kerosene jet fuel, naphtha jet fuel, LPG, kerosene, and asphalt. Moreover, the average yearly imports of each of these products increased in 1973–1974 relative to their levels in 1972, when import controls were still in effect. On the other hand, the United States was a net exporter of petrochemical feedstocks, special naphthas, and lubricating oils in 1973–1974. Petroleum-based waxes showed very little trade in this period, but most indications (for example, table 4.4) suggest that the United States is an exporter under normal circumstances. Road oils and still gas for fuel showed no imports or exports at all in 1973–1974. Petroleum coke, aviation gasoline, and miscellaneous products were exported in substantial quantities without any offsetting imports.[20]

In order to proceed with the interpretation of data on US imports and exports, note should be made of the geographical distribution of the sources of imported refined products. A large part of US refined product imports has historically come from refining operations in Canada, Mexico, and the Central America-Caribbean area. In fact, in 1973–1974 (that is, years substantially free of the effects of import controls and entitlements), over 85% of US refined product imports originated in these regions and over 77% originated in the Central America-Caribbean area alone.[21] To a significant extent, these large market shares of US near neighbors reflect a long-standing bias in US oil import policy toward the encouragement of economic development and friendly relations in the Western Hemisphere. Most notably, from 1959 through early 1973, the MOIP used extra import quota allocations to promote product and, especially, crude oil imports from Canada and Mexico and to spur the development of refining capacity in the Central America-Caribbean area.[22]

The active promotion of near-neighbor production and refining capacity under US oil import policy has restrained US reliance on other major refined product marketing areas such as Western Europe. The possibility exists that the entitlements subsidy could reduce imports of refined products from other than near-neighbor sources such as Europe (with respect to which, following the RAND study, the United States might be modeled as a price taker) to zero and leave the United States as an importer of only near-neighbor refined petroleum. Near neighbors will export to the United States so long as the net price received in the United States exceeds the net price received in the more distant market. This condition will only fail to be satisfied if the entitlements subsidy reduces domestic product prices by more than (a) transportation-and-tariff costs between the distant market and the United States plus (b) the difference between near-neighbor/distant market transportation-and-tariff costs and near-neighbor/US transportation-and-tariff costs.

To illustrate this, let P_{UR} equal the price of products in the United States; P_E, the price in the distant market; T_1, distant market/US transportation-and-tariff costs; T_2, near-neighbor/US transportation-and-tariff costs; and T_3, near-neighbor/distant market transportation-and-tariff costs. For a near-neighbor refiner selling in the United States, the net price received is $P_{UR} - T_2$. Similarly, the net price received in the distant market is $P_E - T_3$. In the absence of an entitlements program in the United States, $P_{UR} = P_E + T_1$. Assume that the United States is a price taker in distant product markets and denote the price in the United States after introduction of an entitlements program as $\bar{P}_{UR}$; then the United States may quit importing from distant sources (for example, figure 4.2). Near neighbors, however, will continue to sell products to the United States unless $\bar{P}_{UR} - T_2 < P_E - T_3$ or, alternatively, $P_E - \bar{P}_{UR} > T_3 - T_2$. Since $P_{UR} = P_E + T_1$ implies $P_E = P_{UR} - T_1$, this last expression implies that near neighbors will stop exporting to the United States only if $P_{UR} - \bar{P}_{UR} > T_1 + T_3 - T_2$.

The importance of the foregoing is the following: In a world in which the United States is a product price taker with respect to distant markets, a decline in imports to zero is not a necessary condition for the entitlements program to have reduced US product prices. Near neighbors may continue to sell to the United States even if the United States stops importation of rest-of-world products upon introduction of the entitlements subsidy. Moreover, if near-neighbor imports do fall to zero, the entitlements program must be reducing US product prices by at least $T_1 + T_3 - T_2$. If near-neighbor imports remain positive, $T_1 + T_3 - T_2$ is a maximum bound on any entitlements-induced fall in domestic product prices. Based on representative markets and transportation-and-tariff costs for 1978, this bound is in the range of 4–5¢ per gallon.[23]

In addition to considerations arising from the geographical pattern of refined product import supply sources, the examination of the pre- and post-entitlements position of the United States in international markets is confounded by special factors that affect imports and exports, but that do not reflect underlying demand-and-supply conditions. For example, despite export controls and the severe discouragement to the exporting of refined products under the entitlements program, the United States has continued to export at least small quantities of most products since the introduction of entitlements. Among the factors that individually or together explain particular instances of anomalous exporting behavior of this type are selective exceptions and exemptions of export sales from the export quotas and/or the discouraging provisions of the entitlements program; imperfect enforcement of export control and/or entitlement regulations; long-term contracts for the export of US products; sales by US refineries in selected geographical markets in which price-making behavior is possible; sales of specialty products or blends in which few non-US substitutes are available and price-making behavior is possible; and short-run market disequilibria in which the price of products paid by foreigners is sufficiently above the domestic price to induce US exports.

Of course, the behavior of product imports at any particular time may also reflect special factors that complicate attempts to assess underlying supply-and-demand conditions. As noted in section 4.3, small amounts of imports may arise as a result of long-term contracts or distress sales by foreign refiners or marketers with short-run transportation or storage constraints. Similarly, distress purchases by domestic firms with contractual supply obligations that cannot be fulfilled in the short run with domestic products may result in imports. Then, too, the typically seasonal patterns of petroleum product demands may result in a situation in which the United States is only a seasonal participant in selected international product markets. Moreover, as discussed in chapter 2, banked cost regulations may encourage imports of higher-priced foreign products by refiners operating at or near ceiling prices. Finally, adjustment must be made in any examination of imports under the entitlements program for the unique position, as previously noted, of refineries located in US territories and possessions. Although sales of products from these refineries are still reported as imports, the entitlements program makes them effectively domestic operations. For all of these reasons, small quantities of observed imports of any particular product should not inspire confidence in characterizations of domestic product markets as markets in which the United States is an importer in the sense implied by figures 4.1 and 4.3.

If the data on product imports are imperfectly revealing and subject to competing interpretations, the issue of the appropriate methodology for analyzing available information necessarily arises. One approach to the data (and the approach to be used here) is to assume that there is an underlying set of supply-and-demand relationships that would clearly make the United States either an importer or (if allowed to be) an exporter of the major refined products were it not for the presence of random, obfuscating special factors. The stochastic nature of the data-generating process implies the relevance of statistical measures of the significance of import or export levels. Unfortunately, use of, for example, tests whether the mean of monthly imports under the entitlements program is significantly different from zero are complicated by at least two problems. First, the applicability of such measures rests on the assumption that the underlying supply-and-demand relationships and the distribution of any random perturbations around the underlying relationships are stable during the entitlements program. Sources of nonrandom change in market conditions include the occasional granting of direct product import entitlements, the program of special tariffs imposed by the Ford Administration in 1975, and seasonality in product demand. Although the last factor is countered by the complementary seasonality of domestic production and inventories, any remaining seasonal patterns in the demand for imports or supply of exports could mean that an accurate characterization of the US markets for refined products should vary according to season; for example, sometimes the United States is an importer of products and sometimes it is not. Second, export controls preclude the observation of positive exports that may otherwise occur as a result of underlying supply-and-demand conditions during the entitlements program. By eliminating observations that might be had from one end of the range of the sampling distribution, export controls tend to bias results toward acceptance of the conclusion that the United States is an importer of refined products. Complications of these sorts are addressed in the examination of individual product markets.

4.4.3 The US Position in International Petroleum Markets

Bearing in mind the foregoing caveats, qualifications, and institutional details, an attempt can now be made to find the proper characterizations of the markets for selected products. This section turns first to the cases of residual fuel, middle distillates, and gasoline and then examines the remaining minor refined products. Throughout, methods and results of tests for the statistical significance of import levels and market shares follow those reported in Kalt (1980).

Residual fuel, middle distillates, and gasoline collectively account for approximately three fourths of the average barrel of refined products consumed in the United States. Moreover, as pointed out above, the United States typically imports these three products in the absence of countervailing regulation. Figure 4.5 shows the monthly market shares of residual fuel, middle distillate, and gasoline imports from November 1972 through January 1980. Figure 4.6 shows the market shares of imports of these products from other than Canadian, Mexican, and Central America-Caribbean near neighbors. Sales to the United States by those refiners that are located in US territories and possessions and that receive the entitlements subsidy for such sales are not included as imports.[24] Market share is calculated as the percentage of domestic consumption accounted for by gross imports.

Residual Fuel Imports Figures 4.5 and 4.6 make it clear that the United States has consistently purchased significant amounts of residual fuel in the international market—both before and after the intro-

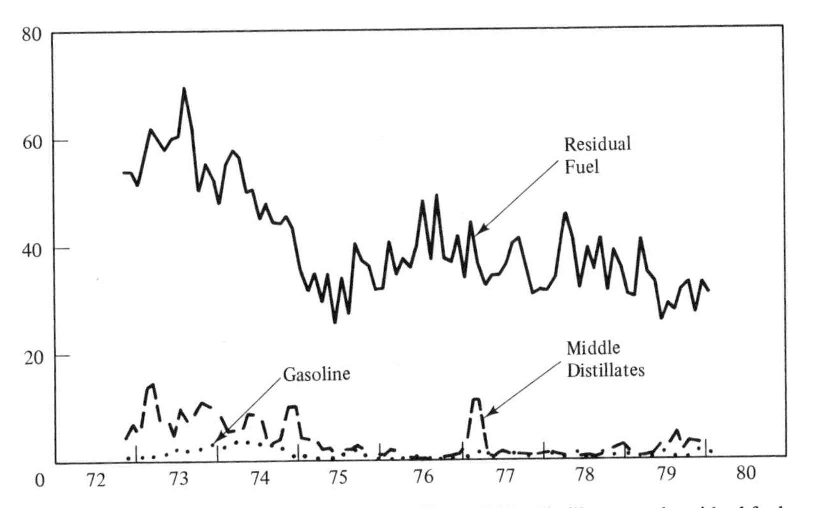

Figure 4.5 Import market shares for gasoline, middle distillates, and residual fuel: November 1972 to January 1980. Excludes imports from foreign sources eligible for the entitlement subsidy. (*Sources:* US Department of the Interior, Bureau of Mines, *P.A.D. Districts Supply/Demand Monthly*; US Department of Energy, *Energy Data Reports*.)

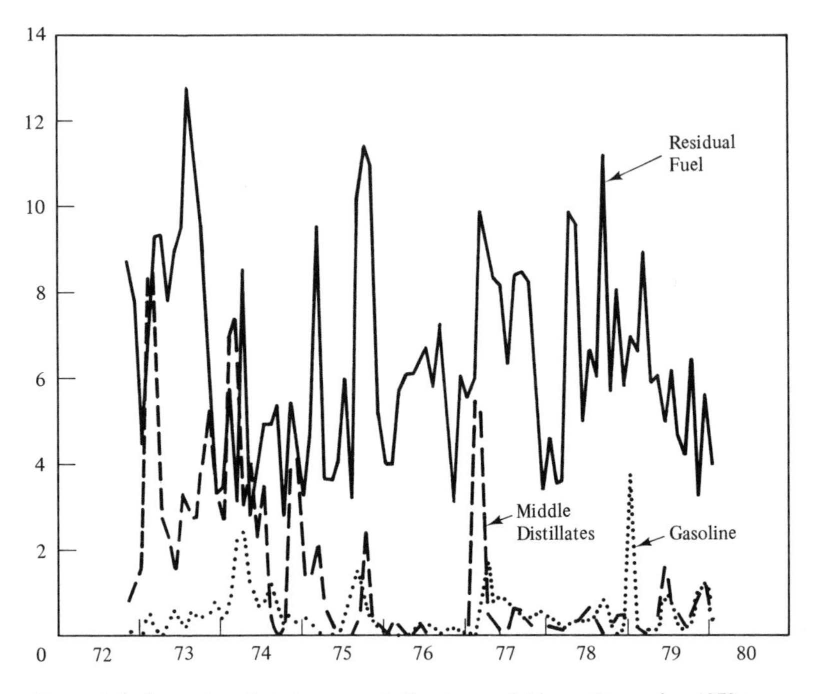

Figure 4.6 Import market shares excluding near neighbors: November 1972 to January 1980. (*Sources:* US Department of the Interior, Bureau of Mines, *P.A.D. Districts Supply/Demand Monthly*; US Department of Energy, *Energy Data Reports*.)

duction of the entitlements program. The monthly market share and level of residual fuel imports in the two years preceding the beginning of the entitlements program in November 1974 averaged 54.5% and 45.0 mmb, respectively. Over the period from February 1975 through January 1976 (that is, when residual fuel imports received no direct entitlements subsidy), the average market share was still significant at 33.1% and average imports were a substantial 24.2 mmb per month. During the periods in which residual fuel imports have been eligible for a direct entitlements subsidy (November 1974–January 1975 and February 1976 onward), the market shares and import levels have averaged 37.7% and 35.0 mmb per month.

The consistent quantitative significance of residual fuel imports clearly suggests support for the assertion that the United States has not been isolated from international residual fuel markets either before or after the introduction of entitlements. This is not to say, however, that the entitlements program has had no effect on the importing of residual fuel. In the analysis of section 4.2, it was found that, under all characterizations of the US product market, the entitlements program implies a decline in the level (and market share) of imports. Accordingly, the market share of residual fuel imports over February 1975–January 1976 was significantly less than in both the period preceding the entitlements program and the period when residual fuel imports received direct entitlements.

Conclusions to be drawn from these data and tests must be subject to the caveats raised in section 4.4.2. The period in which residual fuel imports declined, however, corresponds almost exactly to the period of crude oil tariffs (which could be expected to raise imports of residual fuel). The fact that imports fell despite offsetting tariff policy supports the conclusion that the entitlements program has tended to reduce product imports. At the very least, it can be said that the data of figure 4.5 do not contradict a characterization of the US residual fuel market as one in which the entitlements subsidy to the use of crude oil reduces import levels and import market shares but leaves imports positive and residual fuel prices unaffected (as in figure 4.1). To support this characterization of the residual fuel market, however, is not to argue that domestic prices have, in fact, been unaffected since the beginning of the entitlements program. The provisions that have granted direct entitlements to residual fuel imports during most of the period since November 1974 reduce the private marginal cost of these imports and, within the context of figure 4.1, can be expected to reduce domestic residual fuel prices by approximately the amount of the direct subsidy (a vertical drop in the imported product supply R_i in figure 4.1). Only during the period from February 1975 through January 1976 would it be true that domestic residual fuel prices were unaffected by the entitlements program. Even this statement may have to be qualified, however, by observations on the geographical distribution of import sources.

As the differences between figures 4.5 and 4.6 imply, the majority of residual fuel imports comes from Canada, Mexico, and the Central America-Caribbean nations. In fact, over the two years preceding the introduction of the entitlements program, the market share and level of residual fuel imports from other than near neighbors were significantly greater than zero (implying US reliance on other than near-neighbor international markets), but averaged only 6.8% and 5.5 mmb per month. As expected, these imports declined during the period of the entitlements program in which direct subsidies to residual fuel imports were absent (February 1975–January 1976). Their market share averaged slightly less than 6.4% and their level averaged 4.7 mmb per month. These declines, however, were fairly small and were not of strong statistical significance. Those months in which residual fuel imports have received direct entitlements show a market share for imports from other than near neighbors of slightly more than 6.2% and a corresponding average import level of 5.5 mmb. Except for three months of unusually high imports during February 1975–January 1976, these figures would be significantly higher than in that period.

The one anomalous conclusion suggested by the residual fuel import data of figures 4.5 and 4.6 is that the major part of the entitlements-induced expansion of domestic output and reduction in imports of residual fuel has been borne by imports of residual fuel from near neighbors, rather than other sources. Of the 20.8-mmb decline in average monthly imports from all sources that was observed when comparing the two-year period preceding the entitlements program to the period of no direct subsidization, only 0.8 mmb was attributable to world sources other than near neighbors. On a percentage basis, the near-neighbor reduction in residual fuel imports was 46%, compared to roughly 15% for other world sources.

These observations conflict with an expectation that other than near neighbors should be the marginal suppliers of residual fuel to the United States. This expectation arises, in part, from the transportation cost advantages enjoyed by near neighbors (and see section 4.4.2). Additionally, the price elasticity of residual fuel import supply can be expected to be less for near neighbors than for other sources. The import supply elasticity described by (4.1a) can be rewritten as

$$\epsilon_{\text{ir}} = (\epsilon_{\text{fr}} - \eta_{\text{fr}}) \frac{R_{\text{F}}}{R_{\text{I}}} + \eta_{\text{fr}}. \tag{4.1b}$$

If near-neighbor and rest-of-the-world refining operations are subject to similar technologies and costs, and if at-home demand elasticities are comparable, both ϵ_{fr} and η_{fr} would be approximately the same for near-neighbor and rest-of-the-world sources. Differences in elasticities of import supply would then depend on the ratios $R_{\text{F}}/R_{\text{I}}$. In fact, in 1974 this ratio was approximately 1.5 for near neighbors and 17.5 for rest-of-the-world (free world) sources of residual fuel.[25] Assume that ϵ_{fr} and η_{fr} are constant over relevant ranges; then this suggests that the elasticity of import supply for near neighbors is less than for other sources since ϵ_{ir} is larger when R_F/R_I is larger:

$$\frac{\partial \epsilon_{ir}}{\partial (R_F/R_I)} = \epsilon_{fr} - \eta_{fr} > 0. \tag{4.2}$$

In short, any given change in the price received in the United States could be expected to induce a smaller percentage change in imports of residual fuel from near neighbors than from other sources.

As a conjecture, the failure of the data to reveal this result may reflect a situation in which the comparatively low import levels from other than near neighbors are due to special factors. These could include long-term contracts, distress sales, attempts by US refiners to raise domestic price ceilings (prior to decontrol in mid-1976) by averaging higher-priced foreign products into their product mix, or purchases of specialty product (for example, in terms of sulfur content) from rest-of-the-world refiners. If this is the case, the US residual fuel market under the entitlements program may indeed have been isolated (in the sense implied by figure 4.2 or 4.4) from rest-of-the-world supply sources. The magnitudes of the market shares and levels of rest-of-the-world imports, however, argue against confident assertion of this conjecture.

In general, the data of figures 4.5 and 4.6 make it hard to accept the possibility that US residual fuel prices are totally independent of foreign prices under the entitlements program. Thus, if the United States is a price taker with respect to the international market as in figure 4.1, domestic residual fuel prices can only be reduced by direct subsidization of imports. The extent, if any, to which the United States is a price maker in international residual fuel markets is examined in section 4.5.

Middle Distillate Imports Imports of middle distillates historically have played a smaller role in the domestic middle distillate market than residual fuel imports have played in the domestic residual fuel market—as figure 4.5 indicates. Nevertheless, in the two years preceding the introduction of the entitlements program in November 1974, both the average market share (7.8%) and level (7.4 mmb per month) of middle distillate imports were significantly positive. For the first three months of the entitlements program (November 1974–January 1975), during the severe winter of 1977 (February–March), and again during the middle of 1979 (May–October), middle distillate imports were eligible for direct entitlements. Under this direct subsidization, the mean market share and level of imports were 5.9% and 6.3 mmb per month, respectively. Although these figures suggest fairly substantial US purchases of foreign middle distillates, they are somewhat misleading with respect to the 1979 period of direct subsidization. Taking the two earlier periods of direct subsidization alone, middle distillate imports were more significant in the domestic market, averaging 10.8 mmb per month with a monthly average market share of 9.5%. The much smaller US purchases of foreign middle distillates in the 1979 period (averaging 2.6 mmb per month and a market share of 3.0%) suggests that the 1979 direct import subsidy was not sufficient to overcome a foreign-US price differential and move the United States into the role of an active international buyer. In fact, there is evidence of a sizable price differential in the middle of 1979. Based on weekly data, Caribbean spot prices for middle distillate exceeded New York harbor spot prices by an average of 20.4¢ per gallon over May–October 1979, compared to the direct entitlements subsidy of 11.9¢ per gallon.[26] Later in this section, corroborating evidence from trade press reports is reviewed. These reports indicate little US purchasing of foreign middle distillates during the 1979 period of direct import subsidization.

As expected, both the market share and level of middle distillate imports during the entitlements program (without direct import subsidization) have been significantly less than during both the period before November 1974 and the periods of direct subsidization. The market share from February 1975 through January 1980 (excluding months of direct subsidization) averaged only 1.5%. The level of imports averaged 1.6 mmb per month. The extent of US purchasing in world middle distillate markets is obviously quite low under the entitlements program (without direct subsidization); and the observed average market share and import level are not clearly significant. Indeed, in some months, the market share of middle distillate imports has fallen well below its 1.5% average and been less than 0.5%. Consequently, it is not hard to suspect that the relatively small amounts of imports during the entitlements program reflect the special factors discussed above and that, in fact, the underlying supply-and-demand conditions are such that the United States would export middle distillates if allowed to do so. If this is so, US prices for middle distillates have been held below world prices (as in figure 4.2 or 4.4).

When imports of middle distillates from Canada, Mexico, and the Central America-Caribbean area are excluded from the data on market shares and levels, the weakness of any link between US and world middle distillate markets becomes more apparent. Over the entire period of November 1972–January 1980, more than 65% of middle distillate imports came from these near neighbors. The data recorded in figure 4.6 indicate that the monthly market share and level of imports from sources other than near neighbors were rather insignificant and averaged 3.3% and 2.7 mmb, respectively, in the two years preceding November 1974. In the periods of direct subsidization, middle distillate imports were also fairly insignificant. The market share averaged only 2.2% and imports averaged 2.6 mmb per month. When the 1979 episode is removed from these periods for the reasons noted above, US buying from other than near neighbors appears still to have been small, but more significant (market share of 4.0% and imports of 4.6 mmb per month). Except for the November 1974–January 1975 and February–March 1977 months of direct subsidization of imports, the United States appears to be substantially independent of other than near-neighbor suppliers of middle distillates during the period of the entitlements subsidy to the domestic refining of crude oil.

The entitlements subsidy to crude oil use has been accompanied by significant declines in the market share and level of middle distillate imports from other than near-neighbor sources. From February 1975 through January 1980 (excepting the months of direct subsidization), the market share of these imports averaged a mere 0.4%. Import quantities averaged only 0.4 mmb per month. In fact, figure 4.6 reveals that in 1976, for example, imports of middle distillates from other than near neighbors were zero for half of the months of the year; and, over all of 1976–early 1980, the market share of these imports did not get above 1% except during the period of direct subsidization and December 1979.

When comparing the preentitlements period to the months covered by the entitlements program, the decline in average monthly imports from other than near neighbors (2.3 mmb) might have been expected to be larger than the decline from near neighbors (5.8 mmb); or, at least, the decline in imports from distant sources might have been expected to have exhausted these imports before the development of a decline in imports from near neighbors. As discussed in connection with residual fuel, however, special factors may yield positive, but small, imports from distant markets despite underlying supply-and-demand conditions indicating otherwise. Indeed, the evidence of figure 4.6 indicates that, at the prices prevailing under the entitlements program, the United States does not have a demand for middle distillates from distant markets. The picture of the middle distillate market that emerges under the entitlements program is one in which, except for special reasons and in most periods of direct import subsidization, the United States fails to import from distant markets and may even fail to import from near neighbors.

Gasoline Imports Of the many petroleum products used domestically, gasoline accounts for the largest share of total consumption. Over 1973–1977, for example, approximately 40% of refined petroleum product consumption was in the form of gasoline (compared to 16% and 18% for residual fuel and middle distillates, respectively).[27] Gasoline's role in the refined product import market, on the other hand, is very small. In the two years preceding the introduction of entitlements, the market share of imported gasoline averaged 2.4%; and imports averaged only 4.9 mmb per month. Despite their small magnitudes, these mean shares and import levels appear to have been significantly positive. As expected, however, these figures have been reduced substantially since November 1974. Under the entitlements program (November 1974 onward), the market share of imported gasoline has averaged only 0.9%. The level of imports has averaged 1.9 mmb per month. These values do not appear to have been significant. Moreover, they fail to provide strong support for the contention that the United States is an importer of gasoline under the entitlements program.

When imports of gasoline from sources other than near neighbors are examined, the generally insignificant level of US involvement in the international market is reconfirmed. The mean market share and level of imports were not statistically significant over the period from November 1972 through October 1974. This result may be misleading, however, due to the timing of the loosening and termination of oil import controls, as well as the effects of binding controls on domestic gasoline prices in early 1974. The sharp rise in gasoline imports in March and April 1974 (see figure 4.6), for example, corresponds to a period in which price controls had resulted in shortages of domestically produced gasoline. When these two months are removed and the mean market share and level of gasoline imports from other than near neighbors are calculated over May 1973–October 1974, both the market share and level are significantly positive, although small. The mean market share over this period was 0.6% and imports averaged 1.3 mmb per month. These figures may be compared to an average share of 0.4% and average imports of 0.8 mmb per month under the entitlements program (November 1974 onward). Neither of these figures is substantially different from zero; nor is either significantly less than its counterpart during the May 1973–October 1974 preentitlements period.

While US participation in distant gasoline markets had generally been quite limited since November 1974, Lichtblau (see section 4.3.1) suggests that 1977 presented a different situation and notes that weak foreign demand for gasoline may have lowered foreign prices to levels closer to US prices. If this in fact occurred, it might be expected that the US would return to foreign gasoline markets and resume importation. In support of this possibility, figures 4.5 and 4.6 indicate that the market shares of imported gasoline rose somewhat in 1977 and thereafter tended to decline. For imports from all sources, the mean market share and the mean level were significantly different from zero over 1977; but this was not true for imports from other than near neighbors. Thus, if 1977 saw the United States reentering international gasoline markets, the evidence indicates that this reentry most likely occurred primarily in near-neighbor markets. A somewhat similar pattern seems to have applied to parts of 1979, when binding price controls and the shortages in domestic gasoline markets apparently sent US buyers into near-neighbor gasoline markets. Of course, these patterns of behavior are consistent with the expectation that near neighbors are marginal suppliers of products when the United States moves from a position in which it is not importing to the position of an importer.

The discussion of section 4.4.2 indicated that a substantial decline in US product prices is required to induce the termination of near-neighbor imports—and yet this appears to have occurred in the case of gasoline over much of the period since November 1974. The cases of gasoline imports from the Netherlands Antilles and Venezuela immediately prior to and during the entitlements program are instructive in this regard. These two countries are major Central America-Caribbean area exporters of refined petroleum products. In 1974, for example, they accounted for over 54% of near-neighbor imports of gasoline. Under the entitlements program, gasoline imports from the Netherlands Antilles and Venezuela have been substantially reduced. In fact, gasoline was not imported from these two countries in the majority of the months of 1975–1979. Over the period between the end of import controls and the start of entitlements, neither of these countries had failed to sell gasoline to the United States in even a single month.[28]

In addition to the Netherlands Antilles and Venezuela, Trinidad-Tobago has historically been a major supplier of refined products to the United States. Trinidad-Tobago exports of gasoline to the United States, for example, accounted for approximately 33% of near-neighbor sales to the United States in 1974. Unlike the Netherlands Antilles and Venezuela, Trinidad-Tobago has consistently exported gasoline to the United States under the entitlements program. This country, for example, failed to sell gasoline to the United States in only two months during 1975–1977, which appears to be somewhat anomalous in light of the Netherlands Antilles's and Venezuelan patterns of trade and the geographical proximity of the three nations. Interviews with representatives of the major refining operation (a US-based firm) in Trinidad-Tobago reveal that, although net prices received in the United States have been perceived as less than net prices on sales in distant markets during most of the entitlements period, imports of gasoline from a Trinidad-Tobago subsidiary to the United States have served to satisfy "broader company objectives." These objectives have included adjustment to domestic regulatory policies (such as banked cost regulations) and fulfillment of existing domestic contracts to deliver gasoline. In short, imports of gasoline from Trinidad-Tobago under the entitlements program appear to be due in large part to special factors. Significantly, the same company has been the major importer (until recently) of gasoline from distant sources (such as the Middle East). This was stopped, according to company representatives, because it was recognized that the purchase price (*excluding* transportation-and-tariff costs) of gasoline for export in distant markets exceeded the US pump price.

To conclude, the characterization of the US gasoline market as one in which the United States has been isolated by the entitlements program from world markets appears to be supported by available data on the patterns of import trade. The United States showed some tendency to import gasoline from near and distant sources during the period in which imports were neither controlled under the MOIP nor counteracted by the entitlements program. Except for some return to importing in parts of 1977 and 1979, however, the United States has been completely isolated from international markets since the entitlements subsidy began. By the characterizations of figures 4.2 and 4.4, EPAA/

EPCA regulation would appear to have lowered domestic gasoline prices. Moreover, the general disappearance of even near-neighbor gasoline suppliers from domestic markets indicates the regulation-induced price reductions have been substantial.

Trade Press Reports The foregoing analysis of refined product import data indicates that the only major product for which the United States can be credibly portrayed as an active importer (in the sense of figures 4.1 and 4.3) under the entitlements program is residual fuel. Imports of gasoline and middle distillates generally have been negligible under the entitlements program, although occasional purchases of these products from near neighbors have apparently taken place. The characterization of US refined product markets as markets linked by price-taking trade to world markets appears to stand up under scrutiny only in the case of residual fuel.

Notwithstanding these conclusions, the evidence on refined product imports derived from the Bureau of Mines data used in figures 4.5 and 4.6 is less than perfect. Particularly at the extremely low levels of imports observed for gasoline and middle distillates, these data do not permit clear discrimination between special factor imports and imports that arise from conditions of the type portrayed in figure 4.1 or 4.3. Fortunately, there are market observers and participants who specialize in answering precisely the kinds of questions of interest in this study (for example, Is the United States buying in world markets?; Is US buying affecting world product prices?). Perhaps the most current and authoritative of the publicly accessible specialists of this type is the *Petroleum Intelligence Weekly (PIW)*, which periodically publishes accounts and assessments of price trends and trading patterns in international refined product markets.

PIW has addressed the issues of the US presence in world markets and US-foreign price differences on numerous instances since the introduction of the entitlements program. Table 4.1 summarizes the *PIW* evidence on the US position vis-à-vis foreign refined product spot markets since early 1975. N* indicates that the United States was not purchasing products in any foreign markets; N, that *PIW* reported that the United States was not purchasing products in other than near-neighbor markets, but that no information was available to determine whether products were being imported from near neighbors; N′, that the United States was buying from near neighbors, but not from more distant sources; Y*, that the United States was buying from both near

Table 4.1 *PIW* Accounts of US Positions in International Product Markets[a]

	Jan.	Feb.	March	April	May	June	July	Aug.	Sept.	Oct.	Nov.	Dec.
1975												
Gasoline						N*			Y*	N		
Middle distillates				N*	N*	N*			N*	N*		
Residual fuel				N*	Y*	Y*			Y*	Y*	Y*	N?
1976												
Gasoline					N	Y*	N					N*
Middle distillates		N′			N				N*	N*		N*
Residual fuel	N′				N′			Y*	N′	N′	Y*	Y*
1977												
Gasoline		Y*	Y*	N*	N	Y*?	Y*	N		N*		N*
Middle distillates	N′	Y*	Y*	N*	N′	N′	N*		N*	N*		N*
Residual fuel	N′	Y*	Y*	Y*		Y*	Y*	N	Y*		N	N

1978												
Gasoline			N*	N	N			Y*	N		N*	
Middle distillates	N*	N′	N*	N*		N′	N*	N*	N*	N*	N*	
Residual fuel	N*			Y*	N				Y*	Y*	N′	
1979												
Gasoline					N	Y*?	N*	N	N	Y*	Y*	Y*
Middle distillates	N*			N*	Y*	N′?	N*	N*	N*			
Residual fuel			Y*		Y*	Y*			Y*	Y*		N
1980												
Gasoline	N	N		N	N*							
Middle distillates												
Residual fuel		Y*	N	N′	N′		Y*					

Source: Petroleum and Energy Intelligence Weekly, Inc., *Petroleum Intelligence Weekly*. Reproduced by written permission from *Petroleum Intelligence Weekly*, copyrighted 1974–80 by Petroleum and Energy Intelligence Weekly, Inc.

a. For the meaning of N, N*, N′, and Y*, see the text.

neighbors and rest-of-the-world sources; and ?, that there is some uncertainty as to the interpretation of *PIW* statements. The *PIW* reports come in forms such as*

1. Cold weather on the United States east coast, however, has no appreciable impact on gas oil [middle distillate] prices in the Caribbean or Europe. Overseas gas oil [middle distillate] has been effectively priced out of the U.S. market by lower-cost domestic oil. (28 April 1975, p. 3)

2. So far, elimination of the 60¢ a barrel fee on refined product imports into the United States has failed to stir any rise in sales of European products to the U.S., since prices in the U.S. are still generally too low to encourage it. . . . In early October, in fact, there was even a significant flow of gas oil [middle distillate] to Rotterdam from the Caribbean and Canada which under more normal circumstances would have been sold in the U.S. (13 October 1975, p. 8)

3. The big upturn in U.S. demand for fuel oil [residual fuel] imports is primarily responsible for the Caribbean price surge. . . . The spurt in U.S. fuel demand is also starting to attract European fuel supplies now, due to tight supply in the Caribbean. (22 November 1976, p. 2)

Table 4.1 tells much the same story. Over most of the period of the entitlements program, the United States has consistently imported residual fuel from both near and distant (for example, European) sources. Occasionally, residual fuel imports from other than near neighbors have been uneconomical. In contrast, middle distillate imports from both near and distant sources have been blocked consistently by domestic regulatory policies. The only exceptions appear to be brief periods in early 1977 and mid-1979. Gasoline imports, particularly from other than near neighbors, have also been relatively rare under the entitlements program. *PIW* views the appearance of US buyers in international gasoline and middle distillate markets as newsworthy, and its failure to comment on the US market role (that is, the blanks in table 4.1) apparently indicates the absence of US buyers.

PIW has summed up the case of middle distillates by noting that

U.S. companies in recent years haven't been able to afford gas oil [middle distillate] imports, since spot prices abroad have been too high in relation to low controlled domestic prices . . . In fact, the U.S. hasn't imported any substantial volume in more than five years. U.S. price controls have kept U.S. oil firms out of the world market in recent years, in effect "giving the Caribbean gas oil supply to Europe" as one trader put it. (11 June 1979, p. 5)

*Reproduced by written permission from *Petroleum Intelligence Weekly*, copyrighted 1974–80 by Petroleum and Energy Intelligence Weekly, Inc.

The story in gasoline appears to be much the same. Significantly, those instances when US buyers have appeared in other than near-neighbor gasoline markets (for example, June 1976, the first half of 1977, June 1979, and late 1979) appear to be transitory, disequilibrium periods. Commenting on the June 1976 episode, for example, *PIW* reports that

> a sharp decline in United States east coast prices killed off the burgeoning early June trans-Atlantic gasoline trade almost as quickly as it began. Traders generally don't see any revival of this market for the rest of the summer. U.S. buying of European gasoline was strictly a "nine cargo wonder" in the words of one trader. Even some of those were sold at a loss as U.S. prices sagged while speculative cargoes were en route. (12 July 1976, p. 5)

Regarding the 1977 period (noted by Lichtblau, 1977, as a period in which US-foreign price equality may have been restored), *PIW* comments that

> the big blow to Europe's gasoline expectations came in a sudden plunge in landed United States east coast prices, resulting in big losses on speculative European cargoes and probably killing off hopes for a lively Europe-U.S. gasoline trade for the rest of the summer. . . . At least five unsold cargoes of European gasoline arrived in the U.S. in the past few weeks, with losses running as high as a half-million dollars on some cargoes. Further distress sales could result from disposing of gasoline already prepared to U.S. specifications. . . . It's now clear that European prices this summer will be determined solely by European demand, with no assist from U.S. buying. (16 May 1977, p. 7)

In parts of 1979, US gasoline sellers found themselves subject to binding price ceilings and public pressure to avoid shortages. *PIW* notes that

> although U.S.-grade gasoline costs about 10¢ a U.S. gallon more in Europe than on the U.S. east coast, a number of inquiries for this grade were reported in Europe in early April. Some U.S. companies were even reported willing to buy high-priced European gasoline to roll into their U.S. cost-base. (16 April 1979, p. 6)

Moreover,

> U.S. prices were too low to justify European imports. But oil companies on both sides of the Altantic were under public pressure to reduce shortages, which could force some buying of high-priced spot cargoes. (16 July 1979, p. 8)

In this setting for 1979 gasoline markets and with a direct subsidy for middle distillate imports,

U.S. buyers of light refined products (gasoline and distillate) . . . entered the world market briefly in June but pulled back quickly as domestic prices dipped. This left some speculative traders with unsold oil supplies "on the water" to the U.S. European and Caribbean light product supplies are still uneconomic versus domestic U.S. alternatives (2 July 1979, p. 7)

In late 1979, however,

Some gasoline was reported being imported to the U.S. at a loss to ensure retail supplies, due to public criticism of the oil companies both for last summer's shortages and for "excess" profits. Landed costs at New York harbor last week were about $1.12 a U.S. gallon, versus pump prices as low as $1.00. (12 November 1979, p. 7)

The late 1979 surge in US importing was short-lived, as "Slumping prices and demand in the United States largely dried up buying of European gasoline," (14 January 1980, p. 8), and "Collapse of the U.S. market has driven Mediterranean cargoes into Northwest Europe, undermining prices there" (21 April 1980, p. 7). In fact, by early 1980, the only market in the United States for foreign refined products was in specially prepared residual fuels:

With U.S. markets [for middle distillate and high sulfur residual fuel] saturated, Caribbean supplies are continuing to flow to Europe. . . . The only exception is U.S.-oriented 0.3% sulfur fuel oil [residual fuel]. (26 May 1980, p. 5)

The general picture of US refined product markets under the entitlements program that emerges from the trade press reports summarized in table 4.1 is one in which the United States is isolated from international gasoline and middle distillate markets, but usually tied by trade to international residual fuel markets. In the case of the lighter products, US prices are too low to permit product imports from traditional Central America-Caribbean area and European sources; and the trade press is rife with references to products from these sources being "locked out" of US markets, "abnormal" movements of products from the Central America-Caribbean area to Europe (rather than the United States), and US regulatory policies that have lowered domestic oil prices below world market levels.

Minor Products The refining of crude petroleum typically results in the creation of dozens of marketable products. While the greatest portion of these refined products is accounted for by gasoline, middle distillates, and residual fuel, numerous minor products are produced.

These often serve specialty interests and generally compete with few substitutes. In the case of some of these minor products, the concentration of interests has apparently led policymakers to be concerned with the perceived effects of regulation on their prices. Commercial airlines, for example, have consistently presented organized (and at least partially successful) opposition to proposals for the deregulation of petroleum prices based on their perception of the likely effects of such proposals on the price of jet fuel.[29]

Some insight into the nature of the markets for minor products and, in particular, some insight into the effects of current federal regulation can be obtained through examination of disaggregated data. Tables 4.2A and 4.2B show the market shares of imports of nine minor refined products from all sources and from other than near neighbors during 1973–1979. The market shares of these imports have generally declined since the introduction of the entitlements program. By 1976, the market shares of imported naphtha jet fuel, kerosene, special naphthas, asphalt, and petrochemical feedstocks were all less than 1% (the special case of lubricants is discussed below). In fact, table 4.2B makes it clear that even the small amounts of imports of these products came almost entirely from near neighbors. Only kerosene jet fuel and liquefied petroleum gas had market shares of imports from other than near neighbors in excess of 1% in 1976. The short-lived decline in the entitlements subsidy in 1978 and the disequilibrium in international markets in 1979 may have been the sources of the reappearance of naphtha jet fuel and special naphtha imports in those years.

The important issue to be addressed is whether the imports of minor products are sufficiently large to justify a characterization of the United States as a significant importer under the entitlements program (that is, figure 4.1 or 4.3). Tables 4.3A and 4.3B provide information on the frequency of US participation in international minor refined product markets. The entries in table 4.3A indicate the number of months, in each year from 1973 through 1979, in which the United States did *not* import any minor products. Table 4.3B indicates the number of months in which the United States did *not* import minor products from any distant sources.

Tables 4.3A and 4.3B suggest that the entitlements program has typically isolated the United States from international naphtha jet fuel and kerosene markets. The frequency of naphtha jet fuel imports picked up substantially in 1979—a year when shortages were reported in domestic markets. Still, during 1975–1979, the United States failed

Table 4.2A Import Market Shares of Minor Refined Products from All Sources: 1973–1979[a]
(Percent)

	Naphtha jet fuel	Kerosene jet fuel	Kerosene	Special naphtha	Lubricants	Wax	Asphalt	LPG	Petrochemical feedstocks
1973	13.00	19.39	0.82	0.27	0.10	5.42	4.63	11.68	2.19
1974	6.83	17.29	1.59	2.84	0.54	5.13	6.67	11.51	2.44
1975	1.29	12.66	0.07	0.16	0.25	2.60	3.36	11.23	1.68
1976	0.71	6.91	0	0.15	0.35	2.27	0.22	12.18	0.90
1977	0.01	6.81	0.01	3.15	2.35	NA	0.94	11.35	2.04
1978	2.14	7.58	0	4.10	2.26	NA	0.52	8.69	0.61
1979	3.69	5.90	0.48	8.49	1.93	NA	0.85	13.70	0.62

Source: US Department of the Interior, Bureau of Mines, *P.A.D. Districts Supply/Demand Annual.*
a. NA, not available.

Table 4.2B Import Market Shares of Minor Refined Products from Other Than Near Neighbors: 1973–1979[a]
(Percent)

	Naphtha jet fuel	Kerosene jet fuel	Kerosene	Special naphtha	Lubricants	Wax	Asphalt	LPG	Petrochemical feedstocks
1973	4.06	5.09	0.79	0	0.10	3.69	0.06	0.77	0.42
1974	0.94	4.44	1.50	0	0.28	3.15	0+	0.54	0.57
1975	0.32	3.47	0	0	0.18	1.22	0.01	1.34	0.13
1976	0.02	1.82	0	0	0.07	0.43	0.04	1.56	0.11
1977	0+	2.62	0+	0	2.15	NA	0	1.48	0.21
1978	1.43	2.69	0	2.14	1.98	NA	0	0.75	0.11
1979	1.76	2.39	0.48	5.16	1.56	NA	0	2.22	0.20

Source: US Department of the Interior, Bureau of Mines, *P.A.D. Districts Supply/Demand Annual.*
a. NA, not available.

Table 4.3A Import Patterns of Minor Refined Products from All Sources: 1973–1979[a]
(Number of Months without Imports)

	Naptha jet fuel	Kerosene jet fuel	Kerosene	Special naphtha	Lubricants	Wax	Asphalt	LPG	Petrochemical feedstocks
1973	0	0	2	1	0	0	0	0	0
1974	0	0	1	1	0	0	0	0	2
1975	7	0	9	2	2	0	0	0	6
1976	10	0	12	0	1	0	0	0	6
1977	7	0	9	0	2	NA	0	0	0
1978	5	0	12	0	0	NA	4	0	0
1979	1	0	10	0	0	NA	2	0	0

Source: US Department of the Interior, Bureau of Mines, *P.A.D. Districts Supply/Demand Monthly*.
a. NA, not available.

Table 4.3B Import Patterns of Minor Refined Products from Other Than Near Neighbors: 1973–1979[a]
(Number of Months without Imports)

	Naphtha jet fuel	Kerosene jet fuel	Kerosene	Special naphtha	Lubricants	Wax	Asphalt	LPG	Petrochemical feedstocks
1973	3	0	9	12	4	0	8	0	10
1974	5	0	6	12	0	0	12	0	9
1975	10	0	12	12	3	0	11	0	11
1976	11	0	12	12	4	2	11	1	9
1977	10	0	11	12	4	NA	12	0	2
1978	7	0	12	7	2	NA	12	0	6
1979	2	0	11	6	0	NA	12	0	5

Source: US Department of the Interior, Bureau of Mines, *P.A.D. Districts Supply/Demand Monthly*.
a. NA, not available.

to import any naphtha jet fuel in a majority of the 60 months of the period and only imported from other than near neighbors in 20 months. The United States consistently imported kerosene from near neighbors prior to the entitlements program. Imports from other sources were somewhat sporadic. Still, in half of the months of 1973–1974, the United States imported kerosene from Europe. In the 60 months of 1975–1979, however, the United States purchased kerosene from other than near neighbors in only 2 months; and the United States imported kerosene from near neighbors in only 8 months of the 1975–1979 period. Moreover, exports of kerosene, which are subject to the provisions of the export quota program, have exceeded imports under the entitlements program, despite the fact that crude oil used to produce exported kerosene does not receive the entitlements subsidy. In 1975, for example, imports of kerosene were 45 mb, while exports were 52 mb. Observed import levels must have been due to special factors. These patterns support a characterization of the underlying supply-and-demand conditions in the domestic markets for naphtha jet fuel and kerosene as constituting markets in which the entitlements subsidy to domestic refining has usually driven true imports to zero and reduced domestic prices.

A variant of this conclusion on the effects of the entitlements subsidy appears to apply to petrochemical feedstocks. During 1973–1974, the United States consistently imported feedstocks from near neighbors (primarily Venezuela and Trinidad-Tobago) and occasionally imported from European, Asian, and Middle Eastern countries. On net, however, the United States was an exporter. This pattern apparently reflects the notable lack of homogeneity in the data category of "petrochemical feedstocks"; that is, the United States was an importer of some feedstocks and an exporter of others prior to entitlements. At any rate, in 1975–1976, the United States imported petrochemical feedstocks in only half of the 24 months and only imported from other than near neighbors in four months. The trend in imports was briefly reversed in 1977. The United States imported feedstocks from Europe (albeit in fairly small quantities) in all but two months of 1977. In 1978 and 1979, however, the frequency of imports was back down. As testimony to the insignificance and special factor nature of the observed levels of petrochemical feedstock imports, exports of these products have exceeded imports under the entitlements program—despite the fact that crude oil used to produce feedstocks that are exported is not eligible for the entitlements subsidy. In 1975, for example, exports of petrochemical feedstocks were four times greater than imports.

Special naphthas and asphalt represent cases in which the United States has imported little or nothing from other than near neighbors either before or after the introduction of the entitlements program. Even imports of these products from near neighbors, moreover, may reflect special circumstances in which the United States can hardly be considered to be a price taker and is effectively isolated from world markets. Imports of special naphthas, for example, came entirely from Canada during 1973–1977, except for three months in 1977. These imports, in fact, enter the United States almost exclusively from central Canada via the North Central states. The relatively high transportation costs that Canadian sellers would face in trying to sell in any other market increase the amount by which the price paid by US purchasers can fall without driving imports out of the North Central US market. The only times the United States has purchased special naphthas from other than near neighbors were in the summer of 1978 and parts of 1979. The first episode reflected weak European prices and the temporary decline in the entitlements subsidy, while the 1979 importing from distant sources was in response to domestic refiners' buying of naphthas for re-forming into gasoline during the memorable gasoline shortages. By the second half of 1979, special naphtha imports from distant sources had ceased. In general, the United States has been a net exporter of special naphthas. This was true even in 1978. For numerous reasons, then, observed imports are apparently not representative of underlying conditions in the domestic special naphtha market. With the possible exception of the summer of 1978, figures 4.1 and 4.3 have generally been inapplicable to the special naphtha case.

In the case of asphalt, US imports have historically come almost exclusively into the East Coast from the Netherlands Antilles, Venezuela, and Canada. Under the entitlements program, imports of asphalt from Canada and Venezuela have become fairly insignificant and nothing was imported from Venezuela in ten months of 1976. The Netherlands Antilles, however, has continued to sell asphalt to the United States since the introduction of the entitlements program. This anomaly (that is, Venezuela and the Netherlands Antilles are less than 100 miles apart) may result from special circumstances of the same type as those that have led to consistently positive imports of gasoline from Trinidad-Tobago while other Central America-Caribbean refiners have generally ceased sales to the United States. In short, domestic asphalt markets, as well as special naphtha markets, have apparently been isolated from international trade under the entitlements program.

Tables 4.3A and 4.3B provide no indication that the entitlements subsidy has resulted in the isolation of US kerosene jet fuel or LPG markets from trade with either near neighbors or other sources. Although the market shares of the imports of these products from other than near neighbors are small, the United States consistently imported positive amounts from distant sources during 1975–1979. Through the end of 1978, the United States had received kerosene jet fuel from distant sources in every month since the beginning of the entitlements program. Over the same period, the United States failed to import LPG from distant sources in only one month (September 1976). Representation of the domestic kerosene jet fuel and LPG markets as ones in which the entitlements subsidy has driven out all imports and reduced domestic prices is not supported by the data.

The same conclusion seems to be justified by tables 4.3A and 4.3B for the wax market. This conclusion, however, is probably incorrect. As with kerosene jet fuel and LPG, the United States has consistently continued to import positive amounts of wax from both near neighbors and more distant sources since the introduction of entitlements. The quantities of these imports, however, have been extremely small under the entitlements program. During 1973–1974, wax imports from distant sources averaged 22.7 mb per month. During 1975–1976, these imports averaged only 6.6 mb per month. The insignificance of this quantity is attested to by the fact that more than seven times this amount was exported from the United States to both near and distant markets in 1975 under the provisions of the export license program. This suggests that observed imports do not provide an accurate picture of the supply-and-demand conditions underlying the domestic wax market since the introduction of the entitlements program. It is most likely the case that US supply-and-demand conditions are such that the United States is not an importer of wax. If the crude oil behind exports of wax did not lose the entitlements subsidy, the United States would export more wax under the entitlements program. In the presence of such a regulatory penalty to exports, the entitlements subsidy to domestic refining reduces domestic wax prices.

Lubricating oils receive special treatment under both export controls and the entitlements program. Specifically, lubricating oils are not subject to export quotas, and the crude oil used to produce exported lubricating oils is eligible for the entitlements subsidy (unlike crude oil that can be attributed to other exported products).[30] Given these provisions, domestic refiners have been exporting significant quantities of lubricating oils under the entitlements program. In 1975, for example, the United States exported 9.1 mmb of lubricating oils. Imports from near neighbors and Europe totalled only 0.1 mmb over the same period. Thus any attempt to characterize the underlying supply-and-demand conditions in the domestic lubricating oils market in such a way as to conclude that the United States is a price-taking importer is undoubtedly incorrect. Of course, this is not to say that the entitlements program has lowered domestic lubricating oil prices. Such a conclusion requires that US exporters of lubricating oils not be price takers in international markets. Based on the 43% share of US lubricating oil production in the world market prior to entitlements, the assumption of price taking by US producers as a whole seems unlikely to be true.

Summary By way of summary, the analysis of US importing behavior before and after the introduction of the entitlements program indicates that, in the cases of some refined products, the United States is typically an importer in the absence of regulatory policies which discourage such behavior and that imports have essentially ceased since November 1974. These products include gasoline, middle distillates, naphtha jet fuel, asphalt, and kerosene. For these products, domestic prices have been lowered relative to international prices (as in figure 4.2 or 4.4). Other products appear to have been consistently imported from near neighbors and, most likely, from more distant markets both before and after November 1974. These include residual fuel, kerosene jet fuel, and LPG. It is most likely that these products have not had their prices reduced relative to foreign prices by the entitlements program—except for residual fuel during the periods of direct import subsidization.

In the cases of products such as special naphthas, lubricating oils, petrochemical feedstocks, and wax, the United States tends to be an exporter in the absence of entitlements and export controls. Special naphtha, petrochemical feedstock, and wax exports have been officially blocked and/or discouraged during the entitlements program; and these products can be expected to have had prices below world levels since the beginning of the program. Lubricating oil exports have not been officially discouraged for most of the period since November

1974. Consequently, US prices should be equated to international prices (after appropriate adjustment for transportation-and-tariff costs); and the large market share of US lubricating oil production in world output suggests that the entitlements subsidy to domestic refiners has most likely had a depressing effect on worldwide prices. The possibility of US price making in international lubricant and other world product markets is investigated in detail in the following section.

4.4.4 US Price-Making Power in Refined Product Markets

The United States is the largest single petroleum product consumer and producer in the world. The total combined production of refined products by Western Europe is typically about the same as US production, but US consumption has generally exceeded Western European consumption by approximately 20% in the 1970s.[31] The US shares of world refined product production and consumption are shown on a product-by-product basis in table 4.4. These market shares attest to the potential for US market power in world refined product trading.

Table 4.4 US Market Shares in World Production and Consumption of Refined Petroleum Products: 1974[a]
(Percent)

	Share of world production	Share of world consumption
Gasoline	47.5	48.1
Middle distillates	22.0	23.5
Residual fuel	9.4	17.9
Jet fuel (all types)	44.8	51.5
Special naphthas	5.2	4.5
Lubricating oils	43.1	33.9
Wax	45.7	44.9
Asphalt[b]	41.9	42.9
LPG[c]	11.3	45.2
Petrochemical feedstocks	NA	NA
Kerosene	6.3	7.1
All products	24.7	30.8

Sources: US Department of the Interior, Bureau of Mines, *International Petroleum Annual* (1974); and United Nations (1975).

a. World production and consumption of special naphthas, lubricating oils, wax, and asphalt are exclusive of Sino-Soviet totals. United States production and consumption of each product includes figures for US territorial areas eligible for the entitlements program. NA, not available.

b. Asphalt includes asphalt and road oils.

c. United States LPG production is based on production at refineries. When LPG production from all US sources is included, the US market share of world production is 43.1%.

Notwithstanding the entitlements program, the United States continues to sell or purchase selected products in international markets. Unless trade is directly subsidized, such products are subject to price reductions only to the extent that entitlements-induced changes in US supplies affect world prices. This section estimates the extent of US price-making power in world refined product markets and examines associated price changes resulting from the entitlements subsidy. In keeping with the usual conclusions of standard incidence analysis, the effect of the entitlements subsidy on market prices is found to depend on relevant supply and demand elasticities.

International Trade and the Entitlements Subsidy Representative domestic and foreign markets for an individual refined product typically imported by the United States are shown in figures 4.7A and 4.7B. The US demand for imports is the horizontal difference between domestic demand D_u and domestic supply R_u below the intersection of D_u and R_u at P_U. This excess US demand is shown in figure 4.8 as D_i. The foreign supply of refined products to the United States is shown in figure 4.8 as R_i and is the horizontal difference between D_f and R_f above their intersection at P_F (in figure 4.7B). Transportation costs and tariffs are omitted for simplicity here, but are accounted for explicitly in the empirical analysis below.

The equilibrium world price of the representative refined product is determined in the excess supply-and-demand market depicted in figure 4.8. As shown, this price is P_{UR}, and R_I is supplied to the United States by foreign sources. The introduction of an entitlements subsidy causes the domestic supply of the product to increase. This is shown by the shift from R_u to $\hat{R}_u$ in figure 4.7A. The result of this increase in supply is to reduce the US demand for imports to $\hat{D}_i$. In fact, figure 4.8 makes it clear that the entitlements program is tantamount to a tax on imports of the refined product. This "tax" is approximated by $P_U - \hat{P}_U$, which is the change in the US autarky price. When this tax exceeds $P_U - P_F$, the United States discontinues its importing; and, with export controls in the United States, the domestic price of products falls from P_{UR} to below P_F. When the tax is less than $P_U - P_F$, the United States and foreign price falls to $\hat{P}_{UR}$—as in figure 4.8. If import supply R_i is perfectly elastic as in the RAND Corporation's (1977) model, there is no change in P_{UR}.

The magnitudes of any entitlements-induced reductions in the prices of those products traded in markets that resemble figures 4.7A and 4.8

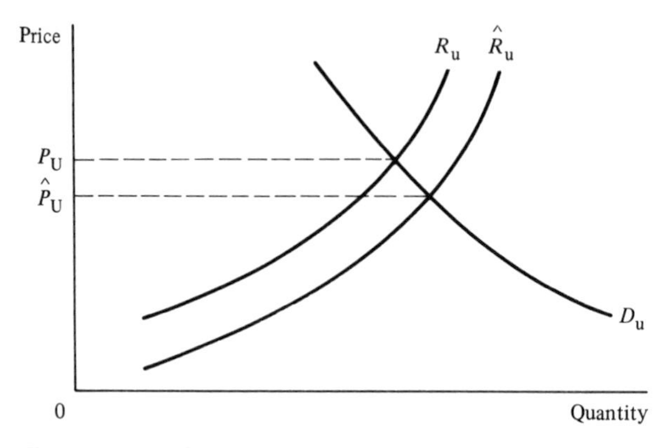

Figure 4.7A US product supply and demand.

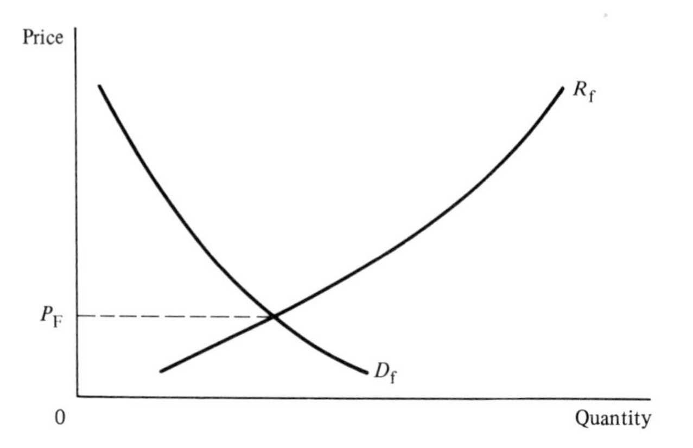

Figure 4.7B Foreign product supply and demand.
Kalt 4.7B

depend upon the price elasticities of import supply and demand, as well as the size of the implicit tax on product imports (that is, $P_U - \hat{P}_U$). This tax, in turn, depends directly on the US own-demand and own-supply price elasticities—as suggested by figure 4.7A. In order to develop explicit relationships relating these various elasticities to the effect of the entitlements subsidy on product prices, it is instructive to look first at import supply and demand.

Without straining the tax analogy, the effect of the implicit tax of the entitlements program on product import demand can be usefully thought of as a net price change (from P_{UR} to $\hat{P}_{UR} + P_U - \hat{P}_U$) that induces a movement along D_i until $\hat{R}_I$ is demanded. From standard incidence analysis (and figure 4.8), this net price change is less than the full tax so long as refined product import supply/demand is not per-

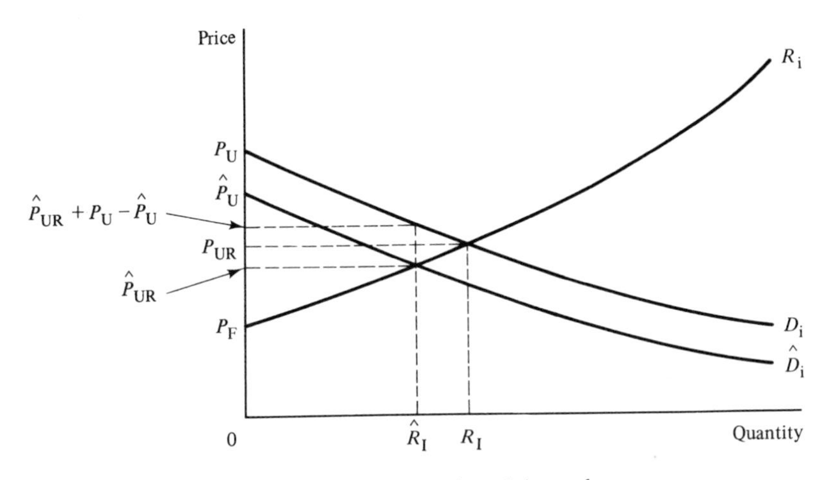

Figure 4.8 International product excess supply and demand.
Kalt 4.8

fectly inelastic/elastic; and the effect of the tax on import demand is partially offset by the induced reduction in the price of imports. As shown, the net price change relevant to import demand is $(P_U - \hat{P}_U) - (P_{UR} - \hat{P}_{UR})$. Noting this permits the elasticity of the demand for imports to be approximated by

$$\eta_{ir} = \frac{\Delta R_I^d}{R_I} \frac{P_{UR}}{(P_U - \hat{P}_U) - (P_{UR} - \hat{P}_{UR})} \leqslant 0, \tag{4.3}$$

where ΔR_I^d is the induced change in import quantity demanded (that is, $\hat{R}_I - R_I$).

When the entitlements program is imposed, suppliers of product imports merely perceive a change in the price they receive. This price change is $\hat{P}_{UR} - P_{UR} \leqslant 0$. Thus, the elasticity of import supply can be expressed as

$$\epsilon_{ir} = \frac{\Delta R_I^s}{R_I} \frac{P_{UR}}{\hat{P}_{UR} - P_{UR}} \geqslant 0, \tag{4.4}$$

where ΔR_I^s is the induced change in import quantity supplied (that is, $\hat{R}_I - R_I$).

A useful relation between (4.3) and (4.4) can be examined by noting that, in equilibrium, $\Delta R_I^d = \Delta R_I^s$ and writing (after simplifying)

$$\frac{\eta_{ir}}{\epsilon_{ir}} = \frac{\hat{P}_{UR} - P_{UR}}{(P_U - \hat{P}_U) + (P_{UR} - \hat{P}_{UR})}. \tag{4.5}$$

Rearrangement of the terms in (4.5) allows the supply-side incidence of the implicit product import tax (that is, the induced decline in product prices $P_{UR} - \hat{P}_{UR}$) to be expressed as a fraction of the implicit tax $P_U - \hat{P}_U$:

$$\frac{P_{UR} - \hat{P}_{UR}}{P_U - \hat{P}_U} = \frac{-\eta_{ir}}{\epsilon_{ir} - \eta_{ir}}. \tag{4.6}$$

This expression, of course, is a standard result of incidence analysis. Its usefulness, however, is limited unless an estimate of the entitlements-induced shift in the demand for imports and, thereby, $P_U - \hat{P}_U$ can be obtained. Figures 4.7A and 4.8 indicate that this shift can be approximated by the change in product price that would result from the entitlements subsidy if the United States were completely isolated from world markets. This change in the autarky price, in turn, depends on the magnitude of the subsidy-induced change in the supply of domestic refined products (the move from R_u to $\hat{R}_u$ in figure 4.7A). Following industry consensus, the RAND Corporation (1977), and available evidence, the elasticity of substitution in refining between crude oil and other factors is taken to be negligible. Consequently, with an input-output ratio of approximately unity, a given crude oil subsidy translates into an equivalent fall in refiners' supply-price schedules. As a result, the subsidy-induced change in the US autarky price and, hence, the induced shift in import demand can be expressed directly as a fraction of the entitlements subsidy (SUBSIDY). Specifically, in the autarky case, the elasticities of domestic demand and supply can be written, respectively, as

$$\eta_{ur} = \frac{\Delta R_U^d}{R_U} \frac{P_U}{\hat{P}_U - P_U} \leq 0 \tag{4.7}$$

and

$$\epsilon_{ur} = \frac{\Delta R_U^s}{R_U} \frac{P_U}{\text{SUBSIDY} - (P_U - \hat{P}_U)} \geq 0, \tag{4.8}$$

where ΔR_U^s is the change in autarky quantity supplied; ΔR_U^d, the change in autarky quantity demanded; and SUBSIDY $- (P_U - \hat{P}_U)$, the net subsidy to domestic refining after taking into account any subsidy-induced decline in product prices. Since $\Delta R_U^d = \Delta R_U^s$ under autarky, (4.7) and (4.8) can be combined to yield

$$\frac{\eta_{ur}}{\epsilon_{ur}} = \frac{\text{SUBSIDY} - (P_U - \hat{P}_U)}{\hat{P}_U - P_U}. \tag{4.9}$$

Rearranging (4.9) gives

$$\frac{P_U - \hat{P}_U}{\text{SUBSIDY}} = \frac{\epsilon_{ur}}{\epsilon_{ur} - \eta_{ur}}, \tag{4.10}$$

which is the subsidy-case analogue to the tax case described by (4.6).

Solving (4.10) for $P_U - \hat{P}_U$ and substituting into (4.6) yields the result of interest:

$$\frac{P_{UR} - \hat{P}_{UR}}{\text{SUBSIDY}} = \frac{-\eta_{ir}}{\epsilon_{ir} - \eta_{ir}} \frac{\epsilon_{ur}}{\epsilon_{ur} - \eta_{ur}}. \tag{4.11}$$

When the United States is an importer of products under the entitlements program, (4.11) is the fraction of the entitlements subsidy borne by product consumers in the form of lower prices. A similar expression applies to the cases in which the United States is a product exporter.[32] Significantly, the subsidy incidence given by (4.11) indicates that this incidence depends on more than the elasticity of import supply—the RAND study notwithstanding. For example, even with a high (but less than infinite) value for ϵ_{ir}, the left-hand side of (4.11) can be large (and approach unity) if η_{ir} is large and/or (4.10) is large.

The Quantitative Significance of US Price Making in World Product Markets Numerical values for (4.11) require estimates of ϵ_{ur}, η_{ur}, ϵ_{ir}, and η_{ir}. Independent estimates of ϵ_{ur} and η_{ur} have been made by numerous analysts. Estimates of ϵ_{ir} and η_{ir} on a disaggregated, product-by-product basis, however, are more difficult to come by. For the task at hand, these elasticities can be written, respectively, as (4.1a) and

$$\eta_{ir} = \eta_{ur} \frac{R_{UI}}{R_I} - \epsilon_{ur} \frac{R_U}{R_I}. \tag{4.12}$$

As the RAND Corporation (1977) study has emphasized, ϵ_{ir} should be adjusted to reflect transportation-and-tariff costs. As it stands, (4.1a) shows the elasticity of the supply of imports with respect to the FAS price. The elasticity of import supply with respect to US prices is approximated by multiplying (4.1a) by the ratio of CIF to FAS prices.[33]

Table 4.5 shows estimates of short- and long-run import and domestic product supply-and-demand elasticities. Estimates of domestic product demand elasticities are from Alt, Bopp, and Lady (1976) and are representative of the magnitudes reported by other investigators.[34] Estimates of non-US product demand elasticities are from Houthakker and Kennedy as reported in Taylor (1977) and are based on data for nine non-US OECD countries. No estimates of the foreign elasticity of demand for jet fuel, special naphthas, lubricating oils, wax, asphalt, LPG, or petrochemical feedstocks are available. For these products, it is assumed that $\eta_{fr} = \eta_{ur}$. Alt, Bopp, and Lady do not provide estimates of η_{ur} for naphtha jet fuel, special naphthas, lubricating oils, wax, asphalt, or petrochemical feedstocks. They do, however, provide estimates for a category of "other" products. Except for naphtha jet fuel, special naphthas, and petrochemicals, the estimates for other products are used for product elasticities not otherwise estimated. The estimated values of η_{ur} for kerosene jet fuel are used to estimate η_{ur} for naphtha jet fuel; and the estimates of η_{ur} for LPG provided by Alt, Bopp, and Lady are assumed to hold for special naphthas and petrochemicals. The short run is taken to be one year, and the long-run estimates are based on the reported limiting values of η_{ur}.

Following the lead of the RAND Corporation (1977) study in its analysis of aggregate import supply elasticities (see section 4.3.2), 1974 values for production, consumption, and international trade flows are used in the hopes of capturing a characterization of the world product market relatively free of effects of the MOIP and the entitlements program. Also after the RAND study, it is assumed that $\epsilon_{ur} = \epsilon_{fr}$ and, in the short run, $\epsilon_{ur} = 0.2$. For the longer run, $\epsilon_{ur} = 0.5$. Data on production, consumption, and trade flows are provided by the Bureau of Mines and the United Nations. Where possible, data for the Sino-Soviet bloc are included.[35] United States territories and possessions qualifying for the entitlements subsidy are treated as part of the domestic market. Data on product-by-product CIF and FAS prices are taken from the Department of Commerce figures; and CIF values typically exceeded FAS values by roughly 7% in 1974.[36] For those products that the United States typically exports (wax, lubricating oils, and petrochemicals), the values of ϵ_{ir} and η_{ir} in table 4.5 refer to the elasticities of US export supply and foreign demand for US exports, respectively.

Table 4.5 indicates that the import supply elasticities faced by the United States are generally large. Only jet fuel and LPG have short- or

Table 4.5 US Power in World Refined Product Markets

	Gasoline		Middle distillates		Residual fuel		Jet fuel		Lubricating oils	
	SR	LR	SR	LR	SR	LR	SR	LR	SR	LR
Elasticity of import demand[a]	−35.30	−70.29	−5.00	−9.56	−0.66	−1.01	−2.77	−4.64	−3.28	−6.26
Elasticity of import supply[b]	54.67	107.82	27.31	58.40	13.10	21.97	3.25	5.80	2.12	4.24
Elasticity of domestic demand	−0.30	−0.51	−0.19	−0.26	−0.22	−0.23	−0.21	−0.21	−0.28	−0.44
Elasticity of domestic supply	0.20	0.50	0.20	0.50	0.20	0.50	0.20	0.50	0.20	0.50
Reduction in price as percentage of subsidy—US participation in world markets	15.57	19.63	7.86	9.22	2.31	3.01	22.45	31.30	16.23	40.38
Reduction in price as percentage of subsidy—no US participation in world markets	39.68	49.75	50.76	65.53	48.19	68.59	48.78	70.42	41.32	53.19

	Wax		Asphalt		LPG		Petrochemical feedstocks		Kerosene	
	SR	LR	SR	LR	SR	LR	SR	LR	SR	LR
Elasticity of import demand[a]	−31.54	−60.75	−19.97	−38.66	−0.31	−0.41	−17.22	−30.91	−3.43	−15.01
Elasticity of import supply[b]	24.22	47.14	28.88	56.21	0.87	1.71	53.71	95.87	49.65	383.98
Elasticity of domestic demand	−0.28	−0.44	−0.28	−0.44	−0.18	−0.18	−0.18	−0.18	−0.17	−1.08
Elasticity of domestic supply	0.20	0.50	0.20	0.50	0.20	0.50	0.20	0.50	0.20	0.50
Reduction in price as percentage of subsidy—US participation in world markets	23.37	29.95	16.89	21.68	13.72	14.16	39.75	55.52	3.49	1.19
Reduction in price as percentage of subsidy—no US participation in world markets	41.32	53.19	41.32	53.19	52.49	73.42	52.49	73.42	54.05	31.65

Sources: Based on data from Alt, Bopp, and Lady (1976); Houthakker and Kennedy, as reported in Taylor (1977); US Department of the Interior, Bureau of Mines, *Petroleum Statement Annual* (1975); and United Nations (1975).

a. Elasticity of foreign demand for US exports in the cases of lubricating oils, wax, and petrochemical feedstocks.

b. Elasticity of US export supply in the cases of lubricating oils, wax, and petrochemical feedstocks.

long-run import supply elasticities that are less than 10; and exports of lubricating oils face a demand elasticity less than 10. The expression in (4.11), however, indicates that the effect of the entitlements subsidy on world prices depends on more than ϵ_{ir}. Accordingly, the last two rows of table 4.5 report estimates of the percentage of the entitlements subsidy that is reflected in lower product prices. When the United States is an active participant in world markets, (4.11) is applicable and is shown in the next-to-last row of table 4.5.[37] When the United States is not a participant in world markets, (4.10) is applicable and is shown as the last row of table 4.5.

The results of table 4.5 should be subject to some skepticism in view of the necessary assumptions concerning data and parameters. Nevertheless, these results are suggestive of the potential for US policies to affect refined product prices. When the United States is a participant in international markets, it shows greatest short- and long-run price-making ability in petrochemical feedstocks, wax, jet fuel, asphalt, lubricating oils, and gasoline. The United States apparently has virtually no international price-making ability in residual fuel or kerosene markets. For most products, when the United States is isolated from international markets, 40% or more of a subsidy to the US refining industry shows up as a price reduction in the short run. In the long run, most products have 50% or more of a subsidy reflected in lower prices.

Table 4.5 leaves unanswered the question, Which measure of subsidy incidence is applicable for any particular refined product? The analysis above has indicated that the United States has apparently been removed from international markets in several products by the effects of the entitlements program. These products include gasoline, middle distillates, asphalt, and kerosene. For these products, the last row of table 4.5 [which estimates $(P_U - \hat{P}_U)$/SUBSIDY] is an upper bound on the fraction of the entitlements subsidy which has been passed on to product consumers: If $P_U - \hat{P}_U$ exceeds $P_U - P_{UR}$ and the United States discontinues imports as a result of the entitlements program, $\hat{P}_U$ becomes the US price, the induced price change is $P_{UR} - \hat{P}_U$, and $(P_{UR} - \hat{P}_U) \leq (P_U - \hat{P}_U)$. Similarly, the next-to-last row of table 4.5 is a lower bound on the fraction of subsidy passthrough for these products. It estimates $P_{UR} - \hat{P}_{UR}$, while the actual price change is $(P_{UR} - \hat{P}_U) \geq (P_{UR} - \hat{P}_{UR})$. Once the United States has been moved out of international markets by the entitlements program, however, the last row of table 4.5

Table 4.6 Reduction in Price of Weighted Average US Barrel of Refined Products—Percentage of Entitlements Subsidy

Assumption on US role in international product markets	Domestically produced barrel[a]		Domestically consumed barrel[a]	
	SR	LR	SR	LR
US trades, selected products[b]	37.25	47.49	32.48	41.08
US trades, all products	13.94	18.77	13.93	17.83
US trades, no products	44.73	58.01	44.73	58.01

Sources: Based on data from Alt, Bopp, and Lady (1976); Houthakker and Kennedy, as reported in Taylor (1977); and US Department of the Interior, Bureau of Mines, *Petroleum Statement Annual* (1975).
a. Individual products in the average domestically produced and consumed barrel are weighted by their 1974 percentage refinery yields and shares in total domestic product demand, respectively.
b. See text on selection of products which the United States trades internationally under the entitlements program.

does estimate the *marginal* subsidy passthrough accurately—assuming data and parameter values are not inaccurate.

In the cases of residual fuel, jet fuel, lubricating oils, and LPG, the United States has apparently remained an active participant in international markets under the entitlements program; and the appropriate measure of subsidy incidence is that reported as the next-to-last row of table 4.5. For those products, such as wax and petrochemical feedstocks, which the United States would actively sell in world markets in the absence of the entitlements program, the appropriate measure of incidence is that shown in the last row of table 4.5. The eligibility for the entitlements subsidy of domestically refined crude oil attributable to exported products acts as a tax on product exports. This tax exactly offsets the export subsidy of $P_U - \hat{P}_U$ implied in figure 4.7A; it reduces the demand price perceived by US exporters by $P_U - \hat{P}_U$, and the domestic price falls by this amount.

Any *aggregate* measure of the incidence of the entitlements subsidy must take into account the interproduct differences in international market participation. Table 4.6 presents three alternative sets of estimates of the weighted average subsidy-induced price decline for the representative domestically produced and consumed barrels of refined products. Each alternative represents different assumptions concerning each product's place in world markets under the entitlements program. The second and third sets of estimates assume that the United States participates in the international market for each product and that the United States is isolated from the international market for each product, respectively. These estimates bound the aggregate percentage reduction in US prices that results under the entitlements program. The first set of estimates, on the other hand, represents the "best guess" as to the nature of US involvement in international markets under the entitlements program. This best guess is based on the decisions made above concerning the applicability of the next-to-last and last rows of table 4.5 to each product. For residual fuel, jet fuel, lubricating oils, and LPG, the next-to-last row is applicable. For other products, the last row of table 4.5 is used in calculating the best guess in table 4.6.

From table 4.6, it appears that, in the short run, domestic refiners lose between 14 and 45% of the per unit entitlements subsidy as a result of the price reductions that arise from the expansion of their collective output. For the long run, the reduction in the price of the domestically produced barrel of products appears to be on the order of 19–58%. The best-guess estimates of the incidence on the price of this barrel are 37% for the short run and 47% for the long run. The average barrel of products consumed in the United States shows a similar pattern of incidence. In the short run, between 14 and 45% of the entitlements subsidy is reflected in lower prices, with a best guess of approximately 32%. In the long run, the range on the incidence of the subsidy is 18–58%, with a best guess of approximately 41%. With these short- and long-run best guesses at the incidence of the entitlements subsidy, the implied reductions in the price of the domestically consumed product barrel range from $0.51 (short run) and $0.66 (long run) in 1978 to $1.65 (short run) and $2.12 (long run) in the first quarter of 1980. The corresponding range for the domestically produced barrel is from $0.59 (short run) and $0.75 (long run) in 1978 to $1.91 (short run) and $5.43 (long run) in the first quarter of 1980.

To provide some sensitivity analysis of the results of table 4.6, estimates of subsidy incidence under an assumption of much greater (long-run) supply responsiveness than embodied in table 4.5 have been made. With greater supply responsiveness, import supply elasticities would increase. Other things being equal, the effect of the entitlements subsidy on refined product prices would then be reduced. On the other hand, greater refiner supply responsiveness increases $P_U - \hat{P}_U$ in figure 4.7A and, hence, in figure 4.8. This translates, other things being equal, into a larger decline in refined product prices. To examine these offsetting factors, a long-run value of $\epsilon_{ur} = \epsilon_{fr} = 99$ was assumed. This very high elasticity is consistent with an international refining industry characterized by approximately constant returns to scale, no fixed factors (for example, land or building permits), and price taking (as a group) in all factor markets. Using this elasticity and assuming that the United States does *not* trade any products in world markets lead to the conclusion that virtually all of the entitlements subsidy—99.6% of the domestically produced and consumed barrel—is reflected in lower product prices. Even if the United States trades *all* products in world markets, 36.2% of the entitlements subsidy is passed through on the domestically produced barrel and 31.7% is passed through on the domestically consumed barrel. These trades-no-products and trades-all-products estimates bracket any best-guess incidence estimate.

The values of table 4.6 suggest that the impact of the entitlements program on domestic refined product prices has not been so insignificant as to justify treating them as zero. The relatively large impact on product prices apparently arise for two reasons: (a) although the United States faces highly elastic supplies of imports of most refined petroleum products, the demands for imports of most products are also extremely elastic and are very responsive to the disincentives to importing that result under the entitlements program; and (b) in the cases of several products, the entitlements program has made the role of foreign markets in determining US refined product prices irrelevant. The results of table 4.6 also imply that an assumption that 100% of the entitlements subsidy is passed through in the form of lower product prices is also unlikely to be accurate. From the values reported here, it appears that refined product producers and refined product consumers are sharing in the wealth transfer carried out by the combination of crude oil price controls and the entitlements program. An attempt will be made to estimate aggregate wealth effects. Before this is done, however, corroborating evidence on the effects of the entitlements program on the levels of refined product prices will be presented.

4.5 US and Foreign Refined Product Prices: Direct Comparisons

The analysis of the possible effects of domestic regulatory policies on the US position in world refined product markets has indicated that it is possible for the entitlements program and export controls to remove the United States from international markets. The possibility that this has, in fact, occurred in the cases of several products is supported by the data on imports discussed in section 4.4 and by the considerable evidence from industry trade publications. If the United States has, indeed, pulled out of world product markets since the beginning of the entitlements program, domestic refined product prices should be observed to be lower than (prospective) landed costs of foreign products.

4.5.1 Selected Data on Foreign and Domestic Prices

Numerous data sets have been examined and numerous comparisons of foreign and domestic product prices have been made in Kalt (1980). In response to the doubts raised by the RAND Corporation (1977) about the reliability of calculated foreign-domestic price differences, these comparisons have attempted to account for, and avoid, the problems of obtaining price quotations from comparable markets and adjusting for tariffs, currency valuations, transportation costs, and differences in contract terms. The resulting comparisons tell a consistent story about the impacts of the entitlements program on domestic product prices. Direct observations of foreign-US price differences indicate that the entitlements subsidy to the domestic refining of crude oil has moved the United States out of international gasoline and middle distillate markets and held domestic gasoline and middle distillate prices well below the levels prevailing in the rest of the world. These conclusions do not hold for residual fuel, which has had its domestic price held below world levels only when imports have been directly subsidized by the entitlements program. The evidence presented in Kalt (1980) in support of these results is summarized here.

United States Central Intelligence Agency (1976) data on tax-exclusive and exchange rate-adjusted consumer prices for seven industrial nations reveal that US price increases in gasoline markets from October 1973 to October 1975 exceeded only those of Canada. Middle distillate price increases in the United States over the same before-and-after OPEC price increase period were smaller than the increases in every country except Canada and Italy. United States Bureau of Mines data (as reported in the *International Petroleum Annual*, 1974, 1976) on tax-exclusive, exchange-rate adjusted retail gasoline prices for thirty-seven non-OPEC countries indicate that only four countries had price increases over July 1973–July 1976 less than the US increases. On a percentage basis, the US gasoline price increase was 41.6 percentage points smaller than the average increase in the non-OPEC countries and was 21.6 percentage points less than the average OECD increase. The US Federal Energy Administration (1976c, 1977b) calculated landed foreign-US wholesale price differences for the three major products and two foreign trading centers (Italy and Rotterdam). The FEA data indicate that domestic gasoline prices averaged 6.2¢ per gallon less than Rotterdam prices and 4.6¢ per gallon less than Italian prices from January 1975 through May 1977. From February 1975 through May 1976 (a period free of direct entitlements subsidization of product imports), the Rotterdam-US difference in middle distillate prices averaged 5.9¢ and the Italy-US difference averaged 5.8¢. Residual fuel price differences over the same period, however, were insignificant and averaged only 1.3¢ for the Rotterdam-US case and −0.5¢ for the Italy-US case. In the handful of months of direct import subsidization covered by the FEA data, the US residual fuel price was less than the Rotterdam price and the Italian price by 4.7¢ and 3.1¢, respectively.

Royal Dutch Shell quarterly indexes (as reported in *PIW*) of its per barrel extax sales revenues (realizations) on the refined product sales of the company around the world during 1972–early 1976 indicate very high initial correlations between US and foreign gasoline and middle distillate realizations. These correlations break down dramatically after introduction of the entitlements program—suggesting the presence, then severance, of tying international trade. United States-non-European residual fuel realizations exhibit a similar, but less pronounced pattern. United States-European residual fuel realizations, however, are highly correlated throughout the period 1972–early 1976 (which was substantially free of direct import subsidization) and show no breakdown in correlation after the start of the entitlements program—suggesting the presence and persistence of tying trade. The general implications of these data are supported by evidence developed by the Ford Administration Task Force on Reform of Federal Energy Administration Regulation (MacAvoy, 1977). This Task Force found that US gasoline and middle distillate spot prices were consistently lower than delivered Rotterdam spot prices in 1975–1976. United States residual fuel spot prices, however, were consistently below Rotterdam delivered prices only during periods of direct import subsidization.

One of the most revealing data sets examined in Kalt (1980) is that provided by the Standard Oil Company of Indiana's (Amoco) *Financial and Statistical Supplement to the 1977 Annual Report* (1978a). Amoco reports realized sales prices for the average barrel of refined products from both its domestic and overseas operations over 1968–1977. It also reports US and overseas sales quantities and US prices for each major product and other products. These data permit the calculation of some particularly enlightening figures.

Table 4.7 Domestic and Overseas Refined Product Prices Realized per Composite Barrel
(Dollars)

	(1) US prices; US barrel	(2) Foreign prices; foreign barrel	(3) US prices; foreign barrel	(4) Difference (3) − (2)
1968	5.75	3.78	4.42	0.64
1969	5.80	3.53	4.46	0.93
1970	5.92	3.74	4.75	1.01
1971	5.29	4.37	5.30	0.93
1972	6.05	4.34	5.23	0.89
1973	7.06	6.05	6.18	0.13
1974	11.97	11.59	11.83	0.24
1975	13.40	13.27	12.68	−0.59
1976	14.41	13.82	12.68	−1.14
1977	15.88	15.08	14.24	−0.84

Source: Standard Oil Company of Indiana (1978a, pp. 20–21).

Because the average US and foreign barrels of refined products do not contain the same proportions of individual products and because individual products have unequal market values, a comparison of actual US and foreign realizations on average barrels of products (columns 1 and 2 of table 4.7) is not especially informative. The Amoco data, however, permit a barrel of refined products with *foreign* proportions to be valued at *US* prices for the individual components. The foreign average barrel can be hypothetically ''produced'' and ''sold'' in the United States. The US value of this domestically produced barrel of refined products with foreign proportions is shown in column 3 of table 4.7. A relevant comparison of foreign and domestic refined product prices can now be made between the actual value of the foreign barrel and the value it would have if it were produced and sold in the United States. The difference between these values is shown in column 4 of table 4.7.

The *pattern* in the differences between the overseas and domestic values of the foreign barrel of products reflects the history of petroleum market regulations. As expected (because the United States was an importer of most products), the prices realized in domestic refining operations exceeded foreign realized prices during the period of US import controls. In the period between the end of import controls in early 1973 and the beginning of the entitlements program in late 1974, domestic-foreign price differences were reduced—although tariff-and-transportation costs kept domestic prices slightly higher than

Table 4.8 The Estimated Effects of the Entitlements Program on US Refined Product Prices
(Dollars per Barrel)

	(1)	(2)	(3)	(4)	(5)
Alternative adjustments[a]					
1975	−2.08	−1.56	−1.51	−1.50	−0.92
1976	−2.55	−2.03	−1.98	−1.95	−1.34
1977	−2.27	−1.76	−1.70	−1.62	−0.97
1975–1977 average as percentage of entitlements subsidy	91.5	71.0	68.9	67.2	42.8

Sources: Standard Oil Company of Indiana (1978a, pp. 20–21); US International Trade Commission (1975); National Petroleum Refiners Association, *Oil Import Digest* (25 August 1974 as modified by Presidential Proclamation 4341, 23 January 1975); and Herold, *Petroleum Outlook* (January 1978, p. 43).
a. Actual US price-delivered foreign price differences from column 1 are adjusted for possible bias in Amoco data. These adjustments, by column number, are (2) 52¢ per barrel overestimate of differences; (3) 57¢ overestimate of differences; (4) bias of same proportion as in 1974; (5) bias of same proportion as in 1973. For complete explanation see text.

foreign prices. Beginning in 1975, the entitlements program apparently reduced the domestic prices realized by refiners (or at least Amoco) to levels below overseas prices. Beginning in 1975, a refiner selling a barrel of refined products with typical foreign proportions of individual products could expect to receive less for that barrel in the United States as compared to the non-US market.

The magnitude of the entitlements-induced reduction in domestic refined product prices can be deduced by noting that, in the absence of the current program, the *domestic* value of a domestically produced barrel with the foreign composition should approximately equal that barrel's actual foreign value (column 2 of table 4.7) *plus* the tariff-and-transportation expenses that would be incurred in delivering a barrel of products to the United States. That is, in the absence of the entitlements program, the *delivered* price of products could be expected to set domestic prices; and it would be incorrect to report the 1975–1977 figures of column 4 in table 4.7 as the entitlements-induced domestic price reductions.

The difference between the actual US price of a domestically produced barrel with Amoco's foreign composition and the delivered foreign price of the foreign barrel during 1975–1977 is shown in column 1 of table 4.8. This difference provides an estimate of the entitlements-induced reduction in domestic product prices if Amoco's data collection and reporting methods are reliable and if the Amoco data represent refiner realizations for foreign and domestic refineries that face the same costs of getting product to and from ports of exit and entry, respectively. Concerning the former qualification, it is worth pointing out that the noted coherency in the pattern of the US-foreign price relationship prior to 1975 engenders some confidence in the Amoco numbers; and there is no obvious reason to believe that subsequent data are less reliable. In fact, oil industry and trade journal opinion supports this conclusion. The *Petroleum Intelligence Weekly* reports that "one large U.S. marketer finds its realization prices are within one-half cent a gallon of Amoco's 1977 sales realization" and that "[Amoco's] product prices are considered 'fairly typical' of the U.S. refining and marketing industry, in the opinion of some major oil companies canvassed by *Petroleum Intelligence Weekly*."[38]

The most relevant international comparison that could be attempted with the Amoco data is between a Western Europe export center and a US East Coast port. The calculation of the delivered price of Amoco's foreign product barrel, however, is based on average foreign realizations (with actual points of production and sale unknown), while the calculated domestic value is based on average values realized across the United States. To some extent, then, both foreign and domestic realizations may reflect intraregional transportation-and-tariff costs. In Amoco's case, only about 9% of its domestic refining capacity is located on the East Coast, while 36% is located on the Gulf Coast and the remaining 55% is located in the Midwest and Rocky Mountain states.[39] Of these regions, the Gulf Coast tends to be a net "exporter" to other regions of the United States, and Gulf Coast realizations could be expected to be slightly less than both the domestic average and East Coast realizations. With respect to the foreign realizations, Amoco's most important refining center is in Great Britain, with major operations also located in Italy and Australia and minor operations located in India and Southeast Asia.[40] The operations in Great Britain come closest to being the appropriate location for the price comparisons at hand, but the direction and size of any bias imparted to estimates of the delivered price of foreign products remain uncertain. If there are no biases in the data, table 4.8 indicates that 91.5% of the entitlements subsidy has shown up as a reduction in the domestic value of Amoco's average product barrel of foreign composition.

A measure of any inaccuracy in the Amoco price data can be had by examining foreign and domestic realizations in a period free of import controls and the entitlements program. In such a period, the delivered price of a barrel of foreign products and the price of a domestic barrel of the same composition should be equal in the absence of measurement error and product market disequilibrium. The closest approximation to such a period is 1973–1974—import controls were declining in effect immediately preceding their official abandonment in April 1973, and the entitlements program did not appear until very late in 1974. In 1973, the delivered price of a foreign barrel exceeded the calculated value of the domestically refined barrel of the same composition by approximately $0.57, and in 1974 the difference was $0.52.

Although the sample size is small, these considerations suggest that the Amoco data tend systematically to overstate differences between foreign and domestic prices. It is most probable, however, that the differences found for 1973–1974 are overestimates of any bias in the Amoco data. Domestic realizations in the last quarter of 1973 and the first half of 1974 reflected the impact of binding domestic product price controls; and price comparisons for most products at this time reveal reported US prices that are much lower than delivered foreign prices (by as much as $19.00 per barrel in middle distillates, for example).[41] Nevertheless, even if the $0.52 per barrel difference is accepted as a bias in the Amoco data, the estimated differences in delivered foreign prices and domestic prices (and the estimated impacts of the entitlements subsidy) are in the range of $1.56–2.03 per barrel over 1975–1977—as reported in column 2 of table 4.8. Alternatively, if it is assumed that there is a $0.57 per barrel bias in the Amoco data that persists throughout 1975–1977, the estimated differences in delivered foreign prices and domestic prices are in the range of $1.51–1.98 per barrel (column 3 of table 4.8). If it is assumed that there is a bias in the Amoco data that persists throughout 1975–1977 in the same proportion as $0.52 bears to the delivered price of foreign products in 1974, the estimated differences in delivered foreign prices and domestic prices are in the range of $1.50–1.95 per barrel (column 4 of table 4.8). If it is assumed that there is a bias in the Amoco data that persists throughout 1975–1977 in the same proportion as $0.57 bears to the 1973 delivered price of foreign products, the estimated differences in delivered foreign prices and domestic prices are in the range of $0.92–1.34 per barrel (column 5 of table 4.8).

The final row of table 4.8 presents estimates of the percentage of the entitlements subsidy that has been passed through in the form of lower prices. These estimates range from approximately 43 to over 91%, depending on the method of adjustment for possible biases in the data. Regardless of how the Amoco data are qualified and modified, however, they consistently reveal that the entitlements subsidy has substantially reduced US refined product prices.

4.5.2 Econometric Evidence on the Effects of the Entitlements Subsidy

Platt's Oil Price Handbook and Oilmanac has been reporting data on the international petroleum industry for several decades. Included in the reported data are time series on refined product prices at major petroleum importing and exporting centers. *Platt's* time-series evidence on the relationship between US and foreign gasoline, middle distillate, and residual fuel prices is examined econometrically in this section in order to account for the effects of changing regulatory regimes and the possibility of a lack of complete comparability in the foreign and domestic markets covered by the *Platt's* reporting service.

The Data The relative abundance of data on, and the importance of, their ports to petroleum trade have dictated the choice of Rotterdam and New York as the markets for examination here. Rotterdam has been called "Europe's oil giant" by the industry press and is both a major refining center and an entrepôt for a "huge" volume of transshipment traffic in crude oil and refined products.[42] Similarly, New York City is an important center for East Coast refined product trade and has historically served as a major port for the landing of both imported products and shipments of domestic products from the Gulf Coast. As noted above, the relevant comparison of Rotterdam and New York prices is between New York prices and *delivered* Rotterdam prices. The *Platt's* data report FOB bulk (barge) prices for Rotterdam and harbor terminal prices for New York. To calculate the delivered price of Rotterdam products, tariff-and-transportation costs are added to the FOB price.[43]

Ideally, a comparison of domestic and delivered foreign prices would involve the price of a gallon of domestic product at the precise point of delivery in New York and under the same terms of delivery as a gallon of Rotterdam product. It cannot be claimed that this ideal has been satisfied by the *Platt's* price series. It is known, for example, that much of the Rotterdam barge traffic moves to inland European markets. Then, too, any New York harbor handling charges are not reflected in the estimated delivered Rotterdam prices. Although a study commissioned by the European Economic Community (1978) has concluded that *Platt's* data are generally reliable, the most that can be expected is that any biases imparted to the computed price differences are constant or random and do not change with changes in the regulatory regimes whose effects are at issue.

The *Platt's* data employed here cover the period May 1968–October 1976 (May 1970–October 1976 for residual fuel). During much of this period, of course, refined product imports were controlled by the Mandatory Oil Import Control Program (MOIP). This program was officially in effect through April 1973, but residual fuel imports into the East Coast were decontrolled in 1966. Import controls on middle distillates were gradually relaxed from 1967 onward; and import controls on all products were relaxed substantially in the months immediately preceding the end of the MOIP. During periods of effective import control, delivered Rotterdam prices could be expected to be exceeded by New York prices since the MOIP tended to prop up domestic prices. This seems to be borne out by the data. Residual fuel, free of substantial import controls, showed domestic prices that were slightly lower than delivered foreign prices in 1970 and the first half of 1971—averaging 1.2¢ per gallon less than delivered foreign prices from May 1970 through May 1971. From June 1971 through April 1973, New York residual fuel prices generally exceeded estimated delivered Rotterdam prices by small amounts—with the differences averaging 0.79¢ per gallon over the period. Except for a period in late 1970 and early 1971 when New York middle distillate prices fell below delivered Rotterdam prices, both middle distillate and gasoline prices in New York consistently exceeded estimated delivered Rotterdam prices until the last few months of the MOIP. For middle distillates this difference was typically in the range of 2–3¢ per gallon. For gasoline, this difference was typically in the range of 3–5¢ per gallon.

From August 1971 through the start of the entitlements program, US prices were subject to several different price control regimes of varying degrees of bindingness. Rotterdam spot prices have not been controlled. United States price control systems might be expected to hold (at least reported) domestic product prices below prices for products traded in the international market. The most memorable period of binding domestic controls covered late 1973 through the first half of 1974, when the OAPEC embargo and cartel-type supply reductions drove both crude oil and refined product prices up dramatically in international markets and domestic price ceilings failed to keep pace with the costs of marginal supplies. During this period, New York gasoline, middle distillate, and residual fuel prices appear to have been held far below Rotterdam prices. In the following analyses, it will be possible to test for periods during which controls were binding.

In the case of all three of the refined products covered by the *Platt's* data, the differences between the New York product price and the delivered Rotterdam price showed moves in the negative direction in November 1974 (that is, when the entitlements program began). Indeed, from November 1974 through October 1976, New York prices of gasoline, middle distillates, and residual fuel were consistently below delivered prices of foreign products. In the cases of gasoline and middle distillates, the gap between domestic and foreign prices appears to have been atypical for the price differences observed in most previous periods (except for the periods of binding domestic price controls). The residual fuel price differences observed under the entitlements program, on the other hand, are not large relative to differences observed in other periods. Conclusions from the patterns in these differences, however, are confounded by the problems of deciphering the independent effects of changing regulatory regimes and random factors. Full analysis of the observed differences in New York and delivered Rotterdam product prices warrants econometric investigation.

The Models The competing hypotheses at issue concern the effects of the entitlements program on domestic refined product prices. Many of the other effects have been easier to discern. There is little question, for example, that the entitlements subsidy has reduced imports of refined products—as the evidence presented in section 4.4 demonstrates. In fact, some of this evidence indicates that the entitlements program has most likely isolated the United States from several international refined product markets. Such isolation is consistent with the data of section 4.5.1, which shows some US product prices to be below foreign product prices. This possibility can be tested econometrically with the *Platt's* data. If it is found that the entitlements subsidy cannot explain observed differences in New York and delivered Rotterdam prices, the hypothesis that US product markets can be characterized by figure 4.1 or 4.3 is supported. On the other hand, figure 4.2 or 4.4 and an entitlements-induced decline in domestic product prices are supported if the entitlements subsidy can explain the observed tendency for New York product prices to fall short of delivered Rotterdam prices after November 1974.

The most straightforward model of the determination of the difference in domestic and foreign refined product prices would be

$$(P_{NY} - P_{ROT})_t = \beta_1 \text{SUBSIDY}_t + \mu_t, \tag{4.13}$$

where P_{NY} is the New York terminal price per gallon of product; P_{ROT}, the delivered price of Rotterdam product; SUBSIDY, the per gallon entitlements subsidy; and μ_t, the stochastic component of $P_{NY} - P_{ROT}$ (which might arise, for example, because of random market disequilibria). If the United States has been isolated from world product markets by the entitlements program, $\beta_1 < 0$ and β_1 estimates the marginal fraction of the subsidy passed through to domestic prices. Otherwise, $\beta_1 = 0$.

Unfortunately, the simple statement of expression (4.13) must be complicated by several factors. As noted, much of the period covered by the *Platt's* data was accompanied by import or price controls; and the relationship between domestic and foreign product prices could be expected to reflect their impact. The oil import control program persisted into 1973. Even prior to the end of the MOIP, however, petroleum products had become subject to price controls (see chapter 1). The Nixon Administration imposed its Phase I economy-wide wage and price freeze from 15 August 1971 through 13 November 1971. During this time, however, petroleum product prices were tending to decline around the world and a freeze on prices at levels prevailing prior to 15 August 1971 apparently left market prices generally unaffected. Phase I controls were followed immediately by Phase II. This regime remained in effect until 10 January 1973 and was aimed primarily at controlling the prices of the largest enterprises in the economy. Phase III was a voluntary program of controls on economy-wide prices, lasting from 11 January 1973 through 13 June 1973. On 6 March 1973, however, Special Rule No. 1 reimposed mandatory controls on firms in the petroleum industry with annual revenues in excess of $250 million. This rule lasted until 13 June 1973 and encompassed the 23 largest oil companies. From 14 June through 18 August 1973 the voluntary aspects of Phase III were abandoned and a mandatory economy-wide control system was reimposed. Phase IV controls were begun on 19 August 1973, but Phase IV petroleum rules were not fully implemented until 7 September 1973. These rules were embodied in EPAA in November 1973 with little substantive change and were in effect during the OAPEC embargo period. Their cost passthrough provisions for refined product sellers eventually made the Phase IV-EPAA refined product price controls nonbinding for most market participants. In their early period, however, they were undoubtedly responsible for the shortages and queuing that occurred in many refined product markets.

In order to account for the several regulatory regimes that came and went during the period preceding the entitlements program, dummy variables are introduced into (4.13). The use of dummy variables to account for the impact of price controls and import quotas is not a first-best solution. It would be preferable to account for the various regulatory programs with continuous, nondichotomous measures of the stringency of these programs, rather than merely their existence or nonexistence. One possible measure of the strength of the oil import control program, for example, would be a series on import ticket values. As discussed in chapter 1, the values of import tickets provide an estimate of the magnitude by which domestic petroleum prices were raised above foreign petroleum prices. Consistent time series on import ticket prices, however, are not publicly available. Similarly, data on the magnitude by which the various price control regimes held domestic prices below market-clearing prices (that is, the degree of their bindingness) are not forthcoming. Some attempt is made below to allow for varying degrees of bindingness during the dramatic period of petroleum shortages under Phase IV-EPAA by ranking relevant months according to reports on the severity of shortages.

The possibility has been noted that there may be systematic, unmeasured differences in New York and Rotterdam quoted prices. These could be the result, for example, of differences in the types of markets in which these prices are quoted, unmeasured transportation, tax, and handling costs, or differences in contract terms. To take account, in some measure, of the possibility that the New York and delivered Rotterdam product prices are not strictly comparable in the manner hypothesized, a constant term is introduced into (4.13). As with the persistent differences in foreign and domestic product prices found in the Amoco data, a similar trait may be present in the *Platt's* data.

As a final modification to (4.13), adjustment must be made for periods during which middle distillate and residual fuel imports were eligible for direct subsidies under the entitlements program. In the early months of the entitlements program, this subsidy was equal to approximately 30% of the per barrel subsidy to crude oil purchases by domestic refiners. This subsidy was suspended by February 1975, but was reintroduced for residual fuel in February 1976 under EPCA and continued at a slightly higher rate through the end of the period covered by the *Platt's* data.

The full-blown expression for the difference in domestic and foreign product prices is

$$
\begin{aligned}
(P_{\mathrm{NY}} - P_{\mathrm{ROT}})_t = {} & \beta_0 + \beta_1 \mathrm{DQUOTA}_t + \beta_2 \mathrm{DPHSI}_t \\
& + \beta_3 \mathrm{DPHSII}_t + \beta_4 \mathrm{DPHSIII}_t + \beta_5 \mathrm{DSPCRULI}_t \\
& + \beta_6 \mathrm{DPHSIIIF}_t + \beta_7 \mathrm{DPHSIV}_t + \beta_8 \mathrm{SUBSIDY}_t + \mu_t .
\end{aligned} \tag{4.14}
$$

In this expression, DQUOTA is a dummy variable which has a value of unity during the period of mandatory oil import controls and is zero otherwise. Similarly, DPHSI, DPHSII, DPHSIII, DSPCRULI, DPHSIIIF, and DPHSIV are dichotomous variables denoting the presence or absence of price controls under Phase I, Phase II, Phase III (voluntary period), Special Rule No. 1, Phase III (freeze period), and Phase IV-EPAA, respectively. The SUBSIDY variable in (4.14) is measured as the per gallon entitlements subsidy to refined product production, net of the special supplemental crude oil import fees imposed by the Ford Administration. A variable DIRECTSUB is introduced below to capture the direct entitlements subsidies to middle distillates and residual fuel imports and is measured in cents per gallon in appropriate months.

If price controls on the refined product in question were binding, the coefficients on the price control dummies should be negative (that is, binding controls hold New York prices below world market levels). The coefficient on DQUOTA should be positive. Of central interest, of course, is the coefficient on SUBSIDY. If it is significantly negative, evidence is provided that the United States has pulled out of world product markets and domestic prices have been reduced to levels below world market levels. If β_8 is not significantly less than zero, marginal changes in the entitlements subsidy have no impact on the US-foreign price differences. This most plausibly arises when the United States remains a participant in world markets after introduction of the entitlements subsidy. In such a case, any negative effect of the entitlements subsidy on domestic refined product prices must arise from US price-making ability in world markets. It is also a theoretical possibility that the entitlements subsidy causes a US withdrawal from world product markets, but that marginal changes in the entitlements subsidy over the observed range have no impact on domestic prices.

The introduction of the entitlements program might, for example, induce an expansion of domestic refining that suffices to push out all imports but leaves the US refining industry operating in a range of completely inelastic supply. A variant of (4.14), in which the presence of the entitlements subsidy is measured by a dichotomous variable, is introduced to account for this possibility.

The selection of appropriate time periods of DQUOTA and the price control dummies is somewhat problematic. In the case of DQUOTA, the MOIP was officially suspended in early 1973. Prior to this date, however, the MOIP restrictions on imports had been significantly relaxed. Residual fuel imports into the East Coast, of course, were essentially decontrolled in 1966. Distillate imports were decontrolled in January 1973. In addition to these moves, numerous special allocations of both crude and product imports were made in the later years of the MOIP. Bohi and Russell (1978, p. 259) conclude that "after 1971, the world price [of crude oil] increased to and surpassed the domestic price (which was being controlled), removing altogether the scarcity value of imports. The quota program became superfluous, but nevertheless remained on the books until May 1, 1973." Judging by the behavior of product imports (as shown, for example, in figure 4.5), foreign products began to flow more freely into the United States in the third or fourth quarter of 1972.

The periods covered by the price control dummy variables are, for the most part, restricted to their official durations. The exception to this is the treatment of DPHSIV. The Phase IV-EPAA refined product price controls (and their EPCA successors) remained in effect for most products well into 1976; and controls on gasoline prices and propane continue to the present. Under Phase IV, however, the banked cost and passthrough regulations were introduced. These regulations eventually allowed ceiling prices to rise above market-clearing levels. In the early months of the Phase IV-EPAA controls, however, uncontrolled crude oil prices were rising rapidly and the OAPEC embargo was begun. During this period, price controls were undoubtedly binding in many domestic product markets. Indeed, queuing at retail outlets began to be reported in September 1973, which was prior to the beginning of the embargo.[44] Data on banked costs, visible queuing in certain markets, and reports of wholesalers and retailers on their ability to fulfill allocation obligations and acquire supplies indicate the Phase IV-EPAA controls were generally binding at least throughout the fall of 1973 and the spring of 1974.[45] Even after mid-1974, however, controls have undoubtedly been binding on some products, to some degree, for some firms, in some markets, at some times. (Alternative treatments of the period covered by DPHSIV are examined below.)

Gasoline Two alternative methods of measuring the effects of Phase IV-EPAA price controls on gasoline are employed in the econometric analysis of (4.14). These are referred to here as model I and model II. In model I, an attempt is made to account for varying degrees of bindingness in Phase IV-EPAA controls. A standard dichotomous dummy variable for these controls implicitly assumes that their bindingness was uniform over the relevant period. Such an assumption, however, is no better than a third-best approximation to experience. In lieu of a continuous measure of the bindingness of gasoline price controls, a second-best approach of ranking relevant months on the basis of reports on the bindingness of controls is adopted in model I.

As noted, reports of shortages in gasoline markets under Phase IV began in September 1973—the month prior to the initiation of the OAPEC embargo. Shortages continued into the winter of 1973–1974 and were apparently most severe in January and February 1974.[46] Data on the percentage of federal allocation requirements actually filled indicates that shortages persisted in some significant degree until May or June 1974.[47] To reflect this pattern of bindingness in model I, DPHSIV takes on a value of unity in September 1973, rises in unit steps to a peak in January and February 1974, and then declines to a value of unity in June 1974. This approach is imperfect in its assumption of a simple step-ranking function, but does attempt to bring applicable external information to bear on the analysis. As will become evident, it turns out that the ranking of shortage months by DPHSIV does not significantly alter the story revealed by the data under examination relative to a standard dichotomous DPHSIV dummy variable. In model II, DPHSIV is represented by such a variable. It has a value of unity from September 1973 through June 1974 and a value of zero otherwise.

In the econometric analysis reported below, all value-based data have been deflated by the Wholesale Price Index for all commodities. The presence of inflation over the 1968–1976 sample period suggests that any nominal difference in domestic and foreign prices in 1968 was not of the same real value as an equal nominal difference in 1976. Notwithstanding this observation, results from Kalt (1980) indicate that estimation of (4.14) with data that were not deflated show no major differences with the findings reported under deflation. Finally, as a middle-ground assumption, DQUOTA assumes a value of unity through the third quarter of 1972. Kalt (1980) has examined the sensitivity of results to changes in this assumption.

Application of ordinary least squares to (4.14) yields low Durbin-Watson statistics and a correspondingly high probability of autocorrelation in the errors of (4.14). The nature of the autoregressive process behind (4.14) can be explicitly analyzed. Following the methods suggested by Nelson (1973) and Box and Jenkins (1976), an ARIMA analysis has been used to examine the ordinary least-squares residuals. This analysis, as discussed in Kalt (1980), strongly suggests that a second-order autoregressive error-generating process lies behind (4.14). Accordingly, the analysis here proceeds under the assumption of second-order autocorrelation. It turns out, as Kalt (1980) indicates, that the basic story told by ordinary least-squares estimation of (4.14) and the estimation of (4.14) under a first-order correction scheme is not substantially altered by this assumption.

Table 4.9 reports estimates of expression (4.14) under an iterative, maximum likelihood Gauss-Newton second-order autocorrelation technique. The results concerning the effect of the entitlements program reported in table 4.9 are found in Kalt (1980) to be fairly insensitive to changes in the period selected for coverage by DQUOTA. As noted, if the oil import quota program raised domestic prices above foreign prices, the coefficient on DQUOTA should be positive. This is clearly supported by the results shown in table 4.9. In fact, the oil import control program apparently raised domestic gasoline prices above foreign prices by approximately 4–5¢ (1967 dollars) per gallon. This information is valuable in its own right since it could be employed to shed light on the wealth and efficiency effects of the Mandatory Oil Import Control Program. Such an effort, however, is beyond the scope of this study.

The regressions reported in table 4.9 are done under two alternative assumptions concerning the effects of the various price control regimes. Specifically, if any phase of price controls was binding on US prices (or, at least, New York Harbor prices), US prices could be expected to be below prices for products in the uncontrolled market for exports and transshipments out of Rotterdam. The dummy variable for such a phase would then have a negative effect on $P_{\mathrm{NY}} - P_{\mathrm{ROT}}$. The first regression for each model imposes no prior beliefs as to whether any particular price control regime was binding or not. The second and third regressions for each model assume that price controls under

Table 4.9 The New York-Rotterdam Price Difference: Coefficient Estimates for Gasoline[a] (t-Statistics in Parentheses; Sample Size = 102)

	DQUOTA	DPHSI	DPHSII	DPHSIII	DSPCRULI	DPHSIIIF	DPHSIV	SUBSIDY	CONSTANT	D-W	$\bar{R}^2$	F-Stat.
Model I	4.683	0.018	−0.227	−0.618	−3.255	−4.989	−3.328	−0.755[b]	−0.768	1.98	0.90	88.69
	(6.64)	(0.014)	(−0.35)	(−0.38)	(−2.36)	(−3.56)	(−11.45)	(−2.66)	(−1.60)			
Model I	4.791	—	—	—	−3.093	−4.831	−3.284	−0.700[b]	−0.922	1.97	0.90	130.50
	(7.39)				(−2.33)	(−3.57)	(−12.00)	(−2.70)	(−1.58)			
Model I	3.871	—	—	—	−4.078	−5.795	−3.556	−1.030[b]	—	1.99	0.90	128.10
	(13.80)				(−3.42)	(−4.76)	(−17.27)	(−6.60)				
Model II	4.876	−0.058	−0.114	−0.795	−2.780	−5.686	−9.215	−0.623[b]	−1.027	1.98	0.87	70.75
	(5.20)	(−0.04)	(−0.13)	(−0.43)	(−1.63)	(−3.28)	(−7.22)	(−1.67)	(−1.15)			
Model II	5.028	—	—	—	−2.511	−5.496	−9.020	−0.560[c]	−1.210	1.97	0.88	104.20
	(5.82)				(−1.58)	(−3.33)	(−7.56)	(−1.65)	(−1.55)			
Model II	3.845	—	—	—	−3.711	−6.645	−10.428	−0.990[b]	—	1.99	0.88	102.50
	(10.48)				(−2.63)	(−4.68)	(−12.31)	(−4.92)				

a. In model I, DPHSIV is a ranking of the months of binding controls according to the reported degree of bindingness (see text for full discussion). In model II, DPHSIV is a standard dichotomous dummy variable denoting the presence or absence of binding Phase IV-EPAA controls.
b. Significantly negative at the 5% level or lower.
c. Significantly less than zero at the 6% level.

Phase I, Phase II, and the voluntary period of Phase III were not binding (that is, $\beta_2, \beta_3, \beta_4 = 0$). This hypothesis is supported by results from the first equation. The coefficients on DPHSI, DPHSII, and DPHSIII are all insignificantly different from zero and contribute little to the explanation of variations in $P_{NY} - P_{ROT}$. The assumption is also supported by the paucity of reports of persistent shortages under the early price control regimes and the absence of widespread, visible queuing. This is not to say, however, that Phases I–III did not produce occasional severe spot shortages of gasoline, but only that there is a high probability that such shortages were not persistently present at the New York terminals sampled by *Platt's*. At any rate, the inclusion or exclusion of the Phase I–III dummy variables makes little difference to the primary task at hand, which is the interpretation of the data evidence concerning the effects of the entitlements subsidy.

The constant term in (4.14) is of interest because it may reveal a systematic difference between what $P_{NY} - P_{ROT}$ is intended to measure and what it actually measures. If the constant is significantly negative, attribution of the observed negative values of $P_{NY} - P_{ROT}$ after November 1974 entirely to the entitlements program would be unwarranted. The exclusion of the constant term is likely to bias the SUBSIDY coefficient toward greater negativity if the constant is, in fact, negative. Such a bias arises because, by constraining the partial relationship between $P_{NY} - P_{ROT}$ and SUBSIDY to pass through the origin in the regression plane when, in fact, it passes through a negative value of $P_{NY} - P_{ROT}$, the relationship is forced to appear more negative than it is. [Note that the time periods covered by the variables in (4.14) are such that SUBSIDY is orthogonal to all of the other variables in (4.14).] If the basic error to be avoided is that of accepting the significance of the entitlements program when, in fact, it is inconsequential, the possibility of systematic negativity in $P_{NY} - P_{ROT}$ should not be rejected readily. In fact, the hypothesis that the constant term is negative cannot be rejected with high levels of confidence on the basis of the results shown in table 4.9.

Under both model I and model II and under both the inclusion and the exclusion of the Phase I–III dummy variables, the null hypothesis that the entitlements program has not reduced domestic gasoline prices below world market levels can be confidently rejected. The confidence levels for the rejection of the hypothesis that the SUBSIDY coefficient is zero range from approximately 6% in the second equation of model II to 0.01% in the third equation of model I.

The summary statistics associated with the estimations reported in table 4.9 indicate that the underlying model of the New York-Rotterdam price difference has fairly impressive explanatory capabilities. In each version shown, approximately 90% of the variation in the dependent variable is explained by the included independent variables. Moreover, the F-statistics leave only a minute probability that the apparent role of these independent variables occurs by pure chance. The explanatory power of (4.14) suggests support for the underlying assumption that New York and Rotterdam markets are linked by trade that tends to remove rapidly any price differences in the absence of intervening regulation.

The numerical value of the SUBSIDY coefficient is at least of as much interest as its negativity. It provides a direct measure of the fraction of any marginal change in the entitlements subsidy that is passed through to the difference in New York and Rotterdam prices. Estimates of the passthrough fraction range between 56 and 99% in table 4.9. In no case does the passthrough fraction differ significantly from 100%. Of the estimates available, that produced in the second version of model I is a good candidate for the appropriate point estimate of the marginal passthrough fraction. The prior information suggesting the general nonbindingness of Phases I, II, and III (voluntary period) price controls, the lack of statistical support for the contrary possibility, the prior information on the degree of bindingness in Phase IV-EPAA controls, and the statistical support for the inclusion of an intercept term make the second version of model I a preferred statement of (4.14). The associated estimate of the marginal passthrough fraction is 70%. The 95% confidence interval for this estimate ranges from 18 to 121%. While the length of the confidence interval does not strengthen trust in the point estimate, the negativity of the effect of the entitlements subsidy on the New York-Rotterdam gasoline price difference is difficult to ignore.

The ability of changes in the entitlements subsidy to cause changes in the difference between US and foreign prices is *prima facie* evidence that the entitlements subsidy has caused domestic prices to move below world levels (and, concomitantly, that the entitlements program has caused a US withdrawal from world markets). As noted in section 4.2 in connection with figures 4.1 and 4.3, so long as the United States participates in world markets and domestic prices are equated to world prices, marginal changes in the entitlements subsidy affect only the level and market share of product imports. In such an environment, the coefficient on SUBSIDY in (4.14) would be expected to be zero.

While the coefficient on SUBSIDY provides evidence that US gasoline prices have been moved below world levels by the entitlements program, it measures only the marginal effect of changes in the subsidy on US prices. It does not measure the amount by which US prices have been held below world levels. One method of determining this magnitude, as well as the magnitudes of the price gaps caused by import quotas and price controls, would be to calculate and compare the mean gaps for each relevant period. Simple examination of the mean New York-Rotterdam price differences, however, overlooks the possibility of a systematic price difference present even in the absence of intervening regulatory policies. To account for this, $P_{NY} - P_{ROT}$ may be regressed as in (4.14) with all independent variables in the form of dummy variables. Coefficients estimated in this manner then represent mean price differences for the periods covered by each regulatory regime *less* the estimated systematic price difference.

Table 4.10 presents estimates of New York-Rotterdam gasoline price differences associated with the relevant regulatory regimes. These estimates are obtained (with adjustment for second-order autocorrelation) with SUBSIDY replaced by a zero-one dummy variable DUMSUB. In table 4.10, estimates are expressed in January 1976 cents per gallon. On the basis of these results, it appears that the entitlements subsidy has caused US gasoline prices to be reduced below world prices by 4–5¢ per gallon. In the "preferred" second version of model I, the point estimate of the entitlements-induced price gap is 4.4¢. The 95% confidence interval for this figure ranges from 3.0 to 5.9¢. By comparison, the entitlements subsidy averaged 4.6¢ per gallon through the sample period. Thus, the results here indicate that approximately 95% of the entitlements subsidy to gasoline has shown up as a decline in US prices relative to world prices.

In summary, the data examined here strongly suggest that, in the case of gasoline, a large portion of the entitlements subsidy to domestic refining has been shifted to product consumers. The *marginal* passthrough fraction appears to have been approximately 70% over the sample period. On average over the sample period, domestic gasoline prices appear to have been more than 4¢ (January 1976 dollars) per gallon below world levels. This magnitude represents an *average* passthrough fraction of roughly 95%. These findings are not inconsistent with evidence reported above, which indicates that US gasoline

Table 4.10 The New York-Rotterdam Gasoline Price Difference—Estimates of Differences Due to Regulatory Regimes: 1968–1976[a] (*t*-Statistics in Parentheses; Coefficients in January 1976 Cents per Gallon)

	DQUOTA	DPHSI	DPHSII	DPHSIII	DSPCRULI	DPHSIIIF	DPHSIV	DUMSUB[b]	CONSTANT	D-W	$\bar{R}^2$	*F*-Stat.
Model I	7.253	−0.028	−0.735	−2.245	−7.152	−10.184	−6.338	−5.046	−0.151	2.00	0.90	92.02
	(5.25)	(−0.01)	(−0.66)	(−0.77)	(−2.86)	(−4.01)	(−11.70)	(−3.29)	(−0.11)			
Model I	7.767	—	—	—	−6.464	−9.494	−6.140	−4.370	−0.822	1.98	0.90	134.30
	(6.26)				(−2.70)	(−3.91)	(−12.26)	(−3.22)	(−0.72)			
Model I	6.948	—	—	—	−7.366	−10.358	−6.383	−5.198	—	1.99	0.90	134.90
	(14.29)				(−3.55)	(−4.89)	(−17.87)	(−7.17)				
Model II	7.089	−0.206	−0.736	−2.858	−6.891	−11.857	−18.688	−5.114	−0.025	2.01	0.88	72.80
	(3.84)	(−0.08)	(−0.51)	(−0.84)	(−2.23)	(−3.65)	(−7.80)	(−2.50)	(−0.01)			
Model II	7.934	—	—	—	−5.605	−10.876	−17.494	−4.087	−1.053	1.98	0.88	106.40
	(4.78)				(−1.95)	(−3.65)	(−7.98)	(−2.26)	(−0.68)			
Model II	6.901	—	—	—	−6.689	−11.862	−18.733	−5.140	—	2.00	0.88	107.00
	(10.81)				(−2.71)	(−4.78)	(−12.68)	(−5.43)				

a. In model I, DPHSIV is a ranking of the months of binding controls according to the reported degree of bindingness (see text for full discussion). In model II, DPHSIV is a standard dichotomous dummy variable denoting the presence or absence of binding Phase IV-EPAA controls.
b. All coefficients significant at the 2% level or lower.

prices are below world levels and that the United States is not an active participant in world gasoline markets under the entitlements program. Termination of federal oil price regulations would have significant impact on domestic gasoline prices and the welfare of domestic purchasers of gasoline.

Middle Distillates The evidence presented in previous sections of this chapter supports the contention that the entitlements subsidy has lowered the domestic price of middle distillates relative to the rest-of-the-world price; that is, SUBSIDY has had a negative effect on $P_{\text{NY}} - P_{\text{ROT}}$. To test this contention econometrically, the basic model developed above for the gasoline case is applied to the *Platt's* middle distillate price data.

For at least several of the regulatory regimes covered by the explanatory variables in (4.14), prior evaluations of their effects on $P_{\text{NY}} - P_{\text{ROT}}$ are clearly identified. Import quotas were imposed on middle distillates until January 1973. These quotas insulated US middle distillate prices from world markets; and DQUOTA can be expected to have a positive coefficient. Phase I price controls were imposed on the entire domestic economy at a time (August 1971) when domestic middle distillate prices were stable and no upward pressure was being exerted by foreign prices. Thus, although controls may have been binding on some market participants, they were not binding on a very widespread basis, and the coefficient on DPHSI may turn out to be insignificant. Similarly, the coefficient on DPHSII is likely to be insignificant. Although there was great concern over the possibility of shortages of middle distillates during Phase II (and some spot shortages did occur), widespread shortages came near but did not materialize.[48] Phase II was followed by the voluntary portion of Phase III and Special Rule No. 1 (which maintained a fringe of uncontrolled market-clearing refiners); and even in the opinions of the price controllers, these regimes allowed markets to clear.[49] Thus, the coefficient on DPHSIII should be insignificant. The coefficient on DSPCRULI should be insignificant or negative, depending on whether the traders sampled by *Platt's* were subject to Special Rule No. 1 or not.

The Phase III freeze begun in June 1973 coincided with a sharp jump in international middle distillate prices. Consequently, controls were binding and produced shortages in middle distillate markets; and in the summer of 1973, truckers and retailers began protesting over the unavailability of diesel fuel. The Phase III freeze was followed by the binding period of Phase IV-EPAA controls. As in the case of gasoline, these controls were most severely binding on domestic middle distillate (especially heating oil) prices in midwinter 1973–1974 and became gradually less stringent over the first half of 1974. To capture this period of binding regimes, a dummy variable, DIIIF-IV, is included in the analysis and has nonzero values from June 1973 through June 1974.[50] DIIIF-IV is expected to have a negative effect on $P_{NY} - P_{ROT}$. Similar to the gasoline case, DIIIF-IV is measured in two alternative ways: in model I, DIIIF-IV is an ordinal ranking rising in unit steps from unity in June 1973 to a peak in December 1973 and then declining; in model II, DIIIF-IV is a standard dichotomous dummy variable.

Finally, a variable DIRECTSUB is included to capture the period of direct import entitlements during November 1974–January 1975. It measures the entitlements subsidy to middle distillate imports in cents per gallon. While the direction of the effect of this subsidy on $P_{NY} - P_{ROT}$ is clear, the significance of its effect is uncertain. Not only was it extremely short-lived, but it also was not adopted (with retroactive applicability to November) until late December 1974. Given nontrivial transport times and uncertainties surrounding the timing of implementation, the direct subsidy to middle distillate imports may not have had time to be reflected in domestic markets.

Table 4.11 reports the estimated effects of the various regulatory regimes on the difference between New York and Rotterdam middle distillate prices. These results are based on an iterative estimation technique that accounts for the possibility of second-order autocorrelation in the residuals. The results of interest (that is, the effect of SUBSIDY) are found in Kalt (1980) to be substantially insensitive to various representations of the time pattern of residuals. The Mandatory Oil Import Program and each of the price control regimes have the expected effect on $P_{NY} - P_{ROT}$. The direct middle distillate import entitlements appear to have had no effect on domestic prices. In model I, the marginal effect of the entitlements subsidy on the New York-Rotterdam price difference is positive and significant at slightly better than the 10% level. Moreover, this measure of the marginal pass-through fraction is within the range of values reported in table 4.5 for the case in which the United States is isolated from world middle distillate markets. In model II, the entitlements subsidy has no significant impact on $P_{NY} - P_{ROT}$; but it is clear that the simple dichotomous dummy variable DIIIF-IV for the period of binding domestic price controls in model II is far from ideal. Although the

Table 4.11 The New York-Rotterdam Price Difference: Coefficient Estimates for Middle Distillates[a]
(t-Statistics in Parentheses: Sample Size = 102)

	DQUOTA	DPHSI	DPHSII	DPHSIII	DSPCRULI	DIIIF-IV	SUBSIDY	DIRECTSUB	CONSTANT	D-W	$\bar{R}^2$	F-Stat.
Model I	6.158	1.314	0.625	−2.215	0.909	−5.210	−0.544[b]	−0.097	−3.690	1.69	0.91	100.54
	(3.69)	(0.56)	(0.37)	(−0.85)	(0.33)	(−9.90)	(−1.39)	(−0.06)	(−2.39)			
Model I	6.355	—	—	—	—	−5.220	−0.531[c]	−0.085	−3.740	1.70	0.91	212.13
	(4.02)					(−10.18)	(−1.43)	(−0.05)	(−2.63)			
Model II	9.689	2.312	2.341	0.150	2.446	−7.905	0.409	1.231	−8.497	1.71	0.86	66.73
	(3.97)	(0.74)	(0.82)	(0.04)	(0.71)	(−2.82)	(0.72)	(0.61)	(−3.90)			
Model II	9.390	—	—	—	—	−7.235	0.307	0.971	−7.795	1.73	0.86	140.80
	(3.82)					(−2.72)	(0.54)	(0.49)	(−3.61)			

a. In model I, DIIIF-IV is a ranking of the months of binding controls according to the reported degree of bindingness (see text for full discussion). In model II, DIIIF-IV is a standard dichotomous variable denoting the presence or absence of binding controls.

b. Significantly less than zero at the 9% level.

c. Significantly less than zero at the 8% level.

results of model II suggest skepticism, model I (particularly the second version) remains the preferred specification for the reasons noted in this and the preceding subsections of section 4.5.2.

Table 4.12 reports estimates of the average cents per gallon difference in New York-Rotterdam middle distillate prices that has been induced by the entitlements subsidy. As in the case of table 4.10, this impact is measured by denoting the absence or presence of the subsidy with a zero-one dummy variable, DUMSUB. A dummy variable, DDRCTSUB, is also included and denotes the absence or presence of the direct subsidy for middle distillate imports. In model II, only DQUOTA and DIIIF-IV appear as significant variables, and both the entitlements subsidy and the direct import subsidy are insignificant with implausible signs. In model I, however, the entitlements subsidy is indicated to have been showing up in middle distillate markets as a significant reduction in US prices relative to foreign prices. The most plausible estimate of this average passthrough is 4.8¢ (January 1976 dollars) per gallon. This is not significantly different from the full per gallon subsidy and clearly supports the conclusion that the entitlements subsidy has reduced US middle distillate prices.

Residual Fuel The evidence from previous sections of this chapter suggests that domestic residual fuel markets have not been isolated from world markets. Consequently, the entitlements subsidy to crude oil use can be expected to have had no significant effect on $P_{\text{NY}} - P_{\text{ROT}}$ for residual fuel. On the other hand, direct import subsidies under the entitlements program can be expected to have lowered the delivered cost of foreign residual fuel and, hence, pushed P_{NY} below P_{ROT}. To examine these contentions, the basic model described by (4.14) is applied to the *Platt's* residual fuel price data.

As noted, DIRECTSUB (covering November 1974–January 1975 and February 1976 onward) and SUBSIDY are expected to have a negative effect and an insignificant effect, respectively, on domestic-world residual fuel price differences. DQUOTA, on the other hand, does not even appear in the residual fuel version of (4.14) because East Coast residual fuel imports were decontrolled in 1966. Of the various price control regimes, Phases I, II, and III (voluntary portion) were not accompanied by a plethora of reports of binding ceilings; and DPHSI, DPHSII, and DPHSIII should have insignificant coefficients. As in the middle distillates case, Special Rule No. 1 may or may not show up as binding in the *Platt's* data. The Phase III freeze and, especially, Phase

Table 4.12 The New York-Rotterdam Middle Distillate Price Difference—Estimates of Differences Due to Regulatory Regimes: 1968–1976[a] (*t*-Statistics in Parentheses; Coefficients in January 1976 Cents per Gallon)

	DQUOTA	DPHSI	DPHSII	DPHSIII	DSPCRULI	DIIIF-IV	DUMSUB	DDRCTSUB	CONSTANT	D-W	$\bar{R}^2$	*F*-Stat.
Model I	4.403	1.136	0.233	−2.782	0.277	−5.632	−5.280[b]	−1.981	−1.771	1.69	0.91	100.54
	(2.49)	(0.05)	(0.15)	(−1.10)	(0.10)	(−10.60)	(−2.59)	(−0.73)	(−1.05)			
Model I	4.873	—	—	—	—	−5.568	−4.830[b]	−1.654	−2.211	1.69	0.91	212.13
	(2.94)					(−10.79)	(−2.53)	(−0.63)	(−1.45)			
Model II	9.258	2.246	2.203	0.132	2.429	−8.138	0.904	1.564	−8.004	1.72	0.86	66.73
	(3.56)	(0.71)	(0.76)	(0.04)	(0.71)	(−2.85)	(0.29)	(0.46)	(−3.33)			
Model II	8.826	—	—	—	—	−7.489	0.164	0.977	−7.186	1.73	0.86	140.80
	(3.38)					(−2.76)	(0.05)	(0.29)	(−3.04)			

a. See note a, table 4.11.
b. Significantly less than zero at the 1% level.

IV-EPAA price controls are associated with fairly strong prior information that they were significantly binding; that is, the coefficients on DPHSIIIF and DPHSIV should be significantly negative.

Table 4.13 reports the estimates of (4.14) for the residual fuel case. An analysis of least-squares residuals in Kalt (1980) suggests second-order autocorrelation, and table 4.13 represents the results of an appropriate estimation technique. Although expected to be insignificant, the early price control regimes show up in both models I and II as having *positive* impacts on the difference between New York and Rotterdam residual fuel prices. The conclusion that controls on domestic prices raise New York prices relative to Rotterdam prices is *a priori* implausible. In the second and third versions of models I and II, prior beliefs concerning the nonbindingness of Phases I–III and Special Rule No. 1 are imposed by dropping corresponding variables. In all versions of models I and II, the entitlements SUBSIDY appears to have no significant effect on $P_{NY} - P_{ROT}$ for residual fuel. This is consistent with the observation that the United States has remained an importer of residual fuel under the entitlements program. In the second and third versions of models I and II, the direct entitlements subsidy to residual fuel imports has a significantly negative impact on $P_{NY} - P_{ROT}$—as expected. In fact, the marginal passthrough fraction of approximately 80% is not out of line with the fraction of roughly 95% predicted by application of (4.6) to the residual fuel import supply-and-demand elasticities reported in table 4.5.

Table 4.14 reports estimates of the average cents per gallon New York-Rotterdam price difference attributable to the entitlements subsidy and direct import subsidy. In general, these estimates are statistically insignificant. In the preferred third version of model I, however, price control regimes that appear to have been nonbinding have been excluded and the average cents per gallon difference attributable to the direct import subsidy is significantly negative. This difference is approximately 84% of the average cents per gallon direct subsidy to residual fuel over the sample period and is consistent with preceding evidence and analysis (for example, table 4.5).

4.6 Conclusion

This chapter has presented evidence on the central question of the effects of EPAA/EPCA regulation on the domestic prices of refined petroleum products. The primary mechanism by which product prices

Table 4.13 The New York-Rotterdam Price Difference: Coefficient Estimates for Residual Fuel[a]
(*t*-Statistics in Parentheses; Sample Size = 78)

	DPHSI	DPHSII	DPHSIII	DSPCRULI	DPHSIIIF	DPHSIV	SUBSIDY	DIRECTSUB	CONSTANT	D-W	$\bar{R}^2$	*F*-Stat.
Model I	3.858 (3.18)	2.345 (3.16)	1.644 (0.92)	−1.412 (−0.72)	0.172 (0.10)	−0.686 (−2.68)	−0.018 (−0.09)	−0.441 (−1.02)	−1.267 (−2.84)	2.11	0.43	8.26
Model I	—	—	—	—	−0.538 (−0.30)	−0.932 (−2.69)	−0.141 (−0.62)	−0.828[b] (−1.55)	−0.313 (−0.68)	1.99	0.36	11.83
Model I	—	—	—	—	—	−0.923 (−2.68)	−0.136 (−0.60)	−0.816[b] (−1.54)	−0.336 (−0.74)	1.98	0.37	16.07
Model II	3.844 (3.02)	2.361 (2.97)	1.649 (0.91)	−1.392 (−0.71)	0.073 (0.042)	−1.928 (−2.18)	0.025 (0.12)	−0.427 (−0.93)	−1.292 (−2.67)	2.13	0.41	7.69
Model II	—	—	—	—	−1.017 (−0.55)	−2.691 (−2.45)	−0.133 (−0.54)	−0.816[c] (−1.49)	−0.322 (−0.68)	2.01	0.35	11.53
Model II	—	—	—	—	—	−2.571 (−2.39)	−0.124 (−0.54)	−0.792[c] (−1.46)	−0.375 (−0.81)	2.00	0.36	15.42

a. In model I, DPHSIV is a ranking of the months of binding controls according to the reported degree of bindingness (see text for full discussion). In model II, DPHSIV is a standard dichotomous dummy variable denoting the presence or absence of binding phase IV-EPAA controls.
b. Significantly less than zero at the 7% level.
c. Significantly less than zero at the 8% level.

Table 4.14 The New York-Rotterdam Residual Fuel Price Difference—Estimates of Differences Due to Regulatory Regimes: 1970–1976[a] (*t*-Statistics in Parentheses; Sample Size = 78)

	DPHSI	DPHSII	DPHSIII	DSPCRULI	DPHSIIIF	DPHSIV	DUMSUB	DDRCTSUB	CONSTANT	D-W	$\bar{R}^2$	*F*-Stat.
Model I	3.863	2.344	1.643	−1.415	0.169	−0.687	0.116	−0.899	−1.264	2.11	0.43	8.30
	(3.25)	(3.31)	(0.92)	(−0.72)	(0.10)	(−2.79)	(0.44)	(−1.17)	(−3.20)			
Model I	—	—	—	—	−0.655	−0.980	−1.107	−0.709	−0.147	1.99	0.37	12.34
					(−0.36)	(−2.88)	(−1.11)	(−0.62)	(−0.31)			
Model I	—	—	—	—	—	−0.895	0.048	−1.480[b]	−0.430	1.98	0.37	16.06
						(−2.59)	(0.17)	(−1.53)	(−0.99)			
Model II	3.829	2.343	1.632	−1.393	0.058	−1.948	0.072	−0.867	−1.274	2.13	0.41	7.77
	(2.96)	(2.84)	(0.89)	(−0.711)	(0.03)	(−2.13)	(0.08)	(−0.90)	(−2.37)			
Model II	—	—	—	—	−1.151	−2.874	−1.105	−0.703	−0.144	2.01	0.36	11.90
					(−0.63)	(−2.65)	(−1.07)	(−0.59)	(−0.29)			
Model II	—	—	—	—	—	−2.738	−1.038	−0.704	−0.211	2.01	0.37	15.88
						(−2.57)	(−1.02)	(−0.60)	(−0.44)			

a. In model I, DPHSIV is a ranking of the months of binding controls according to the reported degree of bindingness (see text for full discussion). In model II, DPHSIV is a standard dichotomous dummy variable denoting the presence or absence of binding Phase IV-EPAA controls.
b. Significantly less than zero at the 7% level.

could be altered under this regulation is the entitlements subsidy to domestic refiners' crude oil inputs. While this subsidy increases the domestic supply of refined products, the magnitude of any concomitant price reductions is an empirical issue. Indeed, it is an empirical issue surrounded by considerable academic and political controversy. A resolution to this controversy has been sought in this chapter and has required an exercise in subsidy-incidence analysis.

Previous studies of the impact of the entitlements subsidy on domestic product prices have been sparse and cursory; and the evidence they have presented has been far from convincing. To remedy these deficiencies, the approach taken here has been to draw out information from a wide variety of sources through a number of techniques. The result is the emergence of a consistent picture of the effects of EPAA/EPCA regulation on domestic product prices.

This picture indicates that domestic refined product prices have been substantially reduced by the entitlements program. In the cases of gasoline, middle distillates, and most minor refined products, the United States has been insulated from competition with unsubsidized foreign products. As a result, the entitlements subsidy operates more strongly than it otherwise would to reduce domestic prices. The econometric study of US and foreign gasoline and middle distillate prices indicates that relatively large and statistically significant fractions of the entitlements subsidy have been showing up as reductions in domestic prices. In those cases in which the United States has remained a refined product importer, the competition of foreign refiners (in the form of relatively elastic import supplies) has restrained any subsidy-induced domestic price decreases. In the most notable such case (residual fuel), however, a direct subsidy to imports is provided by the entitlements program. The econometric analysis of US and foreign residual fuel prices provides statistical (albeit not overwhelming) support for the expectation that the bulk of this direct subsidy tends to be reflected in domestic prices.

The analysis of this chapter makes it clear that the impact of the entitlements subsidy on refined product prices is likely to vary considerably from product to product. As a summary measure, a figure of roughly 40% appears to be a reasonable estimate of the fraction of the entitlements subsidy to crude oil use that has been showing up as a reduction in the price of the typical domestically consumed composite barrel of refined products. A number of factors suggest this figure. A 40% passthrough is in the range of best-guess estimates derived in the analysis surrounding tables 4.5 and 4.6. Significantly, this analysis has taken account of differences across products in the extent of the domestic refining industry's participation and price-making ability in international markets. The results of this approach have been given some corroboration by the econometric examination of US-foreign price differences in the cases of a limited subset of products. The econometric analysis has produced subsidy-incidence estimates for individual products that are not grossly out of line with table 4.5.

Of course, any estimates of the effect of the entitlements subsidy on individual product prices or the price of a composite barrel of products are subject to error. Nevertheless, the important conclusion to be drawn from any passthrough estimate in the neighborhood of 40% is that both refiners and refined product consumers have been realizing substantial benefits from post-embargo federal regulation of the petroleum industry. Aggregation of these benefits, as well as measurement of the total losses of crude oil producers and the net effects of regulation on the allocation of resources is undertaken in chapter 5.

Chapter 5

Measuring Allocative and Distributional Effects

5.1 Introduction

Domestic crude oil price controls have blocked much of the increase in the wealth of crude oil producers that would otherwise have accompanied rising world oil prices since 1973. The value of the inframarginal rents associated with controlled crude oil, but denied to crude oil producers, has been shown in table 2.2 and has averaged approximately $21 billion (1980 dollars) per year during 1974–early 1980. As described in chapter 2, these sums are used to finance the entitlements program—to the benefit of refiners and refined product consumers. The Windfall Profits Tax will soon preempt EPCA crude oil price controls, but the WPT will continue to block large transfers of wealth to domestic crude oil producers. In general, the beneficiaries of petroleum price policy will continue to be the users of crude oil, although the Crude Oil Windfall Profit Tax Act of 1980 makes more of an effort than EPAA/EPCA legislation to direct the sums extracted from producers toward specific interest groups and public projects (see chapter 1).

Chapters 2, 3, and 4 have provided the analytical and empirical settings for estimation of the allocative and distributional consequences of post-embargo petroleum price regulation. This chapter undertakes this estimation. It is found that, as of early 1980, crude oil price controls were preventing crude oil producers from capturing approximately $50 billion (1980 dollars) per year worth of producer surplus. The bulk of this was being directed to refiners and refined product consumers through the entitlements program. Taking account of the incidence of both the entitlements subsidy to crude oil refining and the special entitlements programs, it appears that approximately $32 billion per year was being captured by refiners as of early 1980. Refined product consumers, meanwhile, were gaining from regulation at the rate of roughly $12 billion per year. The allocative distortions of crude oil price controls and the entitlements subsidy were resulting in an annual deadweight loss to the domestic economy that was in the range of $5 billion as of early 1980. At 1980 prices, the Windfall Profits Tax will transfer over $30 billion per year from domestic crude oil producers when it becomes fully effective, but these transfers will continue to be bought at the expense of efficient resource allocation. Preliminary estimates put the supply-side social cost of the WPT at around $2 billion per year.

5.2 Regulation of the Crude Oil Market and Demand-Side Inefficiency

As a method of insulating the direct and indirect users of crude oil from the oil price increases of recent experience, the entitlements program has been remarkably successful and administratively manageable. An endless stream of new rulings and regulations must undoubtedly leave those who are directly involved with anything but the impression that the entitlements program is uncomplicated and administratively tractable. Relative to other governmental subsidy programs of comparable magnitude, however, the entitlements program is accompanied by minimal direct federal control over the disbursement and use of funds. Rather, the bulk of the entitlements program's subsidies is carried on through direct payments between refiners. Moreover, the program's funds are, for the most part, spread across the broad class of refiners and, ultimately, consumers without day-to-day DOE oversight.

Notwithstanding the ingenuity with which the distributional results of the entitlements program are accomplished, the program is not allocatively neutral. As noted in chapter 2, for example, the Small Refiner Bias tends to shift the size distribution of firms in the refining industry toward relative smallness. While this allocative impact and others associated with the special entitlements programs may be nontrivial, primary emphasis here is placed on the allocative consequences of the subsidy to crude oil refining.

5.2.1 Methodology and Measurements

Relative to the unregulated market, the entitlements subsidy, SUBSIDY, lowers the marginal cost of crude oil perceived by domestic refiners—from the world price P_W, to $P_W - \text{SUBSIDY} = P_S$ (see figure 2.4A). The result is an increase in the quantity of crude oil refined. At the margin, subsidized crude oil inputs add P_S to the value of refined product output, as indicated by the (inverse) crude oil demand and marginal valuation schedule D of refiners. To realize incremental added value of P_S, however, the domestic economy pays the world price P_W to foreign crude oil sellers. Assuming that marginal US purchases of crude oil in international markets have no impact on the world price and either that (a) there are no externalities associated with crude oil importation or (b) any externalities that do exist are appropriately accounted for by independent policy measures, the world price of crude oil is the marginal social cost of crude oil to the domestic economy. The difference between this marginal cost and the subsidized marginal cost perceived by domestic refiners is consequently a marginal *deadweight* loss. It is a deadweight loss since the P_W dollars paid to foreign sellers of crude oil for incremental imports exceed the contribution P_S of those imports to the value of the resulting refined products delivered to the domestic economy. In March 1980, for example, marginal crude oil inputs were being acquired from foreign sources for approximately \$37.30 per barrel, but had a marginal value to domestic refiners of only \$32.25 per barrel (the difference being the entitlements subsidy).[1]

The total deadweight loss of the entitlements subsidy is the sum of the differences between P_W and the marginal value of crude oil to domestic refiners at each quantity over the range from the level of crude oil inputs selected in the absence of the subsidy to the level of inputs selected in the presence of the subsidy. If the former level of inputs is denoted by C^* and the latter level of inputs is denoted by C', the deadweight loss of the entitlements subsidy can be expressed equivalently as the difference between the total addition to expenditures on imported crude oil induced by the subsidy, $P_W(C' - C^*)$, and the total value placed on $C' - C^*$ (which is $\int_{C^*}^{C'} D\, dC$). The deadweight loss of the entitlements subsidy is then

$$\text{DWL} = P_W(C' - C^*) - \int_{C^*}^{C'} D\, dC. \tag{5.1}$$

Graphically, this represents area XYZ in figure 2.4A.

An alternative method of measuring DWL is based on the difference between the total amount spent on the entitlements subsidy and the concomitant increase in consumer surplus (or, more properly, "user" surplus) associated with the subsidy-induced expansion of domestic refining. The total amount spent on the entitlements subsidy is SUBSIDY $\cdot\, C' = (P_W - P_S)C'$. The user surplus gain USG can be measured by area $P_W ZXP_S$ in figure 2.4A and expressed as

$$\text{USG} = P_W C^* + \int_{C^*}^{C'} D\, dC - P_S C'. \tag{5.2}$$

The deadweight loss of the entitlements subsidy is then given by

$$\text{DWL} = (P_W - P_S)C' - \left(P_W C^* + \int_{C^*}^{C'} D\, dC - P_S C'\right), \tag{5.3}$$

which reduces directly to (5.1).

Both DWL and USG depend on D and, in particular, on the responsiveness of crude oil demand to the effective price change brought on by the entitlements subsidy. Specification of this responsiveness is complicated by the discrete price change associated with the entitlements subsidy and the relative scarcity of evidence on the price elasticity of demand. Available evidence suggests a price elasticity (one-year) on the order of −0.5; and this value has been employed in previous empirical work by, for example, Bohi and Russell (1978), the RAND Corporation (1977), and Houthakker and Kennedy (1978).[2] Although a sensitivity analysis is undertaken below, a value of −0.5 for the price elasticity of crude oil demand is relied upon here. Moreover, crude oil demand is given a functional form having the property of constant own-price elasticity over the interesting range of prices. In general form, such a demand function can be represented by

$$C = hP^{\eta}, \tag{5.4a}$$

or, in inverse form,

$$D(C) = P = h^{-1/\eta}C^{1/\eta}, \tag{5.4b}$$

where h is a constant and η is the price elasticity of demand.

The h term defines the intercept of the log-linear form of (5.4a) and is given by

$$h = CP^{-\eta}. \tag{5.5}$$

With a given value for η and (5.5) evaluated at observable values C' and P_{s} (which result under the entitlements subsidy) for C and P, the value of h is determinant and can be denoted $\hat{h}$. Then, since the world price P_{w} is observable, the quantity of crude oil C^* that would be demanded in the absence of the entitlements subsidy can be determined directly from (5.4a): $C^* = \hat{h}P_{w}^{\eta}$.

Expression (5.4b) is the explicit form of the demand $D(C)$ in the expressions for the deadweight loss and user surplus gain. Thus, with values on hand for P_{w}, C', C^*, η, and h, the deadweight loss of the entitlements subsidy can be calculated from (5.1). Similarly, USG can be calculated from (5.2).[3]

Table 5.1 presents estimates of the social cost of the entitlements subsidy over the years 1975–1980. The deadweight loss estimates are based on (5.1). The total expenditures on the entitlements subsidy are given by $(P_{w} - P_{s})C'$. The user surplus gain is from (5.2). Additional imports are calculated as $C' - C^*$, and the expenditure on these im-

Table 5.1 The Allocative Effects of the Entitlements Subsidy: 1975–1980 (Dollar Figures in Millions of 1980 Dollars)

	Deadweight loss ($)	Total expenditure on subsidy ($)	User surplus gain ($)	Additional crude oil imports (mmb)	Expenditure on additional imports ($)
1975[a]	1,037.2	18,165.4	17,127.2	490.9	9,601.6
1976	851.8	16,676.6	15,824.8	477.3	8,764.2
1977	654.2	15,449.5	14,795.3	433.0	8,051.9
1978	299.5	10,270.9	9,971.4	304.8	5,285.2
1979	627.4	17,771.6	17,144.2	360.9	9,199.6
1980[b] (Jan.–Mar.)	1,037.8	27,483.1	26,445.3	373.4	14,260.5

Source: Based on data from US Department of Energy, *Monthly Energy Review*.
a. In 1975, the supplementary import fees were imposed. If the marginal social cost of imported crude oil in 1975 had been the refiner acquisition cost *exclusive* of these fees, the deadweight loss of the entitlements subsidy would be estimated at $212.4 million, additional imports would be 222.0 mmb, and expenditures on additional imports would be $4,344.0 million.
b. Annual rate.

ports is $P_w(C' - C^*)$. The level of the demand for crude oil is reevaluated for each year (that is, the h term is allowed to vary from year to year) to account for possible shifts in the demand function. Nominal values have been deflated by the Gross National Product Deflator and expressed in constant 1980 dollars.

The difference between the total amount spent on the entitlements subsidy and the consequent user surplus gain is a deadweight loss in the sense that there exists a hypothetical compensation scheme that could accompany deregulation and leave both crude oil producers (who "finance" the entitlements subsidy through the implicit tax of crude oil price controls) and the beneficiaries of the subsidy better off than they are under regulation. As table 5.1 indicates, the amount spent on the subsidy exceeds the resulting gain to crude oil users. From 1975 to 1978, the deadweight loss of the entitlements subsidy declined from approximately $1,000 million to $300 million per year. This decline reflected the gradual decline in the total rents associated with controlled crude oil and available for the entitlements program (see table 2.2), as well as an increase in the proportion of these rents devoted to special programs such as the Small Refiner Bias, Exceptions and Appeals, and the $0.21 allowance. The downward trend in the social cost of the entitlements subsidy was reversed in early 1979 as a sharp rise in world crude oil prices increased the funds available to the entitlements program by raising the rents associated with controlled domestic crude oil.

The pattern of the entitlements-induced additions to crude oil imports has been similar to the pattern of the deadweight loss. These additions amounted to almost 500 mmb in 1975 and were down to approximately 300 mmb in 1978. By the first quarter of 1980, entitlements-induced imports were up to roughly 373 mmb per year—or roughly 1 mmb per day out of a total daily domestic consumption of crude oil of 14.5 mmb.

The reported estimates of the inducement to imports and associated deadweight loss of the entitlements subsidy are fairly sensitive to the choice of a value for the price elasticity of crude oil demand. To a first approximation, estimates based on an elasticity η' other than -0.5 are in the same proportion to the corresponding estimates in table 5.1 as η' bears to -0.5. In 1978, for example, an elasticity of -0.25 implies a deadweight loss of approximately $151.1 million, or 50.5% of the 1978 figure shown in table 5.1. Additional imports are estimated at 154.5 mmb per year, or 50.7% of the corresponding value reported in table 5.1.

5.2.2 Problems and Qualifications

As an important note to the calculation of the social cost of the entitlements subsidy, it should be pointed out that the deadweight loss of the entitlements subsidy is likely to be underestimated if there are externalities associated with refiners' use of crude oil—particularly imported crude oil. The arguments that have been advanced in favor of a federal regulatory role in the energy sector have uniformly pointed to the possibility of external *costs* associated with crude oil use. Suggested sources of such costs have included national security problems associated with reliance on foreign supply sources, the possibility of macroeconomic disequilibrium induced by an unanticipated increase in the price of foreign oil, and the environmental damage associated with transporting, refining, or burning petroleum. If present, such external costs make the social cost of imported crude oil higher than the (delivered) world price.

Although estimates of the magnitude of any of the possible external costs of crude oil use are unavailable and constitute a task well beyond the scope of this study, the measurement of the social cost of the entitlements subsidy in the presence of such externalities is conceptually straightforward. Consider, for example, a case in which the social

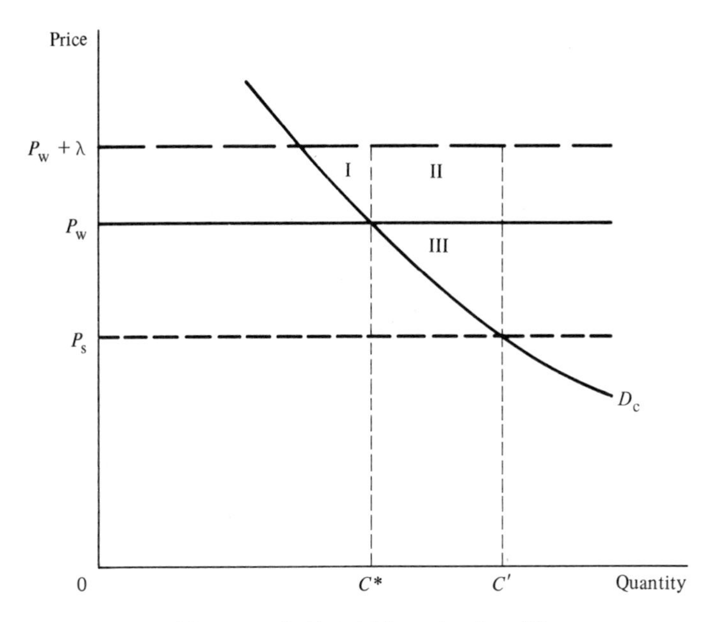

Figure 5.1 Entitlements subsidy with import externalities.

cost of imported crude oil at any quantity exceeds the world price by a constant λ. Graphically, this situation can be represented as in figure 5.1. In the absence of an entitlements subsidy, individual refiners would perceive the price of imported oil P_W as the relevant marginal cost of crude oil inputs and would buy C^*. The domestic economy would incur a deadweight loss equal to area I—and efficiency considerations would suggest the imposition of some method of internalization such as a tariff of λ dollars per barrel. Of course, it might be that the costs of internalizing the externality exceed area I, but, for the purpose of analysis and illustration here, I is treated as the net gain under a least-cost method of internalization. Without such internalization of the externality and with the introduction of an entitlements subsidy, the deadweight loss in the crude oil market would become the sum of areas I, II, and III. The incremental contribution of the entitlements subsidy to this sum is II + III. Area III, of course, is the deadweight loss reported in table 5.1, and area II is given by $\lambda(C' - C^*)$. For example, if there is an external cost of $\lambda = \$1.00$ associated with each barrel of imported crude oil, the roughly 305 million additional barrels of imports induced by the entitlements subsidy in 1978 would have carried an additional deadweight loss of $305 million. The total deadweight loss of the subsidy would then be estimated at approximately $605 million in 1978. In lieu of estimates of λ, policymakers and analysts so inclined could augment the deadweight loss shown in table 5.1 by choosing their own estimates of λ and multiplying by the fourth column of table 5.1.

In a similar vein, the deadweight loss of the entitlements subsidy may be underestimated in table 5.1 if the United States is not a price taker in the world crude oil market. If the US refining industry as a whole has some monopsony power (albeit unexercised) vis-à-vis crude oil imports, the marginal cost of crude oil to the domestic economy exceeds the world price. Marginal crude oil imports would cost the world price *plus* the additional expenditures on inframarginal purchases that result from the induced increase in the world crude oil price. Consequently, the use of the world price as the measure of marginal social cost in the calculations shown in table 5.1 could understate the true marginal burden the domestic economy bears when importing crude oil under the entitlements subsidy. The discussion in section 2.2 suggests that the magnitude of any international price response to the changes in import demand brought on by the entitlements program is most likely quite small. Nevertheless, it is clear that the accuracy of the deadweight loss estimates reported in table 5.1 is dependent on the assumption of a constant world price of crude oil.

5.2.3 Summary

In conclusion, it is obvious that the entitlements subsidy has consequences for the importation of crude oil that are in precisely the opposite direction of policymakers' publicly proclaimed goal of reducing "dependence on foreign oil." The EPAA/EPCA subsidization of crude oil imports through the entitlements program presents domestic refiners with a privately perceived cost of imported crude oil that is less than the value of the resources the economy gives up to acquire such crude oil. The result is an increase in crude oil imports. Taking the world price as the measure of the social marginal cost of imported crude oil shows that the net economic waste associated with the entitlements subsidy is estimated to have varied within the range of $250–1,000 million per year since 1975. These estimates of the net social cost of the entitlements subsidy are likely to be underestimates if there are external costs associated with the importation of crude oil or if incremental US purchases of crude oil on world markets raise world prices.

5.3 Regulation of the Crude Oil Market and Supply-Side Inefficiency

EPAA/EPCA regulation of the petroleum industry has not only induced overconsumption of crude oil; it also has had allocative effects on the production of crude oil from domestic reserves. Drawing on the analysis of chapter 3, this section measures the value of the allocative distortions in domestic crude oil production that are induced by crude oil price controls. In addition, measures of the prospective allocative effects of the Windfall Profits Tax are examined.

5.3.1 Methodology and Measurements

Efficiency in crude oil production requires that, for a given level of demand, the total cost of acquiring oil from foreign and domestic sources be minimized. When the alternative to domestic crude oil is imported oil purchased at a given world price and when there are no externalities associated with imported oil, efficient use of domestic petroleum resources requires the production of all domestic crude oil that can be acquired for a cost that does not exceed the price of foreign oil. This means that, to satisfy the criteria of efficiency, production of crude oil from any domestic field must be carried on until the incremental cost of the last unit of output is equated to the cost of imported crude oil.

The analysis will include considerations of possible intertemporal distortions to crude oil production decisions. But before this is done, the allocative consequences of crude oil price controls can be described by figure 5.2. In this figure, S denotes the schedule of supply prices for a representative producing property. In the absence of crude oil price regulation, producers would face an average and marginal revenue of P_w and would produce Q_u^*. Under EPCA controls, however, a producer with a property classified as lower tier who is able to produce profitably above BPCL (see section 3.3) or a producer with a property classified as upper tier would face a marginal revenue of P_u and would produce only Q_u. The efficiency, or deadweight, loss associated with this production decision amounts to area CDH within any period. A producer with a lower-tier property (under EPCA) or an old oil property (under EPAA) who is unable to produce profitably above BPCL would receive a price of P_0 and would select an output level of Q_0. The deadweight loss associated with this production decision is area BDG. A producer with an old oil property under EPAA or any

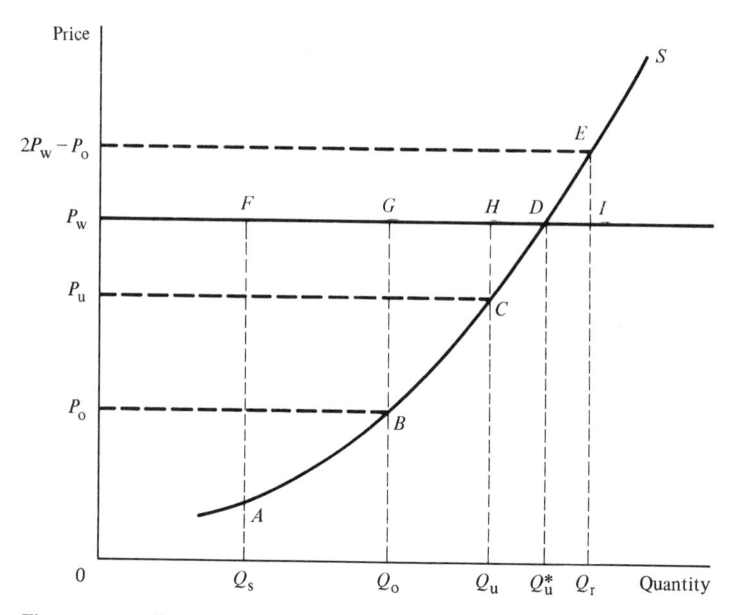

Figure 5.2 Allocative effects of crude oil price controls.

producer with a lower- or upper-tier property under EPCA who finds it profitable to reduce production in order to qualify for stripper prices would produce Q_s. The deadweight loss associated with this production decision is area *ADF*. Of course, for "legitimate" stripper properties, $Q_s > Q_u^*$ and regulations have no allocative impact. Finally, a producer (under EPAA) with an old oil property who is able to produce profitably above BPCL and thereby qualify for the released oil program would produce Q_r. The deadweight loss associated with this production decision is area *DEI*.[4]

As the phased decontrol of crude oil prices proceeds over June 1979–September 1981 and then as the Windfall Profits Tax takes full effect, supply-side allocative distortions will continue to arise. Corresponding to the effective, after-tax prices such as shown in table 3.1, there are production decisions analogous to those described by figure 5.2. Moreover, resulting areas of deadweight loss for each category of oil can continue to be approximated by the difference between the completely unregulated price P_w and the relevant supply price schedule, summed across the range of output foregone as a result of after-tax prices less than P_w. The notable exception to this rather straightforward analysis of the WPT is the case of incremental tertiary oil. As described in chapter 3, a large inframarginal incentive (in the form of a reduction in tax burdens) is provided to those properties otherwise in the WPT's Tier One and Tier Two that are able to increase output above statutorially determined base levels through new tertiary enhanced recovery projects. When information on the extent of these projects becomes available, the allocative consequences of this provision might be roughly measured by asking hypothetically what effective marginal price would have to be offered to producers to induce them to supply the same level of output from tertiary enhanced recovery as they supply in the presence of the WPT inframarginal incentives. This hypothetical price will then provide a location for the supply schedule of tertiary recovery oil and a rough analogy with figure 5.2 could be drawn.

Mathematically, the net economic cost of crude oil price regulation can be expressed as the difference between the regulation-induced change in the expenditures made by the economy to acquire imported crude oil and the regulation-induced savings of domestic resources as a result of reductions in the output of domestic crude oil. Using upper-tier oil as the example, the change in expenditures on imports induced by price controls is $P_{\mathrm{w}}(Q_{\mathrm{u}}^{*} - Q_{\mathrm{u}})$, where Q_{u}^{*} is the output of crude oil that would result in the absence of controls from properties that produce upper-tier oil under controls. Writing the supply price schedule of these properties as $S_{\mathrm{u}} = S_{\mathrm{u}}(Q)$, the savings of domestic resources which arises from the output reduction from Q_{u}^{*} to Q_{u} is $\int_{Q_{\mathrm{u}}}^{Q_{\mathrm{u}}^{*}} S_{\mathrm{u}}\, dQ$. Thus, the deadweight loss associated with upper-tier oil is

$$\mathrm{DWL}_{\mathrm{u}} = P_{\mathrm{w}}(Q_{\mathrm{u}}^{*} - Q_{\mathrm{u}}) - \int_{Q_{\mathrm{u}}}^{Q_{\mathrm{u}}^{*}} S_{\mathrm{u}}\, dQ, \tag{5.6}$$

and analogous expressions pertain to other categories of oil. Taking the price elasticity of supply ϵ_{u} to be constant over the relevant range, the supply price of upper-tier crude oil can be expressed [following (5.4b)] as

$$S_{\mathrm{u}} = g^{-1/\epsilon_{\mathrm{u}}} Q^{1/\epsilon_{\mathrm{u}}}, \tag{5.7}$$

where g is completely analogous to the h term in the constant-elasticity demand function. Substituting for S_{u} in (5.6) and integrating yields the desired measure of $\mathrm{DWL}_{\mathrm{u}}$.[5]

The application of (5.6) is confounded by the potential for divergences between the market price of uncontrolled crude oil P_{w} and the social marginal cost of crude oil to the nation. The implications and possible sources of such divergences will be discussed at length; at this stage, it is only noted that the (delivered) world price of crude oil is being used to measure the marginal social cost of imported crude oil in the regulated and unregulated market settings and that estimates of allocative and distributional effects of controls are sensitive to deviations from this assumption.

Measurement of the allocative consequences of crude oil price regulation must be qualified not only by recognition of the possibility that the marginal cost to the economy of imported crude oil differs from its price, but also by recognition of possible intertemporal effects. In particular, the discussion of chapter 3 indicates that, although the effect of price controls (and, soon, the WPT) on exploration and development activity is negative because controls (and taxes) reduce the value of the stream of returns to be expected from such activity, the effect of controls (and taxes) on the time pattern of extraction from existing wells is ambiguous. The implications for estimation of the allocative consequences of controls include the following: Controls (and taxes) might leave the time path of extraction unaffected and, except for properties unable to cover extraction costs at regulated prices, there might be no allocative distortions associated with output from existing properties. Of course, in any period, controls (and taxes) might result in more or less extraction than the allocative optimum.

The lesson to be drawn is that the estimation of the net social cost of crude oil price controls or the Windfall Profits Tax through the use of expression (5.6) must be subject to caveats. From the results of chapter 3, it appears that, on average, the time path of price controls has been such that the crude oil output reductions implied by a static analysis have been reinforced by the intertemporal aspects of controls. Furthermore, the intertemporal impact of the WPT will depend crucially on the course of world oil prices. At any rate, the estimation of the allocative effects of crude oil price regulation undertaken here follows the methodology laid out by Herfindahl and Kneese (1974) in their important work on the economics of natural resources.

The supply elasticity in (5.6) and (5.7) is the only term that requires external estimation. The considerable efforts by economists toward this end have been described in section 3.5 and are relied upon here. So long as the hypothetical price changes involved in the move from regulated to unregulated regimes would carry the same information about the expected present value of the returns to exploration and development activities as the (typically pre-embargo) price changes examined by those who have estimated supply elasticities, application of (5.6) to supplies from new sources is straightforward. In the absence of evidence to the contrary, the satisfaction of this (admittedly problematical) condition is maintained here (as it is implicitly maintained in Herfindahl and Kneese, 1974). Similarly, (5.6) is applied to supplies from existing reserves under the assumption that the price changes on which elasticity estimates have been based carried the same information on the expected time path of prices as the price changes attendant to the prospective move from the regulated to the unregulated world.

Because different categories of domestic oil are subject to different prices and, presumably, different supply elasticities, the total deadweight loss of federal price regulation must be built up from individual category estimates. The most important categorical distinction is between existing and newly producing supply sources. As discussed in chapter 3, estimates of the elasticity of supply of crude oil from newly developed sources based on data from the pre-embargo era have quite consistently been around unity or slightly less. Although the abundance of exogenous and policy shocks and the scarcity of data points since the OAPEC embargo have so far precluded full-blown reestimations of elasticities, reassessment of matters in the new era of higher world prices has led to a consensus of opinion favoring values in the neighborhood of 0.5. Consensus on the elasticity of supply from existing sources supports considerably smaller values—near 0.1. These estimates of supply elasticities are employed here in the development of "most likely" cases, but cases using other values are also developed to provide information on sensitivity.

In the case of EPAA controls, the task is to estimate the deadweight losses from supply sources producing exclusively old oil and from properties producing released oil.[6] New oil properties were uncontrolled. In the case of EPCA controls, the task is to estimate deadweight losses associated with properties producing exclusively lower-tier oil, existing properties producing upper-tier oil, and foregone supply from newly developed sources. Data are available on total production within various control categories, but they are insufficient to permit determination of, for example, the quantity of old oil produced under EPAA by properties that yielded exclusively old oil versus the quantity of old oil produced from old properties that produced released oil. Such a determination is important because the relevant marginal revenue under controls (and hence the *change* in marginal revenue to which output would respond upon decontrol) is vastly different in the two cases. If all old oil properties produced released oil, the old oil price ceiling did not affect any marginal production decisions under EPAA. On the basis of some scanty data made available to, and reported by, the RAND Corporation (1977), it is estimated in Kalt (1980) that approximately 20% of old oil production came from wells that produced released oil under EPAA. In the absence of sufficient data, it is further assumed, by extrapolation, that 20% of lower-tier output under EPCA has come from lower-tier properties that produce upper-tier oil (that is, lower-tier properties that produce above BPCL).

Once the determination of supply elasticity (for example, ϵ_u) and actual production levels (for example, Q_u) for each control category have been made, the g term in (5.7) can be calculated. Output levels from existing properties under deregulation (for example, Q_u^*) can then be taken from the inverse form of (5.7) (with P_w substituted for S_u), and deadweight losses may be calculated from (5.6). The specification of the allocative effects of price controls on newly developed properties under EPCA, however, is more complicated. Studies that have estimated the supply elasticity of output from newly developed sources have, in fact, estimated the supply elasticity of additions to reserves. As Bohi and Russell (1978) point out, in a long-run (unregulated) market equilibrium, this latter elasticity is equivalent to the long-run elasticity of *output* from additions to reserves; and the base output level from which the output response from new reserves, as implied by the elasticity, to any given price change could be taken is total domestic production from existing supply sources. Consequently, the appropriate technique for estimating the effect of controls on production from new reserves is to first estimate what aggregate domestic production from existing reserves would have been in the absence of controls and then estimate, from that base, the controls-induced change in output from new sources.

Table 5.2 shows the most likely case estimates of the allocative effects of EPAA/EPCA price controls on domestic crude oil. Results are reported on an annual, rather than cumulative basis, and the experiment being performed is a repetitive one of "If price controls are removed at the beginning of year X, what will be the allocative impact within year X?" It is implicitly assumed that explorers and developers appropriately anticipate that year X will bring uncontrolled prices, so that when year X arrives new supply sources are brought into production as desired without a lag.

Table 5.2 indicates that the annual deadweight loss of crude oil price

Table 5.2 The Allocative Effects of Crude Oil Price Regulation (Most Likely Case): 1975–1980
(Dollar Figures in Millions of 1980 Dollars)

	Deadweight loss ($)	Additional crude oil imports (mmb)	Expenditure on additional imports ($)
1975[a]	963.0	99.4	1947.9
1976	1045.5	344.4	6288.3
1977	1213.3	454.6	8443.7
1978	815.7	347.6	6034.1
1979	1851.5	331.2	8443.9
1980[b] (Jan.–Mar.; actual)	4615.9	529.5	20214.3
March 1980[c] (hypothetical decontrol with WPT)	2219.4	486.5	18573.4

Source: Based on data from US Department of Energy, *Monthly Energy Review*.
a. If the marginal social cost of imported crude oil had been the landed cost of imported crude oil *exclusive* of the supplementary import fees imposed in 1975, the deadweight loss of controls would be estimated at $629.2 million, additional imports would be 88.7 mmb, and expenditures on additional imports would be $1,457.2 million.
b. Annual rate based on actual effective prices, after Windfall Profits Tax, as reported in chapter 3.
c. Annual rate based on effective prices, after Windfall Profits Tax and hypothetical complete decontrol as of March 1980, as reported in chapter 3. Estimates do not reflect special WPT provisions for incremental tertiary enhanced recovery oil.

controls has averaged slightly less than $1.8 billion (1980 dollars) since 1975. Under EPAA in 1975, the efficiency loss of controls totalled roughly $1.0 billion. This loss increased with the introduction of EPCA controls, which brought new production under upper-tier ceilings in 1976. Rising domestic ceilings with fairly stable world prices in 1978 reduced the deadweight loss of controls relative to 1976 and 1977. The very steep world price increases in 1979 pushed this loss back up, despite the beginnings of phased decontrol.

March 1980 brought the start of the Windfall Profits Tax and a transition period of overlapping controls and excise taxes. Based on actual after-tax prices, table 5.2 presents estimates of the allocative impacts of federal crude oil price policy as of the first quarter of 1980. The social loss of price regulation at this time was running at an annual rate of over $4.6 billion. To provide some perspective on the impact that the Windfall Profits Tax will have when it becomes fully effective after complete decontrol in the fall of 1981, the last row of table 5.2 reports estimates of the allocative effects of the WPT. These estimates are made under the assumption that after-tax prices received by major and independent producers upon full decontrol are as reported in chapter 3 for March 1980.[7] The estimated deadweight loss of the fully effective Windfall Profits Tax is approximately $2.2 billion per year. This figure is a significant improvement relative to the actual conditions prevailing in March 1980, when crude oil price controls, rather than the WPT, were determining effective prices for several important categories of domestic production. Specifically, the improvement is due to the substantial increases in effective prices for lower- and upper-tier oil that occur when the WPT takes over from EPCA controls. (For example, compare the second and third rows of table 3.1.) Notwithstanding allocative improvements under the WPT, crude oil producers will still be big losers. The $2.2 billion social cost of the WPT will be borne as lost surplus by domestic producers. When this loss is coupled with the estimated direct tax collections of the WPT, producers will have approximately $35.4 billion per year extracted from them by federal regulation.

Importantly, neither the deadweight loss estimate nor producers' tax payments for the hypothetical case of the WPT with no price controls reflect the special incentives provided for tertiary enhanced recovery by the WPT. As noted, the response of supply to these incentives cannot be estimated solely from information on marginal price incentives and marginal costs because of the inframarginal and discrete character of the tax savings attendant to tertiary enhanced recovery projects. The magnitude of the expected supply response is a formidable question in its own right and will require considerable methodological and empirical research efforts that are beyond the scope of the present study. Nevertheless, it can be pointed out that, from an allocative point of view, the most felicitous result of the tertiary recovery incentives would be the case in which these incentives produce the same supply responses that would occur if affected crude oil supplies were not subject to controls or taxes and received a price of P_w. But, even if it were the (extreme) case that *every* property eligible for the tertiary recovery tax benefits would undertake tertiary recovery in the unregulated world, this "most felicitous" case would still leave deadweight social losses from the Windfall Profits Tax running at an annual rate of $0.8 billion (based on March 1980 price incentives for the hypothetical case of WPT with no controls). Moreover, the noted WPT inducements to smallness in tertiary recovery projects and to the creation of projects that would be unable to cover marginal costs at market-determined prices suggest the untenability of the most felicitous case.

Turning to output effects, EPAA/EPCA controls caused the domestic economy to forego the production of an average of approximately 315 mmb of crude oil per year (0.9 mmb/d) during 1975–1979. This resulted in the importation of an equivalent amount of foreign crude oil with an associated average annual increase in the bill for crude oil imports of roughly $6 billion. The bulk of the foregone output has been unrealized production from new reserves. In 1978, for example, less than one third of the controls-induced reduction in domestic output was attributable to existing developed properties. In early 1980, foregone domestic production was running at an annual rate of roughly 530 mmb per year (1.4 mmb/d). In the reported hypothetical case of a fully effective WPT as of March 1980, foregone US output would have been running at a rate of 480 mmb per year (1.3 mmb/d).

Obviously, the estimates of the supply-side costs of EPAA/EPCA regulation and the Windfall Profits Tax depend heavily on the elasticity of the response of domestic crude oil production to price changes. Table 5.3 presents estimates of these costs under alternative assumptions regarding the price responsiveness of domestic production. Whereas the most likely case developed above employed an estimate of 0.1 for the elasticity of supply from existing properties with oil field equipment in operation, table 5.3 presents cases in which this estimate is first halved and then doubled. Similarly, the elasticity of supply from newly developed sources has been halved and doubled relative to the most likely case. Most certainly, the estimates of table 5.3 bound the deadweight loss of crude oil price controls. During most of the 1975–1979 period, the bounds on this loss cover a range of $0.5–2.5 billion per year. Sharp increases in world crude oil prices beginning during 1979 approximately doubled the lower and upper bounds for 1979 relative to 1978. With the full impact of higher world prices, the 1980 (actual case) bounds on the social cost of price regulation rose still higher. The fully effective WPT would have reduced these bounds substantially, primarily by raising lower- and upper-tier prices closer to uncontrolled levels.

As noted in chapter 3, the multitier EPAA/EPCA price control system has tended to reduce the costs of controls relative to a uniform price control system. Assuming the parameters of the most likely case described above, the value of the inefficiency of controls in 1978, for

Table 5.3 The Deadweight Loss of Crude Oil Price Regulation under Alternative Supply Elasticities: 1975–1980 (Billions of 1980 Dollars)

Low case: existing properties' supply elasticity = 0.05

New properties' supply elasticity	1975[a]	1976	1977	1978	1979	1980[b] (Jan.–Mar.; actual)	March 1980[c] (hypothetical)
0.25	$0.5	$0.5	$0.6	$0.4	$0.9	$2.2	$0.8
1.00	$0.5	$0.8	$1.3	$0.7	$1.2	$3.3	$1.9

High case: existing properties' supply elasticity = 0.2

New properties' supply elasticity	1975[a]	1976	1977	1978	1979	1980[b] (Jan.–Mar.; actual)	March 1980[c] (hypothetical)
0.25	$2.0	$1.8	$1.8	$1.3	$3.5	$ 9.6	$2.2
1.00	$2.0	$2.1	$2.5	$1.7	$3.9	$10.9	$3.4

Source: Based on data from US Department of Energy, *Monthly Energy Review*.
a. If the marginal social cost of imported crude oil was the landed cost *exclusive* of the supplementary import fees imposed in 1975, the deadweight loss of controls would be estimated at $0.5 billion (low case) and $2.1 billion (high case).
b. Annual rate. See note b, table 5.2.
c. Annual rate. See note c, table 5.2.

example, would be estimated at over $2.0 billion if all controlled crude oil had been priced at the domestic average. Under the same assumptions as to supply responsiveness, the social loss of the actual multitier controls was reported in table 5.2 to be $0.8 billion. Thus, as a method of extracting crude oil producer wealth, multitier controls have resulted in a less costly distortion of resource allocation than a single-price monopsonistic scheme.

5.3.2 Problems and Qualifications

One of the central goals of this study is to provide quantitative estimates of the effects of current federal petroleum policy. As was the case with the measurement of the demand-side allocative effects of EPAA/EPCA regulation, however, the estimation of the effects of price controls and taxes on crude oil supply is confounded by the possibility of a divergence between the social marginal cost of imported oil and its delivered price. Again, a quixotic search for numerical values of such a divergence is avoided here; and the purpose of this subsection is only to elucidate the directions of any bias in the estimates put forth in tables 5.2 and 5.3.

The implications of a noninternalized external cost to imported crude oil can be illustrated with figure 5.3, which simplifies relative to actual policies by assuming a simple price ceiling (or after-WPT price) of P_u on the domestic supply described by S. With an externality of λ per barrel of imports, the deadweight loss in the unregulated market would

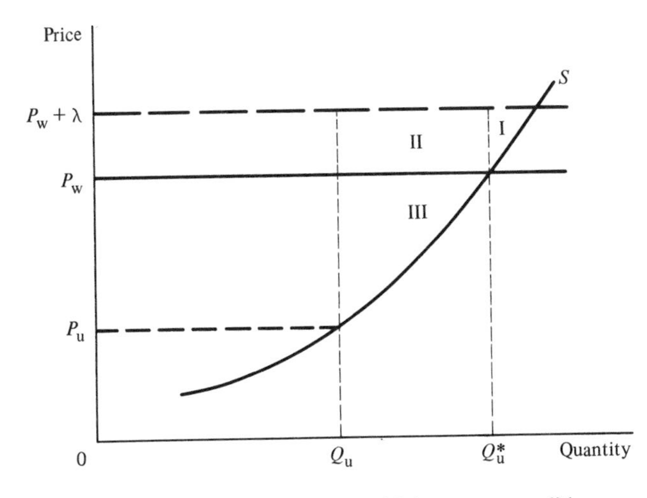

Figure 5.3 Crude oil price controls with import externalities.

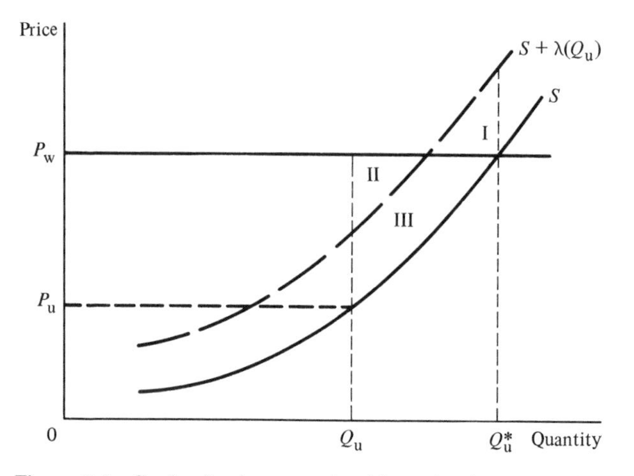

Figure 5.4 Crude oil price controls with production externalities.

be area I. Crude oil price regulation increases the supply-side efficiency loss by II + III to a total of I + II + III. Area III is reported in tables 5.2 and 5.3 and area II is given by λ times the change in crude oil production induced by regulation (that is, $Q_u^* - Q_u$). Thus, with $\lambda =$ \$1.00, for example, the additional 347.6 mmb of imports that table 5.2 reports for 1978 would have a deadweight (externality) loss of \$347.6 million. The total deadweight loss of EPCA controls (that is, area II + III) would then be estimated at close to \$1.2 billion in 1978.

In addition to the possibility that there are noninternalized externalities associated with the importation or domestic use of crude oil, it is also possible that there are externalities associated with the domestic production of crude oil. Such externalities may give rise to domestic production levels that are higher than is optimal. In figure 5.4, unregulated domestic production at a world price of P_w in the presence of a noninternalized marginal external cost of $\lambda = \lambda(Q_u)$ results in an efficiency loss equal to area I. For the sake of simplicity, it is assumed that external damage and crude oil are produced in fixed proportions (so that, for example, an internalizing tax on output of external damage and an internalizing tax on crude oil output are equivalent). The rolling back of effective prices to P_u reduces production; and table 5.2 reports the deadweight loss of controls as area II + III. The true deadweight loss, II, is consequently overstated. Indeed, appropriate selection of P_u could result in II = 0. It is also possible, however, for controls (or taxes) to worsen the allocation of resources, namely, if P_u is so low that

II $>$ I. Regulation represents an improvement under standards of allocative efficiency only when II $<$ I.

Sweeney (1977) has investigated the dynamic implications of externalities in a natural resource market. In the case, for example, of a noninternalized externality associated with resource extraction (for example, environmental pollution), optimal extraction paths depend, other things being equal, on the rate of growth over time of marginal external costs. In general, if noninternalized marginal external costs grow over time at a rate less than the rate of interest, unregulated resource extraction rates will be higher than is optimal in early periods. Intuitively, optimal internalization of a slowly rising external cost (that is, declining in a present-value sense) should cause producers to shift output toward the future in an attempt to avoid relatively high early-period internalized costs. Although a first-best method of internalizing the externality might be to price it directly (for example, with a tax), a price regulation program in which $\dot{\text{GAP}}/\text{GAP} < r$ could lower the early-period extraction rate toward the optimal rate. Alternatively, if noninternalized marginal external costs associated with domestic extraction rise at a rate greater than the rate of interest, unregulated extraction rates will be lower than optimal in early periods; and a price regulation program in which $\dot{\text{GAP}}/\text{GAP} > r$ could hypothetically improve resource allocation by raising early-period extraction rates toward optimal levels.

Noninternalized externalities might be associated with the importation of crude oil, rather than the domestic production of crude oil. Following Sweeney (1977), if the marginal external costs attendant to imports rise over time more rapidly than the interest rate, unregulated early-period extraction rates will be less than optimal, early-period imports will be too great, and too much domestic crude oil will be saved for future use. If the marginal external costs associated with imports rise more rapidly than the interest rate, unregulated early-period domestic crude oil output will be too great, early-period imports will be too small, and too little domestic crude oil will be saved for future use. Intuitively, if the externalities associated with imported oil are rising rapidly, the efficient use of domestic resources requires a policy of "Import now and produce at home later." Price controls or excise taxes on domestic crude oil could serve as such a policy. In either case—marginal external costs rising less rapidly or more rapidly than the interest rate—domestic consumption of crude oil will be too great in every period, since in every period refiners would see a private cost of crude oil that is less than the social marginal cost. An entitlements subsidy magnifies this misallocation of resources by pushing the private cost of crude oil further below its social cost.

5.3.3 Summary

Subject to the foregoing qualifications, it appears that crude oil price controls have resulted in a deadweight loss to the US economy. These controls have caused the domestic economy to hand over more dollars to foreign crude oil suppliers than has been necessary to satisfy any given level of domestic demand. In particular, there has consistently been a range of output from domestic sources foreclosed by price controls but available at marginal costs below the marginal cost of imported crude oil. Although the efficiency loss of this foreclosure has been less under multitier controls than under a uniform price system, estimates of the most likely average annual supply-side waste of EPAA/EPCA regulation over 1975–1979 are in the range of $0.8–2.0 billion. By early 1980, the waste of controls was approaching $5 billion per year. Although commonly viewed as a means of offsetting the distributional effects of crude oil price decontrol, the Windfall Profits Tax perpetuates multitier price regulation and apparently will produce allocative inefficiencies of the same order of magnitude as those produced by EPAA/EPCA price controls.

5.4 Administrative Costs of EPAA/EPCA Regulation

The analysis so far has focused on those social costs of federal petroleum price regulation that arise as the direct effects of this regulation on crude oil production and refining decisions. Other costs, however, are not insignificant. In particular, substantial costs are associated with regulatory administration, enforcement, and compliance.

Discussions of the costs and benefits of regulatory policies often overlook the costs of executing such policies. These costs arise because both governmental and private resources, which might be productively devoted to other tasks, must be applied to the problems of carrying out any regulatory program. In the case at hand, the appropriate treatment of the administrative costs of regulation is not self-evident. On the one hand, it might be argued that although the net effect of petroleum price regulation is some allocative inefficiency, the deadweight losses of regulation might be even larger if, at the margin, fewer resources were devoted to carrying out the program. On the other hand, relative to the unregulated world, resources so devoted can be viewed as wasted if their contribution to the value of society's aggregate output is negative. Since the effects of regulation are being compared to the unregulated case throughout this study, this latter view is adopted here.

The administrative costs of regulation can be thought of as the transaction costs incurred to bring about the transfers accomplished by post-embargo federal petroleum policy. Whereas the transaction costs associated with an efficiency-improving move have to be subtracted from the total gains from trade to arrive at the net social gains of such a move, the transaction costs associated with an efficiency-worsening move increase the absolute value of the resulting negative gains from trade. Unfortunately, a detailed study of the transaction costs of carrying out EPAA/EPCA regulation is confounded by the variety of forms, sometimes unmeasurable, that these costs take. Nevertheless, evidence on some of these costs is available and suggestive of the magnitude of the resource expenditure involved.

5.4.1 Private-Sector Costs

The administrative burden of petroleum price regulation falls directly on two groups: private sector participants in the regulated market and general taxpayers in the public sector. In the private sector, regulation-related costs fall proximately on the regulated industry, but may ultimately be borne in greatest part by consumers of petroleum products. The private sector burdens of regulation arise from (a) the administrative costs of maintaining records and reporting required information (that is, the so-called paperwork burden of regulation); (b) the disruption of normal business practices by the mandated need for federal approval; and (c) the shifting of the arena in which competitive viability is determined from the marketplace toward the political and bureaucratic realm.

Evidence on some part of the first of these costs during fiscal year 1977 was released by a special Presidential Task Force in 1977 (MacAvoy, 1977). In fiscal year 1977, the federal energy regulatory compliance program imposed record keeping and reporting requirements on over 300,000 firms. These firms are conservatively estimated to have had to file over 1,000,000 periodic, one-time, and occasional reports. The value of the man-hours devoted to this effort is estimated to have been approximately $160 million. In addition to the actual filing

Table 5.4 Estimated Federal Budget for Regulatory Administration: 1974–1980[a] (Millions of 1980 Dollars)

1974	71.0
1975	62.8
1976	125.0
1977	133.8
1978	160.0
1979	196.2
1980	198.3

Source: US President, *The Budget of the United States Government* (1976–1980).
a. Estimated under the following assumptions: (a) One half of Energy Information Administration (under DOE) and "energy information and analysis" category (under FEA) budget authorities are associated with petroleum price, allocation, and entitlements regulations. (b) All of Economic Regulatory Administration (under DOE), except "coal utilization," "utility programs," and "emergency preparedness" categories, and all "regulatory programs" category (under FEA) budget authorities are associated with price, allocation, and entitlements regulations. (c) The proportion of ERA (under DOE) and FEA administrative expenses associated with price, allocation, and entitlements regulation is the same as the proportion that the budget authorities of included categories [see (a) and (b)] bear to the authorities of all nonadministrative categories. Fiscal year data converted to annual figures assuming expenditures allocated evenly across each fiscal year.

of reports, administrative costs are incurred to maintain records, monitor regulatory compliance, and ensure adequate representation in the attorneys' arena of exceptions, appeals, and litigation. The Presidential Task Force estimated that the sum of these costs and direct reporting expenses for fiscal year 1977 approached $650 million (1980 dollars).

In addition to the Presidential Task Force estimates of private sector administrative costs, rough estimates of these costs may be indirectly derived from evidence collected by the Center for the Study of American Business (CSAB). Weidenbaum (1979) reports that this Center has culled from the literature the best available estimates of the administrative costs of numerous different types of federal regulation and has found that, on the average, each dollar of federal expenditure on regulatory administration is accompanied by a private sector expenditure of $20. If private administrative costs of federal regulation of the petroleum sector are not atypical of the private administrative costs of federal regulation in general, this 20-to-1 multiplier is applicable in the present context. Using this multiplier and estimates of the federal budget associated with EPAA/EPCA price, allocation, and entitlements regulations (table 5.4), it is estimated that the private sector has incurred regulatory administration and compliance costs that have averaged $2.7 billion (1980 dollars) during 1974–1980. These costs reached almost $4 billion in 1980. Of course, the rather crude methods by which these estimates have been made argue against great confidence in them. Nevertheless, they (as well as the Presidential Task Force estimates) are suggestive of the size of private resource expenditures involved in carrying out the regulation of petroleum prices. What is most striking about the size of these expenditures is that they are of the same order of magnitude as the deadweight allocative losses reported above.

In addition to creating the burden of paperwork, record keeping, and a legal staff, the presence of regulation in petroleum markets has had widespread effects on normal business practices. In many cases, the alteration of these practices has the effect of distorting the cost-minimization efforts of profit-seeking firms. Numerous aspects of refinery operations, for example, including capacity changes, shutdown for maintenance, and sale of a refinery, require either a regulatory approval or clarification of regulatory obligations since refining operations have pricing, entitlement, and buyer-supplier obligations associated with them. Similarly, transactions involving crude oil producers require extensive certification as to quantity and applicable price ceilings; and changes in many aspects of their buyer-supplier relationships with pipelines and refiners are subject to lengthy review, exception, and appeal processes. These regulation-related impediments to doing business raise the costs of transactions and management in petroleum markets.

Some of the costs of carrying out current petroleum price regulations arise through the effects of these regulations on the forms of productivity and rivalry conducive to firms' survival and profit opportunities. As a result of the regulatory process, the viability of a firm depends relatively more than it otherwise would on its ability to "produce" favorable regulation and rulings, as opposed to its ability to produce crude oil, refined products, or marketing services. Some of the costs of the implied changes in firms' production techniques may be captured in the Presidential Task Force and CSAB estimates of the paperwork burden (for example, the staff assigned to handle regulatory paperwork and monitoring may also engage in some lobbying), but many of these costs are likely to be quite subtle and unmeasured. A premium can be expected to be placed on management personnnel, for example, who know relatively more about getting along with, and influencing, regulators than about producing products for their customers.

In general, the presence of a regulatory mechanism that has the ability to shift large amounts of wealth among competing interest groups and that is subsidized by the general taxpayer can be expected to lead interested parties to devote more resources than they otherwise would to lobbying, campaigning, and appealing for favorable regulatory decisions. Indeed, Posner (1975) has pointed out that (a) the economic rents at stake in the typical regulatory setting attract resources into efforts by affected parties trying to capture or avoid the loss of such rents; and (b) consequently, what appear to be pure transfers may be completely transformed into the consumption of scarce resources. While the incentive described by the first observation is undoubtedly operative in the case at hand, there are factors that work against the strong version of the second observation. Significantly, most of the important transfer-affecting regulatory decisions that might be objects of lobbying efforts involve strong public goods attributes. An increase in lower-tier ceiling prices or an increase in the entitlements subsidy, for example, would both benefit and harm broad classes of actors; and a clear free rider effect can be expected to operate on individual actors so as to reduce the commitment of resources to the efforts noted by Posner. Furthermore, even in the absence of this free rider effect, the commitment of resources to influencing the regulatory process may be restrained to the extent individual actors perceive themselves as individually unable to affect such decisions.

While it is not likely that all of the transfers available under federal oil price regulation have been exhausted by attempts at their capture, the foregoing is not meant to imply that no resource-using dissipation has taken place. Far from it. Induced distortions to the normal behavior of market participants are undoubtedly subtle and widespread in the current environment. They are also undoubtedly costly in terms of the resources they consume.

5.4.2 Public-Sector Costs

The resources consumed in the governmental sector must also be considered as costs of petroleum price regulation. Responsibility for this regulation lay primarily with the Federal Energy Office/Administration until 1977, at which time the Department of Energy was created. Neither FEA nor DOE have been inconspicuously small bodies. The Department of Energy, for example, has an estimated budget for fiscal 1980 of $10 billion and employs over 20,000 people.[8] Of course, not all of the resources of DOE or FEA have been allocated to the administration of petroleum market regulation. Unfortunately, precisely isolating the allocation of expenditures within these agencies is not possible. Nevertheless, under the assumptions it describes, table 5.4 reports reasonable figures for the federal budget devoted to petroleum price, allocation, and entitlements regulation during 1974–1980. The budget allocations to EPAA/EPCA policies under study here have typically been in the range of $100–200 million (1980 dollars) per year and have averaged $135 million per year. The Windfall Profits Tax will transfer substantial regulatory responsibility to the Department of the Treasury and, in particular, the Internal Revenue Service. The budgetary impact of this transfer remains to be seen.

5.4.3 Conclusion

The private- and public-sector administrative costs of regulation are the transaction costs incurred in bringing about the reallocations of resources and wealth that occur under federal policy. Many of these costs, particularly in the private sector, are extremely difficult to quantify. Among the unmeasured costs are distortions to normal business practices and alterations in the types of activities and inputs that firms select. The measurable costs of administration and compliance in the private sector most likely have been running at over $650 million per year and may be in the range of $2–4 billion per year. In the public sector, annual expenditures on EPAA/EPCA controls and entitlements have averaged well over $100 million since 1974.

5.5 Distributional Consequences of Petroleum Price Regulation

While the allocative effects of federal price policies hold substantial interest for the economic analyst, participants in the regulated sector presumably attach primary importance to the distributional consequences of regulation. In broad terms, three groups have had a direct distributional stake in post-embargo regulation of the petroleum industry. These are crude oil producers, intermediate users of crude oil (notably refiners), and final consumers of the products made from crude oil. Crude oil price controls have extracted producer surplus from the first of these groups, while the entitlements program has distributed the "proceeds" from controls among the second and third groups. The Windfall Profits Tax promises to follow this general outline in its distributional effects, although beneficiaries among crude oil users will be more specifically targeted and the producers of crude oil substitutes are slated to receive some of the WPT revenues (see chapter 1).

5.5.1 Inter- and Intragroup Interests

For the most part, the distributional interests of crude oil producers, refiners, and consumers are in conflict. They are in competition for shares of the large increase in crude oil producer surplus attendant to rising world oil prices in the 1970s. In the absence of controls, excise taxes, and entitlements, these rising prices would confer wealth gains on domestic owners of crude oil bearing lands and crude oil-producing companies. The users of crude oil, on the other hand, would experience wealth losses in the form of negative income effects (in the case of those who demand refined products) and reduced producer surplus (in the case of those who provide nonpetroleum inputs to the domestic refining-distribution-marketing industry). Refiners and consumers have an interest in blocking the gains of producers; but once these gains are blocked, refiners have no obvious interest in sharing the benefits with consumers and vice versa.

The distributional impact of rising oil prices differs within groups. There are, for example, interproducer and interregional differences in the distribution of the various categories of crude oil (old, new, lower, upper, stripper, and so forth). These differences impart differences in attitudes toward crude oil price controls and excise taxes. Especially in the presence of a regionally based political system, these differences are not inconsequential in policy debates. Moreover, inequalities in the extent to which crude oil producers are vertically integrated into refining can lead to conflicting views over the propriety of regulation. As the mix of operations tilts more toward refining and less toward production of price-controlled crude oil, the likelihood that an integrated company has been a net loser from post-embargo federal oil price regulation is reduced.

Even when degrees of integration are held constant, interrefiner conflicts over the desired scope of regulation are present. Of course, the depressing effects of rising product prices on the quantity of products demanded by consumers tend to subject refiners to some of the incidence of rising crude oil prices. This incidence takes the form of reduced capacity utilization and the writing down of asset values. This implies losses for all refiners. But, given the entitlements program, refiner interests may diverge over issues such as the Small Refiner Bias and the numerous other special entitlements programs. Increases in the Small Refiner Bias, for example, reduce the entitlements subsidy to large refiners and imply a conflict between small and large firms. The Windfall Profits Tax is introducing further heterogeneity to refiners' interests through programs such as the subsidy to synthetic fuel production, which is typically the province of large refiners.

The distributional impact of rising oil prices or policies designed to offset rising oil prices are also unevenly felt by the consumers of petroleum products. For example, while there appears to be little variation in the proportion of income spent on energy across income classes, there is considerable variation within income classes.[9] This variation depends on factors such as urban/rural demographics, cold/warm climates, and the availability and price of public transportation. Moreover, because petroleum may be consumed indirectly in the form of products produced in the industrial, commercial, agriculture, and transportation sectors, the incidence of petroleum product price changes on the ultimate consumer is a (positive) function of the proportion of the budget allocated to energy-intensive goods and services and the inelasticity of demand for such goods and services. Of course, to the extent that the incidence of price changes is not borne by ultimate consumers, it is felt as changes in the fortunes of stockholders and input suppliers in the industrial, commercial, agriculture, and transportation sectors. In general, firms and industries in these sectors are more adversely affected by rising oil prices, the more limited are the possibilities for substitution between energy and other inputs in the production processes available to them.

Finally, differences across intermediate and final consumers in the types of petroleum products consumed give rise to divergent interests in the policy arena. The greater the direct special entitlements subsidies to heating oil and residual fuel imports, for example, the less funds there are for the entitlements subsidy to crude oil refining and, hence, gasoline consumption. The domestic geographical patterns of fuel use by type, as well as the patterns of climate, demographics, and business location, result in regional differences in the distributional impact of rising prices and regulation.

5.5.2 Estimates of Distributional Impact

Bearing in mind the simplifications inherent in characterizing the distributional interests in regulation as a producer-refiner-consumer trichotomy, table 5.5 summarizes the distributional consequences of EPAA/EPCA controls and entitlements during 1975–1980. For illustration, table 5.6 presents the estimates of these consequences in complete detail for 1978 and portrays the method by which the figures reported in table 5.5 have been calculated. All figures are in constant

Table 5.5 Estimated Distributional Effects of Controls and Entitlements (Most Likely Case): 1975–1980[a]
(Billions of 1980 Dollars)

	Crude oil producers	Petroleum refiners	Petroleum product consumers
1975	−23.9	+15.0	+6.9
1976	−18.9	+10.2	+6.8
1977	−18.7	+10.4	+6.4
1978	−14.3	+8.5	+4.7
1979	−32.6	+21.8	+8.3
1980[b] (Jan.–Mar.)	−49.6	+31.7	+12.2

Source: Based on data from US Department of Energy, *Monthly Energy Review*.
a. Most likely case based on 40% passthrough of entitlements subsidy, allocative effects of the entitlements subsidy as described in table 5.1, and allocative effects of EPAA/EPCA controls as described in table 5.2.
b. Annual rate.

Table 5.6 The Distributional Effects of EPAA/EPCA Regulation (Most Likely Case): 1978[a]
(Billions of 1980 Dollars)

Policy	Crude oil producers	Petroleum refiners	Petroleum product consumers
a. Entitlements subsidy	−10.0	+6.0	+4.0
b. Small Refiner Bias	−1.0	+1.0	0
c. Exceptions, appeals, and miscellaneous[b]	−1.0	+1.0	0
d. Product entitlements	−0.4	−0.2	+0.6
e. Strategic Petroleum Reserve[c]	−0.1	0	+0.1
f. 21¢ refiner allowance	−0.7	+0.7	0
g. Deadweight loss of subsidy	−0.3	0	0
h. Deadweight loss of controls	−0.8	0	0
Total	−14.3	+8.5	+4.7

a. Most likely case of 40% passthrough of entitlements subsidy, allocative effects of the entitlements subsidy as described in table 5.1, and allocative effects of EPAA/EPCA controls as described in table 5.2.
b. Includes estimates of miscellaneous special allocations of entitlements, such as those made to imports of certain petrochemical inputs and the use of California crude oil, and estimates of refiner gains from imperfect equalization of average crude oil costs under the entitlements program.
c. It is assumed that in the absence of the subsidy to the Strategic Petroleum Reserve, equivalent revenues would be raised through taxes on the general public. See text for further discussion.

1980 dollars. Costs of regulation associated with administration, enforcement, and compliance have not been included. Inclusion of these other costs, were they all measurable, would increase the losses of crude oil producers and reduce the reported gains of refiners and refined product consumers.

In the absence of crude oil price controls, crude oil producers would have sold their output at world prices in 1978. As shown in table 5.6, the cost of controls to producers in 1978 was $14.3 billion. On the output actually produced and sold under controls, producer surplus of approximately $13.5 billion was lost. In addition to the $13.5 billion lost on actual output, crude oil producers lost the surplus on the output foregone as a result of controls. This, of course, was the deadweight loss of controls in 1978—$0.8 billion. The $13.5 billion lost producer surplus on 1978 output was the rent associated with controlled crude and available for the entitlements program. The bulk of this, $10.0 billion, was transferred to crude oil users. The total amount spent on the entitlements subsidy was $10.3 billion, but crude oil users failed to capture $0.3 billion—and this amount was the demand-side deadweight loss noted in section 5.2. Of the total amount available to the entitlements program in 1978, $3.2 billion (that is, $13.5 − $10.3 billion) was used to finance the special entitlements programs such as the Small Refiner Bias and the $0.21 allowance to refiners.

Of the $3.2 billion spent on special programs in 1978, approximately $1.0 billion was granted to small refiners under the Small Refiner Bias; approximately $1.0 billion was distributed to refiners through Exceptions and Appeals and miscellaneous special grants of entitlements; another $0.4 billion was spent on the direct subsidization of certain refined product imports (primarily residual fuel imports); approximately $0.1 billion was used to fund the Strategic Petroleum Reserve; and approximately $0.7 billion was granted to refiners of controlled crude oil through the $0.21 allowance. Most of the benefits of these special programs are pure transfers to domestic refiners. Product entitlements, such as those available to East Coast residual fuel imports, however, directly reduce the marginal cost of imports and are reflected in lower prices. Based on the incidence measures implied by table 4.5, lower prices for residual fuel resulted in a consumer-user surplus gain of approximately $0.6 billion in 1978. Of this sum, $0.4 billion was directly financed by the entitlements program, while most of the remaining $0.2 billion was transferred from domestic refiners of residual fuel and a small deadweight loss was incurred. Finally, determination of the incidence of the $0.1 billion subsidy to the Strategic Petroleum Reserve is particularly problematical. Were it not financed out of the rents associated with controlled crude oil, this subsidy might be financed through taxes on the general public, refined product consumers, refiners, or crude oil producers. As a likely case, table 5.6 assumes the subsidy would be raised from the general public. If requisite taxes were roughly proportional to income and energy consumption is roughly proportional to income (see above), the incidence of the subsidy can be approximately assigned to product consumers.

The $10.3 billion difference between the $13.5 billion available to the entitlements program and the amount spent on special programs was spent on the entitlements subsidy to crude oil use in 1978. As noted, $0.3 billion of this was a deadweight loss, but $10 billion accrued to crude oil users. These users, however, include both refiners and product consumers; and, as chapter 4 indicates, their division of this $10 billion is not self-evident. Based on the results of chapter 4, it appears likely that approximately 40% of each dollar of the entitlements subsidy accrues to refined product consumers, while 60% is retained by the refining industry. Accordingly, tables 5.5 and 5.6 use these estimates of the subsidy incidence.

As Carlton (1978) has demonstrated, there is a one-to-one correspondence between the deadweight loss of an input price distortion measured in the input market (as it has been here) and the deadweight loss of such a distortion measured in the output market when the elasticity of factor substitution between the input in question and other inputs is zero. At least over the period under consideration here, with refinery capacity and refining technology substantially given, this condition on factor substitution is unlikely to have been violated. With deadweight losses equated, the combined producer and consumer surplus gain from a subsidy as measured in the output market (that is, the difference between the total spent on the subsidy and the deadweight loss) corresponds to the aggregate user surplus gain measured in the input market (area $P_w ZXP_s$ in figure 2.4A). To a first approximation, the division of this gain between consumers and producers is then proportional to the estimated per dollar passthrough. Under the most likely case of a 40% passthrough, the refining sector held onto approximately $6 billion of the $10 billion user surplus gain from the entitlements subsidy in 1978 and product consumers captured the remaining $4 billion. When the distributional effects of the special pro-

Table 5.7 Bounds for Distributional Effects of EPAA/EPCA Regulation on Refiners and Consumers: 1975–1980
(Billions of 1980 Dollars)

Subsidy incidence on consumers	Petroleum refiners		Petroleum product consumers	
	20%	60%	20%	60%
1975	+18.4	+11.6	+3.5	+10.3
1976	+13.4	+7.0	+3.6	+10.0
1977	+13.6	+7.4	+3.2	+9.4
1978	+10.8	+6.8	+2.4	+6.4
1979	+25.1	+18.5	+5.0	+11.6
1980[a] (Jan.–Mar.)	+37.0	+26.4	+6.9	+17.5

a. Annual rate.

grams are added to those of the entitlements subsidy, the total gain of the refining sector under EPAA/EPCA regulation was $8.5 billion in 1978. The total gain of refined product consumers was $4.7 billion.

To bound the estimates of distributional effects, table 5.7 presents calculations of the total distributional impacts on refiners and consumers under assumptions of a 20% and a 60% passthrough of the entitlements subsidy. These bounds are as suggested by the results of chapter 4. Even with a subsidy passthrough of only 20%, refined product consumers have been capturing an average of $4.1 billion per year under controls and entitlements. Refiners, meanwhile, have been capturing an average of $19.7 billion per year. With 60% of the entitlements subsidy borne by product consumers, the average annual transfer to consumers has been $10.9 billion, and the annual transfer to refiners has averaged $13.0 billion.

The preceding estimates of the transfers accomplished under EPAA/EPCA controls and entitlements are subject to some qualifications. First, the public (that is, tax) and private burdens of regulatory administration have not been taken into account. The incidence of the tax burden depends upon the incidence of federal expenditure and revenue policy. Assuming, as in Arrow and Kalt (1979), that the removal of oil price regulations would be accompanied by tax reductions or increases in benefits from other expenditures that are proportional to income and noting the approximate proportionality of expenditures on energy to income leads to the conclusion that the EPAA/EPCA tax burden could be approximately represented as a burden on refined product consumers. The apparent distributional gains of these consumers shown in table 5.5 would then have to be reduced by an average of $0.1 billion in each year (that is, by the values reported in table 5.4).

The incidence of the private-sector administrative costs depends on the split of these costs between producers and refiners and on the extent to which costs proximately borne by refiners are ultimately borne by refined product consumers. Any costs borne proximately by crude oil producers are not passed downstream since crude oil price ceilings are not adjusted for cost changes. Arrow and Kalt (1979) take three fourths to be a reasonable estimate of the fraction of refiners' administrative costs ultimately borne by consumers. Assuming (in the absence of relevant data) that the total private costs of regulatory administration are divided evenly between the crude oil production and refined product production-marketing sectors and taking $1.0 billion (that is, a middle ground between the Presidential Task Force and CSAB estimates) as an estimate of the level of these costs, a three-fourths passthrough would reduce the estimated gains of refiners and consumers reported in tables 5.5–5.7 by approximately $125 million and $375 million per year, respectively. The losses of crude oil producers would be increased by $500 million per year (that is, half of $1.0 billion). Of course, these adjustments to the above tables are only partial adjustments since they exclude the unmeasurable costs of carrying out and adapting to current regulation that were discussed in section 5.4.1.

A second important class of qualifications to the estimates of distributional effects concerns the possibility of divergences between private and social marginal costs. A noninternalized externality associated with crude oil imports, for example, might impose costs on the general populace. It was noted above that although the entitlements subsidy to imports worsens the allocation of resources in the presence of such external costs, the effects of crude oil price regulation on the time pattern of domestic production may be either efficiency improving or efficiency worsening. Thus the net effect of post-embargo oil price policy may be either efficiency improving or efficiency worsening. If regulation, on net, magnifies the inefficiency of preexisting external costs, the apparent distributional gains of refiners and petroleum product consumers may be partially or totally reversed and/or the distributional losses of crude oil producers may be expanded, depending on the incidence of the deadweight losses associated with the offending externalities. Regulation might, for example, exacerbate some national security threat from import dependence; and the diminution in the value of the national security presumably is a cost to all citizens. On the other hand, if the net impact of oil price regulations is efficiency improving, opposite distributional effects result. Efficiency improvements as a result of regulation must accrue to the individuals in the economy, and every individual falls into one or more of the categories of producers, refiners, or consumers.

These general observations are somewhat divorced from the real world context of US energy policy. Yet they raise important questions for policymakers concerned with the distributional effects of energy regulation. They also raise important questions for the economic analyst who, by training, is more likely to be concerned with resource allocation. The most interesting issue is whether the allocative effects of oil price regulation are really as contrary to criteria of optimality as this chapter has indicated or whether there is is some more or less sophisticated allocative case for post-embargo policy that is overlooked by the rather straightforward approach of sections 5.2 and 5.3. Accordingly, attention is now turned toward consideration of the major arguments that have been, or might be, advanced in favor of a federal regulatory role in the crude oil market. Noneconomic, ethical arguments concerning justice in the distribution of property rights to income and action have been summarized and evaluated by Arrow and Kalt (1979). In this study, it is left to the reader to decide on the basis of tables 5.5–5.7 or other considerations whether current regulation is justified on grounds of fairness. Following the classic efficiency-equity distinction, the concern here is with efficiency-based justifications for regulations.

5.6 Resource Allocation in the Crude Oil Market and the Rationalization of Post-Embargo Policies

Recent public and academic debates over US energy policy have produced a number of candidates for the role of resource misallocation that might serve as an efficiency-based justification for regulatory intervention. For the most part, attention has been focused on the possibility that a noninternalized external cost in the form of an overdependence on foreign crude oil (and, in particular, Arab crude oil) is producing a net economic loss for the country. One of the possible sources of such an overdependence—threats to national security—has been touched on in the preceding section. Other possible sources include domestic macroeconomic disruption from a foreign supply interruption and US monopsony power in the world crude oil market. In addition to overdependence arguments, noninternalized environmental externalities, as noted above, have been advanced as reasons for policy action. Finally, an argument may be made for *increased* oil imports based on national defense interests in the Middle East. These various arguments will be considered in turn. Other rationalizations for EPAA/EPCA regulation (indeed, an unlimited number) may exist, but attention here has been restricted to the major classes of plausible arguments.

In the following discussions, the paucity of data-generating, cost-internalizing market transactions and the absence of a convincing methodology for quantifying or even modeling the objectives, constraints, and parameters that control, for example, the behavior of nation-states in international political affairs preclude quantitative estimation of allocative and distributional implications. Still, it is appropriate to identify these implications and round out the picture of US petroleum policy developed thus far.

5.6.1 National Security and Import Dependence

National security interests (broadly defined) have played a role in US oil import policy for several decades. Historically, a primary concern has been the possibility that loss of control over foreign petroleum supplies during time of war may significantly weaken US military defense capabilities. In the present-day context, the overriding concern is that foreign oil reserves (Middle Eastern reserves, in particular) may be seized by an Eastern bloc country or a regime unfriendly to the Western bloc. Once seized, the wealth and monopoly power attendant to control of these reserves could be used in the international political arena to exact concessions from the Western bloc and woo Third World and nonaligned countries. In a similar vein, the vulnerability of the international oil transport system to a hostile naval power has also been a source of insecurity.[10] Finally, the rise of the oil exporting countries to positions of independent political power raises the possibility that the threat of price increases and embargoes will be used to induce domestic petroleum consumers and refiners to press for changes in US foreign policy on such issues as the Arab-Israeli dispute.

In the argument based on national security, an externality is seen as arising from the public good characteristics of national defense. The collective nature of national defense benefits leaves individual consumers of imported crude oil with little or no incentive to internalize the deleterious effects on the national security of increases in their purchases of foreign crude oil. While the theoretical possibility of external national security costs for overdependence on imported oil is clear, measurement of their magnitude is problematical. It is not obvious, for example, whether the national security threat of, say, a 1-mmb/d foreign supply reduction (for example, through an embargo) is greater when the United States is importing 8 mmb/d (as in 1978) than when it is importing 4 mmb/d. The marginal national security externality at actual import levels may be small or zero.

Notwithstanding measurement problems, and granting the existence of a national security externality, the policymaker interested in improving resource allocation would certainly find the entitlements subsidy counterproductive. If oil import dependence threatens national security, it makes little sense to subsidize imports. On the supply side, however, the efficiency-improving policy response in the presence of import externalities is not so obvious and depends on the rate of change over time of such externalities. If marginal national security external costs are seen as rising relatively rapidly over time (that is, faster than the interest rate), a price control or excise tax program which shifts domestic production toward the future (as EPAA/EPCA controls have tended to do) might improve allocative efficiency. The contrary implication is drawn if these externalities are rising relatively slowly. If the future for imported oil looks better than the present, it is appropriate to use domestic resources in the present and rely more heavily on imports in the future.

The time path of any national security problem posed by import dependence could be debated interminably without resolution. Nevertheless, certain relevant observations may be made. The increasing integration (primarily through international capital markets) of OPEC members into the Western economic system, the development of new petroleum reserves in friendly hands (for example, Canada, the North Sea, and Alaska), and the availability of backup technologies (such as coal gasification and liquefaction) suggest declining real national security externalities. On the other hand, evident increases in political instability in important oil-exporting countries such as Iran, recent moves of the Soviet Union in the Persian Gulf area, and uncertainty about the long-term balance of power between the Western and Eastern blocs suggest a less sanguine view of the likely course of these externalities.

5.6.2 Macroeconomic Stability and Import Dependence

Insecurity in the supply of crude oil from foreign sources may threaten the domestic macroeconomic process of income generation as well as national security. The source of such a threat is the possibility of abrupt and unanticipated supply reductions, such as might occur under an embargo or unfriendly takeover of an oil-exporting country. Unanticipated shocks to the supply (and hence the price) of crude oil may touch off what Leijonhufvud (1971) has called a "coordination failure." The resulting macroeconomic disequilibrium may then produce a misallocation of resources and reduction in aggregate income. As Tolley and Wilman (1977) note, an externality can be present because of the public-goods aspects of economy-wide coordination in the production and consumption plans of economic agents. In the presence of such publicness, importers of crude oil have little or no incentive to internalize the effect of a change in their importing behavior on the likelihood of a macroeconomic disruption. Moreover, problems of excluding free riders from the use of information on the likelihood of either an embargo or the probability of unfriendly political changes militate against production of the information that might be put to use in minimizing the impact of a supply shock. Of course, overdependence externalities are present at the margin only if the adjustment problems of the economy, when faced with a foreign supply shock, are a function of the level of imports.

If marginal external costs associated with macroeconomic adjustment problems are positive, the implications to be drawn for the assessment of oil price regulation are the same as in the case of a national security externality: The entitlements subsidy moves the economy further away from an efficient allocation of resources, and the effect of crude oil price controls on allocative efficiency depends on the rate of change of the stringency of controls, the elasticity of supply from new reserves, and the magnitude of marginal noninternalized costs. If, for example, these marginal external costs are expected to rise rapidly in the future, a controls program that shifts domestic production away from the present and toward the future may be consistent with allocative efficiency.

Notwithstanding this last theoretical possibility, it is not hard to believe that certain aspects of post-embargo regulation have exacerbated the macroeconomic adjustment problems associated with foreign supply shocks. The adjustment problem is a consequence of the scarcity of information that might permit the synchronization and validation of agents' plans. One of the functions of markets is to economize on the scarce commodity "information"; and changes in prices and profits in unregulated markets serve as the informational signals and inducements that allow producers and consumers to adjust their behavior rationally when shocks to their environment occur. Regulatory distortions to prices and profits can result in informational distortions.

Consider, for example, the effects of EPAA/EPCA regulation on private incentives to stockpile as insurance against increases in the price of foreign crude oil. In an unregulated market, the prospect of a supply cutback induces stockpiling by those who specialize in assessing the likelihood of such events. This inducement arises because stockpiles of petroleum fetch increased prices during foreign supply reductions. At least in the case of short-term supply reductions, such behavior has the beneficial effects of increasing supplies, moderating price increases, and smoothing out the shock over the adjustment period. While it might be possible to capture uncontrolled prices on stockpiles of uncontrolled crude oil, the effect of refined product price controls (which tend to become binding at times of sharply rising world oil prices—see chapter 2) and export controls is to reduce incentives to stockpile both crude and products. This is reinforced by the buy/sell program and the entitlements program, which act as nonmarket devices for allocating property rights in crude oil.

5.6.3 Controls, Entitlements, and World Oil Prices

If increments to the US demand for imported crude oil (such as those induced by controls and entitlements) tend to bid up world crude oil prices, the marginal burden that the domestic economy bears when importing crude oil is understated by the world price. In the simplest case in which the US refining industry as a whole has some monopsony power, albeit unexercised, vis-à-vis crude oil imports and faces an upward-sloping supply schedule P_W^s with a price elasticity of ϵ_W ($0 < \epsilon_W < \infty$), the marginal supply price MP_W^s is given by

$$MP_W^s = \frac{(\epsilon_W + 1)}{\epsilon_W} P_W^s > P_W^s.$$

In this situation, it is possible to improve the allocation of domestic resources by having the United States exercise its monopsony power through reductions in crude oil imports to the point where the marginal value of crude oil to the domestic economy is equated to MP_w^s (rather than P_w^s, which is the case when refiners individually behave as price takers).

Although the framework above is analytically tractable, the monopolistic character of OPEC—or, at least, Saudi Arabian—supply precludes the treatment of possible US price-making ability with a simple upward-sloping supply schedule. Furthermore, as the discussion of section 3.2 indicates, the *magnitude* of changes in world crude oil prices induced by US oil price policies most likely have been quite small. Nevertheless, the *direction* of the effects of these policies on world oil prices is probably positive. As demonstrated in Kalt (1980), domestic price regulation has the effect of increasing the US demand for crude oil imports and making it less elastic in the relevant range. Thus monopolistic suppliers could be expected to respond to regulation-induced increases in the demand for their product by raising prices. If this effect is significant, the deadweight losses of regulation have been larger than estimated in this chapter. Moreover, the apparent distributional gains of refiners and consumers have been overestimated because induced world price increases offset the entitlements subsidy. The apparent distributional losses of crude oil producers would be overestimated since the estimate of the world price that producers would receive in the absence of regulation is too high.

While it is clear that the exercise of US monopsony power—if it exists—could reduce world prices relative to the case in which each refiner behaves individually as a price taker, the exercise of this power in the presence of monopoly on the supply side of the market constitutes conditions of bilateral monopoly. Under such conditions, the price-quantity solution is *a priori* indeterminant within the range between the pure monopsony and pure monopoly solutions. The bargain struck under conditions of bilateral monopoly is dependent upon the strategies and relative negotiating strengths of buyers and sellers. Even in the absence of international political and military considerations, specification of an optimal strategy for a monopsonistic United States represents an unfinished task. Suffice it to note that, because measures such as the entitlements subsidy and crude oil price controls increase the demand for imported oil and make it less elastic, post-embargo oil price policies in the United States are unlikely to have been part of an optimal strategy.

5.6.4 Environmental Externalities in the Crude Oil Market

Environmental externalities in the crude oil market may arise at the production, refining, and/or transportation stages. Notable forms of environmental damage from crude oil production include ocean pollution from offshore oil rig operations and accidents; air pollution from gas flaring and from burnoff or evaporation following blowouts; and soil and water pollution from emissions of nondegradable oil and dissolved solids (that is, brine). The major environmental damage in the transportation of crude oil arises in connection with the necessarily water-borne transport of oil imports, although pipeline operations are also subject to accidents and pipeline pumping stations typically emit pollutants. Environmental costs in crude oil refining include water pollution from discharge of dissolved salts, suspended solids such as sludge oil, and nondegradable oils and phenols; air pollution from nitrous oxides, sulfides, and hydrocarbon particulates released during processing and burnoff; and soil pollution from disposal of solid wastes such as sludge and residual chemicals and catalysts.

Each of these forms of pollution in the crude oil market might be a likely candidate around which in-depth analyses of both the effects of externalities on resource allocation and remedial market and regulatory mechanisms could be built. Such investigations, however, would go beyond the issues at hand—which are the allocative consequences of post-embargo oil price regulations in the presence of noninternalized environmental externalities. Some of the implications of controls and entitlements in this context have already been touched on. For example, with noninternalized environmental costs arising from crude oil production (but none arising from refining or transportation), the allocative consequences of the entitlements subsidy are as presented in the straightforward analysis of section 5.2.2. Crude oil price controls (or excise taxes), on the other hand, have ambiguous effects. They could conceivably improve the allocation of resources by, for example, reducing early-period extraction (and thereby the generation of environmental damage) when marginal external environmental costs are not rising at a rate in excess of the rate of interest. Such a solution to the externality problem, however, would only represent a first-best solution in the unlikely case in which the only way to vary the output of external costs is to vary the rate of crude oil production. When this case does not obtain, a first-best solution would involve some form of charge for the production of environmental damage (for example, a direct effluent fee, or the creation of liabilities and tradeable rights in pollution), rather than controls or taxes on the production of crude oil itself. The former alternative permits and encourages greater producer flexibility, in terms of selected extraction and abatement technologies, when adjusting to internalized environmental damage costs.

Noninternalized transportation and refining environmental externalities (as well as externalities in end use markets) cause a divergence between the social and perceived private costs of *using* crude oil. As a result, the users of crude oil could be expected to overconsume crude oil, relative to a standard of allocative efficiency, in the absence of a property rights or regulatory regime that promotes internalization. In such a setting, the entitlements subsidy worsens resource allocation by lowering the private costs of using crude oil even further below the social costs. Again, however, the effects of crude oil price controls and excise taxes are ambiguous and depend on the time path of the offending externalities. If, for example, it were the case that the only domestic environmental damage from crude oil use is associated with the transportation of *domestic* crude oil (for example, tanker accidents involving imported oil only occur in foreign waters, but domestic pipelines frequently rupture), crude oil price regulation hypothetically could be used to vary the time path of imports so as to raise imports in earlier/later periods according as domestic marginal external costs of transportation rise relatively slowly/rapidly. To make this point, however, is not to say anything about the practicability of such a convoluted, *n*th-best method of regulating the production of environmental damage.

Numerous direct methods of regulation or internalization through markets, in fact, exist to deal with environmental damage in the crude oil market. Many of these methods have grown out of the environmental movement of the 1960s and 1970s. Governmental units with responsibility for environmental controls over crude oil production, transportation, and refining now range from local land use commissions and state environmental quality boards to the federal-level Environmental Protection Agency and Department of the Interior. On an international scale, the accords coming out of the Law of the Sea Conference and the oil spill liability provisions promoted by the International Energy Agency have (at least potentially) remedied some of the problems of indefinite, distortive, and conflicting liability assignments in ocean-borne oil transportation.

Although assessments of the impact of current environmental policies have produced responses ranging from ineffectiveness to overzealousness, a reliable empirical measurement of the net consequences of these policies is not available. In the crude oil market, the net effect of federal air- and water-quality standards, as well as other regulatory impediments to offshore development, refinery expansion, pipeline construction, deep-water port siting, and well drilling have certainly restrained environmental pollution. The case is not strong that oil price policy is complementary to the existing independent environmental policies. It is hard to rationalize post-embargo price regulation on the grounds of environmental externalities.

5.6.5 Domestic Oil Price Regulation as Foreign Aid

The preceding subsections have addressed the question of the allocative consequences of domestic oil price regulation in the context of noninternalized external *costs*. The entitlements program in particular does not fare well against standards of efficiency in this context, and crude oil price controls and excise taxes yield ambiguous judgments. The analysis leading to these conclusions, however, may be being conducted in the wrong context. Specifically, several considerations suggest a line of reasoning in which domestic oil price regulation is seen as consistent with US national security and foreign policy interests in aligning oil exporting countries, particularly Arab states, with the Western bloc. Although the language of policy debates generally maintains the impression that foreign oil producers constitute threats against which the United States should protect itself through isolation, a strong case can be made that these producers are potential allies whose political fealty should be encouraged and, if need be, purchased. Post-embargo US petroleum price policy has been broadly consistent with this conception of US security interests.

Historically, concern over national security externalities due to import dependence has involved the perception of oil-exporting countries as small, relatively weak states unable to wield significant power in international politics, but subject to military interdiction by unfriendly Eastern bloc countries. According to such a viewpoint, oil-exporting nations sell the United States crude oil, but have limited ability to "sell" the United States active political support in worldwide power struggles. This latter situation has apparently been altered in the 1970s as oil-exporting countries, and especially the members of OPEC, have collectively emerged as an international political force with power independent of the traditional Eastern and Western blocs. In light of the emergence of a group of nations with potentially independent international political power, the postwar history of US-Western European and US-Japanese relations suggests that the United States generally pursues policies of economic integration rather than economic isolation.

The importance of friendly Middle East-US and Arab-US relations in US foreign policy is evidenced daily; and Kissinger (1979) provides a defense for recent policy. These states not only possess large reserves of petroleum, but also play a pivotal role in questions concerning Israel and, perhaps most important, form a strategic buffer between the Soviet Union and Africa. Evidence of active US attempts to attract the allegiance of Arab and Middle Eastern states is not lacking and includes sales of military arms to the two largest OPEC producers (Saudi Arabia and, presumably with regret, Iran); economic aid and technological support given to Egypt as inducement to the expulsion of Soviet "advisors"; subsidized development loans; and grants of money, equipment, education, agricultural goods, and technology through normal foreign aid channels.

In the oil arena, evidence of US support of the interests of Middle East producers *qua* producers is no less abundant—notwithstanding a domestic political rhetoric of import independence and opposition to the OPEC cartel. The State Department, in particular, has exhibited a willingness to foster these interests. State Department support of OPEC, especially in the crucial period of transition to a more or less stable cartel in the early 1970s, is well documented.[11] Moreover, the State Department has been one of the more vociferous proponents of the view that the energy crisis is "real" (in the sense that developments in the last decade reflect a previously drastic underestimation by the world oil market of the true scarcity of petroleum, rather than the exercise of monopoly power by OPEC). After some initial enthusiasm, the State Department has consistently dragged its feet on any plans to weaken the OPEC cartel or exercise countervailing monopsony power.[12]

The State Department is not the only source of implicit or explicit federal support for the interests of oil-exporting countries. United States energy policy, in general, exhibits such support. Not only have entitlements and crude oil price controls tended to increase the domestic demand for crude oil imports and make it less elastic, but the United States has also been notable for its inability (or unwillingness) to engage in any suggested cooperative actions with other importing nations designed to strengthen their joint bargaining power. In fact, in 1977, former Secretary of Defense and then Secretary of Energy James Schlesinger outlined official Administration energy goals and promised that nothing would be done to break OPEC.[13] Also in 1977, the Carter Administration announced an interest in filling the Strategic Petroleum Reserve with Saudi Arabian crude oil, thus offering the Saudis the opportunity to increase their sales without having to lower their price.[14]

If it is in the national security interest to court the international political support of the Middle Eastern oil-exporting countries by selling them arms or granting them foreign aid, it seems plausible that their support could also be encouraged by buying the products they produce. Indeed, if the political support of oil-exporting countries is a positive function of their exports of crude oil to the United States (that is, if their willingness to align themselves with the United States is a positive function of their "dependence" on the US and Western bloc market), crude oil imports into the United States carry external national security *benefits*. These external benefits would not be optimally internalized in the unregulated market because of the public-good aspects of national security. With the possibility of free riding, individual importers would have little or no incentive to take into account the impact of their imports on the national security. Hence, the perceived private marginal cost of imported crude oil would *exceed* the social marginal cost. The entitlements subsidy lowers the private marginal cost of crude oil and, optimally designed, might remedy the implied misallocation of resources. Similarly, if marginal external benefits from crude oil imports are falling or rising relatively slowly over time, a price regulation program that encourages early-period imports could improve resource allocation.

Consider the following question: Other things being equal, would Saudi Arabia be *more likely* or *less likely* to support US foreign policy objectives in the Persian Gulf by, for example, allowing the establishment of a US military base on Saudi soil if the United States accompanied its request for such a base with an announcement that all domestic oil price regulations were being eliminated? Since US oil price regulations tend to bolster the demand for imported crude oil, it is reasonable to answer that Saudi Arabia would be less likely to support US foreign policy objectives. Thus Middle Eastern imports can carry national security benefits, and it is possible that the import dependence fostered by post-embargo policies has improved allocative efficiency by "producing" additional national security. The deadweight losses reported in tables 5.1 and 5.2 might then be interpreted as the price paid for an otherwise underprovided public good; and the reported distributional gains of refiners and consumers would have to be increased, while the distributional losses of crude oil producers would have to be reduced. Of course, economists might prescribe a more direct method of purchasing "national security." But in a time when the vast majority of the public views Arab oil producers as ruthless villains and is uninformed on matters of national security, it may be politically infeasible to make more outright payments to OPEC governments.

5.6.6 Summary

An attempt has been made to rationalize oil price regulation on allocative grounds. The intent of the analysis offered here has been to be suggestive rather than definitive; and certainly there is room for more research. As a general observation, US oil price policies do not fare very well when measured in the context of arguments for intervention based on purported external costs to dependence on imported oil. These policies have been increasing the domestic demand for imported crude oil. Only with some strong (and as yet unsupported) assumptions about the time pattern of dependence externalities can even part (that is, controls and excise taxes) of post-embargo policy be rationalized. It is even more difficult to rationalize oil price regulation on the grounds of environmental resource misallocations. The existence of independent mechanisms for the remedy of environmental problems and the long distance between first-best remedies and the actual provisions of oil price policy render the representation of petroleum price regulation as a form of environmental policy an intellectual contrivance.

The rationalization of post-embargo federal regulation of the petroleum sector that is most consistent with the effects of this regulation characterizes US oil price policy as foreign aid designed to "buy" national security from oil-exporting countries. A conception of US security interests that sees those interests being served through actions that curry the favor of Middle Eastern countries looks favorably on the increases in oil import demand that have been fostered by controls and entitlements. Whether or not such a conception is correct is undoubtedly subject to debate—a debate most fruitfully carried on by those with expertise in international political and military affairs, as well as economics. Such a conception is offered here as only a general conjecture.

5.7 Conclusion

This chapter has provided quantitative estimates of the allocative and distributional effects of federal petroleum price regulation in the post-embargo era. Measurement of these impacts is confounded by the potential for intertemporal regulatory distortions in the crude oil market, possible externalities and resource misallocations even in the absence of federal price regulation, and a lack of data on certain aspects of regulation (such as a part of the costs incurred by the private sector in order to play the regulatory game). The implications of these complications have been discussed at length. Even in their presence, however, the basic conception of post-embargo regulation developed in this study does not appear to be threatened.

Post-embargo regulation of the petroleum industry blocks much of the transfer of wealth from domestic crude oil users (both refiners and consumers) to domestic crude oil producers portended by rising world oil prices. In the most likely case, crude oil price ceilings have denied producers \$14–50 billion (1980 dollars) per year since 1974. Since November 1974, the rent that would otherwise be associated with refiners' ability to buy inframarginal price-controlled crude oil at a lower price than the price of marginal uncontrolled crude oil has been divided among refined product producers and refined product consumers by the entitlements program. The primary tool of this program has been the entitlements subsidy to domestic refiners' use of uncontrolled crude oil. In the most likely case, regulation has allowed refiners to capture \$9–32 billion (1980 dollars) per year since 1975, while refined product consumers have realized gains of \$4–12 billion per year.

The gains of refiners and consumers at the expense of producers have not been allocatively neutral. In particular, crude oil price controls have resulted in a deadweight loss on the supply side of the domestic crude oil market. Even incorporating intertemporal distortions does not alter the conclusion that controls appear to have substantially reduced domestic crude oil production relative to the unregulated environment. In addition, the entitlements subsidy has resulted in an overconsumption of imported crude oil relative to the unregulated market. This has been accomplished by lowering the private cost of imported oil to a level below its price. In the most likely case, the sum of the deadweight losses on the supply and demand sides of the domestic petroleum market have been in the range of \$1–5 billion per year.

When measurable private- and public-sector costs of administering oil price regulations are taken into account, at least another $1 billion per year is added to these losses from controls and entitlements.

When fully effective in October 1981, the Windfall Profits Tax will use a system of excise taxes, rather than price controls, to continue the extraction of wealth from crude oil producers. Based on 1980 prices for uncontrolled crude oil, the WPT will capture roughly $35 billion per year from domestic crude oil producers. The associated distortion of producers' price incentives will apparently generate supply-side deadweight losses on the order of $2 billion per year. On the demand side, the allocative effects of the WPT will not flow from a single major program such as the entitlements program, but will arise from a pork barrel of projects targeted for receipt of tax revenues. Beneficiaries from WPT-funded projects will vary from refiners producing oil substitutes and businesses investing in oil-saving heating and cooling systems to low-income heating oil consumers and the customers of mass transit systems. In general, a substantial portion of the distributional gains from the Windfall Profits Tax will, as under EPAA/EPCA regulation, accrue to users of crude oil.

To note that there are deadweight losses from post-embargo federal petroleum regulation is to note that there exists, at least hypothetically, a trade between crude oil producers, on the one hand, and the beneficiaries of regulation, on the other, that could leave all parties better off. Given the property rights assignment implicit in the regime of controls and entitlements, for example, such a trade would involve crude oil producers "buying" the right to sell their crude oil at uncontrolled prices from refiners and refined product consumers. The potential gain from such a trade would be the deadweight losses from controls and entitlements. The failure to realize this gain implies either (a) a short-run disequilibrium in which interested parties search for appropriate terms of trade or await the appearance of a clever entrepreneur or (b) a long-run equilibrium in which real world political transaction costs prevent the hypothetical trade from ever being made. Less tautologically, it is important to recognize that the "market" in which efficiency-improving trades of this sort might take place is a market in which property rights are not immutable. Rather, they may be altered or transferred without the unanimous voluntary agreement of affected parties. In this market, which is the federal political system, the persistent struggle over oil pricing and the almost daily marginal changes in regulation suggest that the consequences of post-embargo policies are not representative of a static equilibrium. Instead, they appear to reflect an environment in which voters and political entrepreneurs continue to see advantage in bringing about redistributions of property rights and wealth. Understanding this environment, and the behavior of the actors in it, would seem to be an important objective—particularly so in view of the importance of energy matters in the minds of the public, policymakers, and academics. While the bulk of this study has dealt with the *consequences* of post-embargo regulation of the petroleum industry, the analysis now turns toward the *causes,* in the political-economic sense, of this regulation.

Chapter 6

The Political Economy of Federal Petroleum Price Policy: An Analysis of Voting in the US Senate

6.1 Introduction

Previous chapters have focused exclusively on the consequences of recent federal regulation of the petroleum industry. This regulation has been found to be an effective mechanism for blocking or, at least, forestalling much of the intra-US transfer of wealth from intermediate and final users of crude oil to producers of crude oil that has been made possible by international energy developments of the last decade. Crude oil price controls have served as the primary tool by which this transfer has been restrained, while the entitlements program has ensured that refiners and consumers share the distributional benefits of price regulation. Soon, the Windfall Profits Tax and its associated expenditure programs will pick up where controls and entitlements leave off. To be sure, the distributional consequences of post-embargo oil policy have not been realized without allocative effects; but the resulting deadweight losses apparently have not been overwhelming relative to the size of the wealth transfers that have been, and will continue to be, achieved.

The availability of relevant data and tried-and-tested methodologies facilitates investigations of the *effects* of current petroleum industry regulation. If the paucity of research is any indication, however, explanation of the *causes* of this regulation may be a more challenging task. Numerous theses have been advanced, of course, as to why the domestic petroleum industry should or should not be regulated. Nonnormative explanations of why, in fact, the industry is regulated, however, have not been rigorously pursued. This chapter presents an attempt to remedy this deficiency through an analysis of recent voting in the US Senate.

The results of this chapter support the conclusion that the bulk of post-embargo regulation of the petroleum industry has been explicitly designed, at the expense of allocative efficiency, as a mechanism for the redistribution of wealth from crude oil producers to crude oil refiners and refined product consumers. Specifically, it is found that Senate voting behavior is significantly related to the wealth interests of these constituencies and that the policymaking process has been captured to some degree by private interest groups. These results tend to support the view (for example, Wright, 1978) that the domestic energy crisis is a crisis over wealth distribution, rather than a crisis of market or institutional failure. Also important, however, is the finding that some form of ideological, public-interested behavior, which cannot be explained entirely by the petroleum-related economic interests of constituencies, has also played an important role in Senate voting and policy formation.

6.2 Theories of Regulation

The behavior of individuals in their capacities as political actors and the consequences of this behavior for the formation and outcomes of political action are increasingly coming under the scrutiny of economic theorists. Hopefully, this has generated enough insights to allow the avoidance of "ad hocery" and provide a consistent theoretical structure for explaining real world governmental policies. The economic theories of governmental regulation typically employed (in more or less explicit form) in this task may be classed into two broad types. The first, which might be called *allocative*, represents regulation as a mechanism by which undesirable allocative consequences of market failure are eliminated. Allocative theories of regulation are closely associated with traditional welfare economics and have their theoretical bases in public-good theories of the state, which view governmental activity as the means by which collective ends are pursued when such ends are unattainable from the invisible hand. The second type of theories of regulation may be termed *distributive*. These revisionist theories emphasize the intentional effects that governmental regulation can have on the distribution of wealth, as well as on the allocation of resources. They have their theoretical bases in a theory of the state that sees government as a vehicle for the institutionalized redistribution of wealth.

Allocative theories of regulation have often served as normative standards of political action and, with less success, as models of the *outcome* of the political process. Such theories have little value as models of the political, nonmarket *behavior* of acting individuals. Their weakness stems primarily from their theoretical underpinnings and, specifically, their use of either (a) a *deus ex machina* view of government; (b) methodological collectivism; or (c) a model of individual behavior contrary to the choice-theoretic paradigm on which the rest of economics is founded. Examples of the first weakness can be found in numerous introductory economics texts, where governmental intervention is often invoked as the cure-all for market failure. The second weakness pervades much of the early literature on the social welfare function and "society chooses" explanations of governmental policy. The fundamental shortcoming of both the *deus ex machina* and the "society chooses" approaches is that they use conceptual constructs that have no empirically observable referents. There is, for example, no observable entity denoted by *society* having the attributes of choice and action assigned to it by methodological collectivism. Instances of the third weakness are represented by what Buchanan (1954) has called the "bifurcated man," that is, the (usually implicit) assumption that individuals maximize their own utility in their market behavior, while acting to maximize social utility or promote economy-wide efficiency in their political behavior. These three approaches to nonmarket, political behavior have proven so counterintuitive and counterfactual that allocative theories of regulation have failed as modes of positive scientific analysis.

Distributive theories of government presuppose a model of behavior in which the representative individual undertakes and promotes governmental actions for the purpose of improving that individual's own well-being. That is, each individual is taken to be interested in his or her own stake in the distribution of (broadly defined) income and has no direct interest in economy-wide allocative efficiency. This approach is found in both the broad theories of collective choice associated with Olson (1971), Buchanan and Tullock (1965), Downs (1957), Becker (1958), and Breton (1974) and the somewhat narrower theories of regulation associated with such analysts as Stigler (1971), Posner (1974), Jordan (1972), and Peltzman (1976). The methodological individualism (that is, choice-theoretic approach) of these theories is in agreement with the method of analysis that distinguishes economics from other social sciences—and that economists would at least like to believe has led to the relative success of economics in the understanding of social phenomena.

Rational economic behavior may lead individuals to use the political process for the provision of economic goods that the market system fails to provide or provides at greater cost due to, for example, free rider and exclusion problems. Substantive models based on the distributive approach, however, have emphasized that the coercive powers of government provide a means by which nonunanimous redistributions of wealth (which the market system, by definition, does not offer) can be accomplished. Indeed, distributive theories of government have often been interpreted—either gladly or disparagingly, depending on the interpreter's politics—as embodying an indictment of the government as a tool for coercive wealth redistribution that seldom (and only by chance) enhances economic efficiency or some measure of social welfare. Such an indictment, however, is equivalent to criticizing the market economy because firms seek to maximize profits rather than economy-wide efficiency or social welfare. When certain conditions are present, such as competition or the absence of transaction costs, the result, albeit unintended by any individual market participant, is the satisfaction of economy-wide efficiency standards.

Similar reference to an invisible hand can be made in analyzing governmental activity. Competition of the type described by Tiebout (1956) between governments or competition of the kind envisioned by Downs (1957) and Demsetz (1968) between individuals or groups of individuals seeking to become the government can be expected to counteract the deleterious impact that individual welfare maximization may have on economic efficiency. Even in the absence of competition, a monopolistic, wealth-redistributing government may achieve its ends efficiently (for example, through lump-sum transfers) if transaction costs (broadly defined) are zero—just as a monopolistic market firm may promote efficiency through a combination of marginal cost pricing and perfectly discriminating extractions of consumer surplus. In short, distributive theories of governmental regulation do not provide *a priori* information on the efficiency effects of such regulation. In fact, their most significant contribution does not concern the allocative effects of regulation, but is methodological: No individual actor in the political arena seeks economy-wide allocative efficiency as an end.

The foregoing should not be interpreted as precluding all possible forms of public-interest theories of regulation. Distributive, self-interest approaches to the analysis of political action do not rule out policies *explicitly* designed to serve economy-wide "public" interests. Specifically, the economic model of individual choice does not restrict individuals to the maximization of merely their own pecuniary incomes. To represent individuals in both their market and political roles as narrowly egocentric maximizers, unconcerned about the effects of their behavior on other individuals, is to overlook the many instances of altruism (that is, interdependent utilities) that observation and intuition reveal.

It might be thought that a self-interest, distributive approach to governmental action that admits models of regulation that are both egocentric (that is, emphasize wealth redistribution) and altruistic (that is, emphasize public interest) is uninteresting. Broadly conceived, however, a self-interest, distributive approach to governmental action should be viewed as a restatement of the utility-maximizing paradigm of microeconomics. And, in fact, such an approach has the same epistemological role as that paradigm; it can serve as an analytic axiom devoid of interesting empirical content in itself, but providing a conceptual framework by which empirical observations and events can be scientifically comprehended, analyzed, and predicted. Just as the utility-maximizing paradigm of microeconomics becomes useful only when specific, empirically refutable goals and constraints are examined, so the application of the paradigm in the political context becomes interesting only when motives and constraints are invoked.

Most substantive distributive models of regulation have emphasized the ability of regulated industries to capture the agencies that regulate them and to turn regulation in the industries' favor at the expense of the general public and allocative efficiency. Examples of these capture models include Stigler (1971) and Jordan (1972). Recently, Peltzman (1976) has proposed a more general capture theory that stresses the pecuniary interests that multiple groups have in regulation. Faced with many different constituencies, all-or-nothing solutions seem unlikely. Rather, regulators have incentives to "sell" their output (that is, wealth redistributions) to more than one constituency and to allow each to capture a portion of the agency's output. Capture models of both the single- or multigroup types typically give regulators objective functions such as vote maximization, budget maximization, or "squawk" minimization; and they subject regulators to the constraints implied by the self-interested behavior of legislators and, ultimately, voters.

Capture models of regulation have a lengthy list of commissions, agencies, legislative acts, and regulatory proceedings to point to for supporting evidence. In order to salvage some vestige of a public-interest explanation of regulation from this evidence, variations have been added to a straight public-interest approach. One variation sees regulation as originally designed to serve a general public interest, but eventually captured by the regulated industry. Another variation holds that regulation and regulators attempt to serve a public interest, but are frustrated in this effort by inadequate analytic expertise or limited access to information. While these modified public-interest interpretations may fit the facts of particular cases, for the most part, they leave unanswered the question, What preferences and conditions would allow publicly interested behavior ever to become infused in the regulatory process? Capture models of regulation handle this question by excluding pursuit of public-interest objectives from the preferences of policymakers and their constituents.

Probably the most basic testable proposition of the capture models of regulation is the (sometimes implicit) assertion that the altruistic, publicly interested behavior of individual actors in the public arena is such an insignificant factor in explaining the regulatory process that it is empirically dispensable. That is, such behavior is seen as adding so little to our understanding of regulation as to make the use of egocentric, wealth-maximizing capture models more appealing to the positive scientist. Stigler (1972) has noted the possibility of altruistic motives in political behavior that take the form of a sense of civic duty, that is, a duty to serve the (presumably personally defined) interests of the public. Stigler classifies such motives as "consumption motives" and distinguishes them from the self-interested "investment motive" of increasing one's own wealth. He asserts, "The investment motive is rich in empirical implications, and the consumption motive is less well-endowed, so we should see how far we can carry the former analysis before we add the latter" (p. 104). Of course, the validity of this opinion must ultimately stand or fall on the results of a cross-case assessment of the relative merits (in terms of the breadth and ease of application of testable implications) of models that exclude publicly interested behavior versus models that include it. In any particular case of regulation, however, it may be that the altruistic objectives of relevant actors are important determinants of the regulatory process and outcome. In fact, as the followng section indicates, several examinations of the causes of post-embargo federal energy policy report evidence that some form of public-interested behavior is the *only* important explanatory factor.

6.3 Ideology and Interests in the Formation of Energy Policy

In examining the causes of current US energy policies, it is tempting to infer the goals of policymakers and their constituencies from the effects of the policies adopted. This is particularly true with petroleum policy since the distributional impact is large and fairly clear, while efficiency-based or general public-interest arguments for the important elements of post-embargo policy are apparently not easily supported. Previous chapters of this study would strongly suggest that petroleum product consumers and crude oil refiners have captured federal regulation at the expense of crude oil producers.

From time to time even policymakers acknowledge redistributive goals in US petroleum policy. The *National Energy Plan* (US President, 1977b) proposed by the Carter Administration in 1977 advocated the continuation of crude oil price controls. The plan argued (with emphasis from the original),

> The fourfold increase in world oil prices in 1973–74 and the policies of the oil-exporting countries should not be permitted to create unjustified profits for domestic producers at consumers' expense. By raising the world price of oil, the oil-exporting countries have increased the value of American oil in existing wells. This increase in value has not resulted from free market forces or from any risk-taking by U.S. producers. *National energy policy should capture the increase in oil value for the American people.*[1]

In addition to the interesting exclusion of domestic crude oil producers from the class of American *homo sapiens*, the early Carter Administration proposals with respect to petroleum policy were notable for their lack of success in Congress. The most important petroleum-related proposal, for example, would have had the federal government end the entitlements program and directly capture the inframarginal rents associated with price-controlled oil through a "crude oil equalization tax." The proceeds of this tax were then to be distributed directly to consumers through income tax rebates. The clear threat that this proposal posed to refiners' pecuniary interests in the entitlements program seems to have contributed significantly to their strong opposition and ultimate success in blocking the crude oil equalization tax. Moreover, although consumers' inability to capture all of the wealth at stake through passage of this tax contradicts a simple single-group capture model of regulation, it accords quite well with the prediction in Peltzman's (1976) multigroup formulation, which argues that regulation is used to spread out and buffer the income effects of exogenous shocks. If nothing else, previous chapters of this study have shown that the entitlements program and crude oil price controls have been effective buffers against the international oil price increases of recent years; and the Windfall Profits Tax, which has been one of the few major Carter Administration energy proposals actually enacted into law, clearly will continue the process of spreading the producer surplus created by rising world prices among the major participants in domestic oil markets.

Based on the effects of the regulation of petroleum markets, then, there is little to recommend other than a straightforward egocentric, multigroup capture model of the causes of post-embargo federal petroleum policy. Thus it comes as a surprise that empirical studies that have examined the political process by which recent petroleum-related and other energy policies have been shaped have concluded that the characteristics, particularly the pecuniary interests, of policymakers' constituencies have had virtually no impact on legislator behavior and policy formation. These studies, by Mitchell (1977, 1979) and Lopreato and Smoller (1978), report that the overriding determinant of energy policy is the personal political ideologies of policymakers.

6.3.1 Results of Other Studies

Mitchell (1977, 1979), analyzed natural gas deregulation votes in the House of Representatives in 1975–1976 and attempted to explain voting as a function of each representative's political party, whether the representative was from a net gas-producing or gas-consuming state, the presence or absence of substantial shortages of natural gas in the representative's state, the extent to which gas is consumed for heat in each state, certain demographic characteristics (rich/poor, blue collar/white collar, urban/rural), and a measure of the ideological beliefs of each representative. Only the ideology of each congressman, as measured by the Americans for Democratic Action (ADA) liberalness rating scale, had significant predictive power. Mitchell concludes that representatives not only base their votes entirely on ideology but that the deregulation vote is ideologically more polarizing than the average rated issue. These conclusions, he argues, are not atypical of energy politics in general.

Support for Mitchell's results is provided by Lopreato and Smoller (1978), who examine voting behavior on a wide range of energy-related issues in the House of Representatives in the Ninety-Fourth Congress. To create an energy voting scale, Lopreato and Smoller employ the percentage of votes by each representative that agree with the positions taken by the majority of the Texas delegation, which are interpreted as generally favoring energy producers. This scale is then examined (in stepwise regression) as a function of political party affiliation, net state energy production, total state energy production, district demographic characteristics, regional dummies, and a measure of ideology based on the conservativeness ratings of the Americans for Constitutional Action (ACA). This variable is by far the most important determinant of voting behavior, although regional dummies, party, per capita income, poorness, and ethnicity (specifically, blackness) show slight statistical significance. None of the energy-related measures of constituent interests are found to be significant; and Lopreato and Smoller report that three out of every four energy votes can be predicted solely on the basis of ideology. They also find that the House voted slightly more conservatively on energy issues than on other key issues.

The finding that the pecuniary "investment motives" of legislators' constituents are having no influence on the formation of US energy policy would appear to challenge economists' standard conception of causal forces in the political process. Indeed, both Mitchell and Lopreato and Smoller recognize that their empirical results are at odds with the expectation that elected representatives vote the interests of their constituents. Both argue, however, that the economist's standard conception of the link between constituent interests and the behavior of policymakers may yet be saved. Mitchell asserts that "congressmen vote their ideologies as opposed to their constituents' interests [because] either they or their constituents do not know their interests."[2] Lopreato and Smoller argue in a similar vein that "the complexity and recentness of the energy issue" have forced congressmen to fall back on their political ideologies as guides to decisions.[3] In a world in which information on the relation between constituent interests and the consequences of policy proposals is scarce, legislators may quite rationally consult the dictates of an ideology as a shortcut to the service of their constituents. In this view, ideology plays the same role in the economic theory of the political process that managerial rules of thumb play in the theory of the profit-maximizing firm.

While the analogy between managerial rules of thumb and political ideologies is appealing, its applicability to the cases at hand is given little support by the results of Mitchell and Lopreato and Smoller. If ideology is a proxy for constituents' pecuniary interests, the implied collinearity between ideology ratings and measures of those interests should make it difficult to separate econometrically the effects of interests and ideology. This separation, however, has been made; and legislator ideology stands out clearly as an independent explanatory factor. Unless legislators have mistakenly, nonrandomly, and persistently followed the wrong ideological rules of thumb when voting on energy policy, it would seem that the nature, sources, and consequences of ideological behavior warrant closer investigation.

6.3.2 Generalizing Models of Regulation: The Role of Ideology

The discussion to this point has left the concept of ideology unexamined. In the present context, an ideology may be viewed as a more or less consistent set of normative statements as to the best or preferred state of the world. Such statements are ethical in the sense that they are considered prescriptively applicable to the broad class of all actors, rather than merely the actor proposing them. This observation highlights the essentially altruistic nature of ideological goals. Indeed, it is reasonable to interpret ideologies as prescriptions for making the world "a better place," that is, prescriptions for fulfilling their proponent's conception of the public interest.

Of course, ideological behavior in the political arena need not be altruistic. As noted, one source (or explanation) of ideological behavior may be found in the egocentric "investment motives" of political actors: Ideologies can serve as shortcuts to narrowly self-serving political decisions in a world of scarce information. But in addition to this source of ideological behavior, ideology may enter the political arena directly through the altruistic "consumption motives" (that is, preferences in the direct utility function) of constituents and policymakers.

Constituent Ideology It would not be a violation of the economic model of choice behavior to find that the individuals who make up a policymaker's constituency have ideological goals distinct from their more narrowly defined economic interests. Indeed, this is suggested by observations of successful charities, a good part of the time and money spent on political campaigns and lobby organizations, and voter turnouts in the face of an apparent "paradox of voting."[4] Thus narrowly egocentric policymakers could be observed pursuing ideological goals even in the absence of the information deficiencies implied by Mitchell and Lopreato and Smoller. Of course, this observation, by itself, says nothing about how such a consideration could be worked into economic theories of regulation or what the resulting implications might be.

The infusion of altruistic ideological behavior into the economic theory of regulation is certainly contrary to the spirit, if not the formal exposition, of capture models of the political process. But—and this is testimony to the generality of Peltzman's work—the ideological public-interest goals of constituents are fairly easily encompassed by Peltzman's (1976) multigroup capture model of regulation. In a Peltzman-type model, policymakers seek to maximize the support of voters by transferring dollars between competing constituencies. It does not seem to be a substantive change in this model to allow one group of constituents to seek to transfer dollars to another group. Such altruism could be viewed as demands for dollar transfers to the altruistic group, which in turn uses the governmental regulatory apparatus as a substitute for private institutions and engages in charitable giving to the groups that are the objects of its altruism.

Even in the absence of explicit mathematical treatment, it is not hard to speculate on some of the implications of altruistic regulatory wealth transfers. It is plausible, for example, that such transfers are normal goods—some weak evidence for this will be presented. Consequently, altruistic regulatory transfers could be expected to decrease as the extent to which they are financed through taxes on the altruistic group increases. They should also decrease as the costs of mitigating the opposition of taxed outsiders increase, as the costs of organizing the altruistic and the recipient group increase, and as the bureaucratic costs associated with reaching intended recipients increase. Most significantly, Peltzman's conclusion that the regulatory process is likely to serve the interests of multiple groups does not appear to be upset by the inclusion of a group of altruistic individuals pursuing their own interpretation of the public interest. After all, ideologues may be treated as just one of the various groups seeking to capture part of the regulatory output.

It appears to be somewhat more difficult to incorporate constituent ideological interests into a Peltzman-type model when such interests are not income or distributional specific. Altruistic, public-interested political behavior need not be concerned with the (re)distribution of income. The opposition of ideological libertarians to nonmarket solutions to monopoly problems, for example, apparently is not based on a preference for monopolists' wealth and welfare over consumers' wealth and welfare (or even libertarians' egocentric hopes of becoming monopolists themselves). Nevertheless, this opposition is altruistic and based on the libertarians' conception that the world is a better place, and a broad public interest is served, if voluntary exchanges between consenting individuals are unimpeded by government (see Nozick, 1974). This view belies a concern for the *process*, as distinguished from the *outcome* (in terms of the allocation of resources and the distribution of wealth), of social interaction within the economic system. The model of regulation that overlooks this concern, fails to account for the "income" that individuals receive from their ideologies, and represents constituent interests solely with the dollar transfers at stake would overestimate the likelihood of antitrust intervention in a society of ideological libertarians.

Perhaps contrary to the operating assumption of economists in their roles as policy advisors, the foregoing indicates that there is not a one-to-one correspondence between policies designed to serve the public interest and policies designed to promote allocative efficiency. Or, more properly, the appropriate notion of allocative efficiency must be sufficiently broad to include among the goods demanded in the economy the values that individuals place on processes of social interaction. Note, also, that ideological preferences, such as a distaste for intervention in a system of voluntary trade, or, the opposite, a distaste for systems of market exchange (because bargaining brings out the worst in the human spirit?) are not irrational—no survivor test will weed them out. There is no such test for consumption motives. Foregone output resulting from acceptance of monopoly or opposition to the use of markets is merely the "price" paid for the corresponding "output" of ideological rectitude. Of course, even with the broad concept of efficiency suggested by this terminology, the link between efficiency-enhancing policies and promotion of the public interest is tenuous, at best, in a world where efficiency-enhancing policies are seldom accompanied by compensation. The pursuit by constituents or their representatives of such policies for the purposes of furthering the public interest then requires both altruism *and* utilitarianism of either the "interpersonal comparisons" or "equal marginal utility" variety. The latter restriction considerably narrows the class of ideologies for which efficiency and the public interest are synonymous.

Whether or not their conceptions of the public interest are consistent with efficient resource allocation, constituents with process-specific ideological objectives might be as likely to capture the policymaking process as any other interest group. Although the specifics of a Peltzman-type capture model of regulation would apparently have to be altered to incorporate such objectives, some general conclusions seem secure. Pursuit of objectives of this sort should be affected in straightforward ways by the determinants of the "price" of ideological action, such as the costs of organizing, mitigating opposition, and passing on to others administrative and bureaucratic burdens. The result of these factors should be a political process in which ideological, public-interest considerations by utility-maximizing constituents have a positive impact on the form and content of regulation and an equilibrium in which some positive amount of regulatory "output" is devoted to such constituents.

The Policymaker's Ideology In addition to the influence of constituents' ideological interests, ideology may enter the political process through the policymaker's own maximizing behavior. Indeed, if the public statements of most policymakers were fully believable, one would conclude that the *only* motive for their actions is steadfast adherence to the ideology that each holds as the certain path to the best of all possible worlds. On the other hand, economics generally harbors skepticism for all-or-nothing solutions; and constituents' interests, as well as policymakers' nonideological goals, should be expected to have at least some influence on policymakers' actions.

It might be fatuously proposed that policymakers do pursue only their own ideologies but that there is a fortuitous correspondence between these ideologies and constituent interests that makes the role of a policymaker's ideology in any particular policy decision undetectable. The actual source of a discernable role for a policymaker's ideology, however, is the independence that policymakers have from their constituencies. The market system analogue to this independence, of course, is the separation of ownership and control in the corporate firm.

Both the private manager and public policymaker can be expected to maximize their own welfare while on the job subject to the constraints imposed by their employers (that is, stockholders/constituencies) and competitors. In the absence of legal constraints on profits and in the presence of either a costless and continuous market for corporate control or perfect competition between managers, the private manager will maximize his or her own welfare by maximizing the profits of the firm. If these conditions are not met, however, and some on-the-job consumption can be "purchased" by the manager, profits may not be maximized. Analogously, the public policymaker may pursue his or her own interests independent of, but perhaps not inconsistent with, constituents' interests if able to do so.

Most models of regulation leave no room for public-interested, ideological behavior on the part of policymakers. By endowing policymakers with objective functions such as maximizing the vote, rather than their own well-being, such models preclude behavior that is not directly controlled by constituent interests. To be sure, elected officials face direct control by constituents through the voting booth, as well as more continuous pressure from constituents who have some ability to affect such things as the pleasantness of the work day and future nongovernmental employment opportunities. Nevertheless, the nature of the political process suggests that there is substantial separation of ownership and control that can be expected to result in policymaker independence, or ''shirking''—as Alchian and Demsetz (1972) call it. In the case of US Senate seats, for example, the ''market'' for control is hardly costless or continuous. Rather, it meets once every six years, and the costs of organizing and promoting takeover moves are substantial—and made more so by the nontransferability of constituents' ownership shares (that is, votes). This nontransferability precludes takeover organizers from using the difference in organizer and general shareholder (that is, voter) expected outcome, as the basis for takeover bids and requires takeover organizers to place greater emphasis on voter education and mitigation of opposition. Takeover attempts occur, moreover, in an environment in which voters' incentives for participation and acquisition of information are blunted by their inability to register demand intensities directly and the inherent public good characteristics of collective decisions (noted by Olson, 1971).

The ''market'' for control of the political process is not only temporally discrete and plagued by high transaction costs. It also presents voters with lumpy choices. Over the course of, say, six years, numerous issues are examined and decided upon by policymakers without direct voter approval. These issues are not perfectly anticipated and have uneven impact among constituents. As a result, the voter must assess policymakers on the basis of an overall judgment of their expected net impact on the voter's welfare. In fact, the absence of anything like the homogeneous interests of private shareholders in the profits of their firm implies that voters will sometimes be made worse off by the decisions of even those policymakers whom they support.

In addition to notable costliness and discontinuities, the market for political control may be characterized by less than perfect competition. Competition among potential policymakers seeking constituent support could be expected to constrain the degree to which policymakers are able to pursue their own interests at the expense of their constituents' interests. The successful office seeker, however, is a natural monopolist unless constituent influence is completely independent of the geographical boundaries that define the officeholder's jurisdiction. Still, as Stigler (1972) has pointed out, the Demsetzian (1968) competition for the market, which can, at least hypothetically, control the noncompetitive conduct of the private natural monopolist, is potentially applicable to the political process. Several observations, however, cast doubt on the workability of such competition in the political arena. Research by Abrams and Settle (1978), for example, suggests that recently adopted campaign finance laws and ballot status requirements create legal barriers to entry into the political process. More fundamentally, the individual office seekers who might, under the pressure of competition, strike a no-shirking contract with their constituents are to some degree the enforcers of such contracts; and the provision of enforcement is unlikely to be characterized by perfect competition. As Hirshleifer (1976) has pointed out, the individuals who are the government at any moment do not have to take the rules and institutions of the political process as exogenous. The competition to provide and enforce such rules and institutions is ultimately taking place in a Hobbesian world governed only by the "law of the jungle," in which no property or constitutional rules are defined. If nothing else, the inequality of endowments in this context suggests imperfect competition.

Imperfect political competition, the temporal discreteness of voter assessments of policymakers, the problems of registering demand intensities with votes, and the free rider incentives that arise out of the collectiveness of policing and voting all imply imperfect policing of the tendency of policymakers to maximize their own interests. Consequently, with imperfect foresight of the issues to be faced during a policymaker's tenure and a scarcity of information on public policy questions, voters may rationally base their decisions on policymakers' ideologies. By selecting policymakers for whom ideology is a good, voters may hope to place themselves in the position of benefiting, on net, from the shirking of policymakers who maximize their own interests through pursuit of ideological rectitude. Hence the meaning of statements such as "I voted for candidate X because he thinks like I do" or "I voted for candidate X because, in general, his ideology will put me on the winning side." As a corollary, constituents have an interest in electing policymakers who pursue more ideological rectitude as the separation of ownership and control increases. To some degree, policing and policymaker ideology are substitutes from the point of view of constituents. Indeed, it is illustrative that the ready supply of procedural safeguards in the US Senate makes it far easier for a shirking senator to be removed, censured or otherwise penalized for nonideological shirking (such as junketeering, office mischief making, or bribe taking) than for voting his or her own conscience (read ideology).

While each voter may hope that, on average, the ideological shirking of political agents will be consistent with the voter's own pecuniary and ideological interests, the separation of political ownership and control implies that a policymaker's actions on any single issue or set of issues may not be in agreement with voter interests. The extent to which the ideological shirking of policymakers runs counter to constituents' interests can be expected to increase as the trust of voters increases. Policymakers can enhance their brand names by acting in accord with constituent interests. But legal and other impediments to converting those brand names into pecuniary gain, as well as the lack of a market in which ideological rectitude can be purchased, lead to instances of brand name consumption by policymakers. Behind a shield of anonymity, one member of the US House of Representatives describes this process with an apparently poor understanding of its genesis, but a clear grasp of its implications: "It's a weird thing how you can get a district to the point where you can vote the way you want to without getting scalped for it. . . . If they trust you you can vote the way you want and it won't hurt."[5]

Summary In summary, there is good reason to expect policymakers to engage in some degree of shirking by pursuing their own interests. Among these interests are altruistic ideological concerns for self-defined concepts of the public good. The ability of policymakers' own ideologies to affect the regulatory process consequently can not be dismissed *a priori*. Similarly, constituents themselves can be expected to have both egocentric economic objectives and altruistic ideological interests. The possibility that the latter goals can "capture" part of the process also can not be dismissed *a priori*. A general model of regulation admitting public-interest policies originating among both policymakers and constituents must allow for legislators maximizing their own interests, voting behavior that depends on more than personal economic gain, and some separation of policymaker conduct from constituent interests. The extent, if any, to which ideological conceptions of the public interest affect policy formation in any particular case is an empirical issue. This issue will now be studied in the case of Senate voting on petroleum policy.

6.4 Political Support for Petroleum Price Regulation in the US Senate: Study Design

Much of post-embargo US petroleum policy has been made at the legislative, rather than bureaucratic, level. In the process of adopting and amending the components of this policy, numerous roll call votes have been taken. These constitute a promising source of revealing observations on legislator behavior. The votes considered here all concern crude oil price controls and/or the entitlements program. Voting on the recently adopted Crude Oil Windfall Profit Tax Act of 1980 will have to wait for more in-depth study. Notwithstanding some basic consistency (in the form of multitier producer price incentives and the distributional fates of crude oil suppliers and users) over time in post-embargo oil price policy, careful examination of voting on the WPT may produce additional insights.

Data considerations (primarily the relative abundance of statewide data) have led to the selection of US senators as the objects of investigation in this study. In the following empirical analysis, the objective is to examine the determining factors in Senate roll call voting on petroleum policy and, in particular, the roles of ideology and constituent interests. As noted, the ideology revealed in voting by each senator is itself likely to depend to some degree on constituent interests, as well as characteristics of the senator. For econometric purposes, this suggests two concerns: the explanation of energy voting and the explanation of ideology.

6.4.1 Voting on Petroleum Policy

The discussion in section 6.3 makes it clear that senators are likely to have substantial independence in their decision making on any single vote or any set of votes on a single issue. Consequently, senators are simply viewed here as utility maximizers rather than vote maximizers, jurisdiction maximizers, or maximizers of some other specific objective function. Senators are taken to have a wide variety of objectives. These may include, for example, increasing personal wealth, maximizing popularity, securing the easy life, and/or consuming the utility income of ideological rectitude. This formulation means that senators may engage in direct catering to constituents' interests as well as consuming the perquisites of office and voting that does not serve the interests of constituents. Many types of shirking such as junketeering, nepotism, and expense account padding are not available in the act of voting. Most pecuniary objectives, such as postsenatorial employment possibilities or bribe solicitation, are a function of some constituency's interests, and direct financial gains are limited by conflict-of-interest standards. These considerations suggest that the most important objective a senator might pursue in the act of voting independently of (but perhaps not inconsistent with) constituents' interests is ideological rectitude.

Of course, notwithstanding Mitchell and Lopreato and Smoller, senators are not expected to cast their votes on an issue as important as energy policy without some consideration of the interests of their constituents. In the case of petroleum price control and entitlement regulations, the economic interests of all constituents fall into one of three classes: crude oil production, crude oil refining, or petroleum product consumption. As already detailed, crude oil producers clearly have suffered a substantial loss of wealth under current price controls. The refining industry, however, has received a significant subsidy under the entitlements program. Moreover, if any of the entitlements subsidy has been reflected in prices of refined petroleum products that are lower than under deregulation, consumers of refined products have shared in the benefits of the entitlements program.

The ideological implications of post-embargo policy are not so easily identified. Possible ideological groupings include liberal/conservative, egalitarian/voluntarist, conservationist-environmentalist/pro-domestic production, interventionist/libertarian, and populist-consumerist/pro-big business. Although selection of an appropriate distinction is problematical, the positions taken by the Americans for Democratic Action (1976, 1977) on recent energy issues seem to include each of the left-hand sides of the above dichotomies in the ADA definition of liberal. This definition is accepted here and, in the present context, the ADA notion of liberalness is unfavorable to crude oil producers and favorable to post-embargo petroleum industry regulation.

Voting on individual issues is observed as a dichotomous variable reflecting a pro or con position and suggests the applicability of other than ordinary least-squares estimation techniques. As a measure of senatorial positions on petroleum policy, a variable PROCRUDE is constructed from dichotomous observations on votes to reflect the frequency f_i with which a senator casts a vote favorable to crude oil producers. For a senator voting a pro-crude position on r_i out of n_i opportunities,

$$f_i = r_i/n_i. \tag{6.1}$$

Viewing this frequency as a sample probability, our task is to explain the probability of a pro-crude vote as a function of relevant variables. This frequency, however, is bounded by zero and unity. One method of handling this boundedness is through the logit transformation—and the logit technique is employed here.

Following Zellner and Lee (1965), the logit transformation of (6.1) for estimation purposes can be written in stochastic form as

$$\ln\left(\frac{f_i}{1 - f_i}\right) = \ln\left(\frac{r_i}{n_i - r_i}\right) = \beta_0 + \beta_1 X_1 + \beta_2 X_2 \cdots + \epsilon_i, \tag{6.2}$$

where the X_i are explanatory variables and an asymptotically unbiased estimator of the variance of ϵ_i is given by

$$\hat{\text{Var}}(\epsilon_i) = \frac{1}{n_i f_i(1 - f_i)} = \frac{n_i}{r_i(n_i - r_i)}. \tag{6.3}$$

The subscripts in expressions (6.2) and (6.3) indicate heteroskedasticity. With weights defined by the reciprocal of the square root of (6.3), application of weighted least squares to (6.2) yields parameter estimates with the desirable asymptotic properties.[6]

When a senator casts no votes that favor crude oil producers ($r_i = 0$), the right-hand side of (6.3) is undefined. When a senator casts all votes in favor of crude oil producers ($n_i - r_i = 0$), both the right-hand side of (6.3) and the left-hand side of (6.2) are also undefined. Since these cases in fact arise, some adjustment is necessary. Gart and Zweifel (1967) report that a good estimator (in terms of small-sample performance) of (6.2) is

$$L_i = \ln[(r_i + 1/2)/(n_i - r_i + 1/2)]. \tag{6.4}$$

A good complementary estimator of the variance of the logit function is found by Gart and Zweifel to be

$$V_i = 1/(r_i + 1/2) + 1/(n_i - r_i + 1/2). \tag{6.5}$$

These estimators are used here.

The PROCRUDE variable is defined by L_i in (6.4) and varies positively with r_i. Thirty-six roll call votes in which crude oil producer interests were clearly delineated and in which refiner and consumer interests were not in the same direction as producer interests were selected from a much wider sample of petroleum-related votes. (A descriptive list of the selected votes is provided in appendix 6.A.) Generally, measures that would tend to raise the price of domestic crude oil were taken as favorable to the economic position of crude oil producers; measures tending to raise the marginal costs of refining, either through prospective effects on the price of crude oil or the size of the entitlements subsidies, were taken as unfavorable to the economic position of refiners and consumers. The bulk of selected votes took place in 1975–1976, with a few in 1973–1974 and 1977. Although all senators did not vote on all 36 measures, differences in sample size are accounted for by the n_i terms in (6.4) and (6.5). The lower range of PROCRUDE is set by Clark of Iowa, Hathaway of Maine, Jackson of Washington, and Williams of New Jersey. Each of these senators voted on all 36 issues, but cast no pro-crude votes (resulting in PROCRUDE = −4.29). The upper range of PROCRUDE is set by Bartlett of Oklahoma, who cast 35 out of his 36 votes for the pro-crude position (PROCRUDE = 3.16).

The basic measure of ideology used here is the frequency with which a senator voted the ADA liberal position on an ADA-selected sample of (coincidentally) 36 nonenergy issues in 1975–1976.[7] This measure is precisely analogous to (6.1). On the basis of the frequency of pro-liberal votes, a variable PROADA is calculated according to (6.4). The lower range of PROADA is set by Thurmond of South Carolina and Stennis of Mississippi, who both took the conservative position on each of the 35 issues on which they cast votes (PROADA = −4.26). The upper range of PROADA is set by Clark of Iowa, who cast 36 out of 36 votes for the liberal position (PROADA = 4.29).

Selection of measures of crude oil producer, refiner, and consumer interests should be made so as to reflect cross-state differences in the political influence of each group within states. It seems plausible that a senator who is constrained to respond to the interests of constituents must concern himself with the overall importance of producers and refiners (for example, in terms of employment generated, value added, or total revenues) to his state's economy. These types of measures are used here.

To capture the relative importance of crude oil production in each state, as well as the additional revenues at stake in moving domestic crude oil prices to world levels, crude oil interests are measured as the product of state crude oil production in 1975 and the difference between the average landed cost of foreign oil (of comparable quality) and the state's average price of crude oil in 1975, expressed as a fraction of state personal income. This variable is denoted CRUDE.[8] CRUDE is highest for states such as Wyoming, Texas, Louisiana, New Mexico, and Alaska. It has intermediate values for states such as California whose production is substantial but whose economy is very large. CRUDE is lowest in those states that have no crude oil production, including Maryland, Massachusetts, New Jersey, Idaho, Iowa, Hawaii, Oregon, Maine, North Carolina, South Carolina, Minnesota, Washington, Vermont, New Hampshire, Georgia, Wisconsin, Rhode Island, Delaware, and Connecticut.

Refining interests in each state are measured by the value of shipments in the refining sector in 1975, expressed as a fraction of state personal income. Post-embargo regulation of the petroleum industry has distinguished between small and large refiners in significant ways. Most important, the Small Refiner Bias in the entitlements program grants substantial extra subsidy to refiners with less than 175,000 b/d capacity (see chapter 2). To account for the possibility of different interests and political power, refining interests are expressed in two variables: REFINE for the value of large refinery shipments and SMALLREF for the value of small refinery shipments.[9] The values of REFINE are relatively high for states such as Texas, Montana, Oklahoma, Kansas, and Mississippi, and low for states without refining industries, such as South Dakota, Massachusetts, Idaho, Maine, South Carolina, Nevada, Vermont, and Connecticut. SMALLREF is relatively large for states such as Delaware, Wyoming, Kansas, Texas, and Oklahoma and is small for the states noted that have no refining.

The interests of consumers are measured by total state expenditures on energy in 1974, expressed as a fraction of state personal income. Two factors support the choice of expenditures on all forms of energy, rather than petroleum alone, as the basis for a measure of consumer interests. First, as developments since the 1973–1974 Arab embargo have indicated, changes in petroleum prices tend to significantly affect the prices of other forms of energy due to the relatively high degree of substitutability in the aggregate. Second, significant differences in energy prices exist across states and the best available data come in the form of a composite state-by-state energy price index. Expressing total energy expenditures as a share of state personal income is appealing since it is expected that, other things being equal, the elasticity of demand for energy, and hence the magnitude of changes in consumer surplus due to any regulation-induced price changes, will vary positively with the share of energy in total expenditures. Of course, the distributional interests of consumers are tied directly to any such changes in consumer surplus. The consumer interest variable is denoted ENERGYUSE.[10] It is relatively high for states such as West Virginia, Louisiana, Maine, Rhode Island, Alabama, and Massachusetts and is relatively low for states such as North Dakota, Washington, California, and Colorado.

The preceding variable definitions can now be summarized in an explicit statement of the estimation model of primary interest. The energy voting form of (6.2) is

$$\text{PROCRUDE} = \beta_0 + \beta_1\text{CRUDE} + \beta_2\text{REFINE} + \beta_3\text{SMALLREF} + \beta_4\text{ENERGYUSE} + \beta_5\text{PROADA} + \epsilon. \qquad (6.6)$$

If senatorial ideology has a significant effect on petroleum policy voting, the coefficient β_5 of PROADA should be significantly negative. If the distributional interests of constituents have a significant effect on Senate voting, the coefficient β_1 of CRUDE should be unambiguously positive and the coefficient β_3 of SMALLREF should be unambiguously negative. The coefficients of REFINE and ENERGYUSE are somewhat ambiguous. If the entitlements subsidy to refiners' crude oil use at the margin has been passed on completely in the form of lower petroleum product prices (Montgomery, 1977), then the coefficient β_4 of ENERGYUSE should be negative and the coefficient β_2 of REFINE should be insignificantly different from zero. If none of the subsidy has been passed on (RAND Corporation, 1977), the coefficient of REFINE should be negative and the coefficient of ENERGYUSE should be insignificantly different from zero. The evidence presented in chapter 4 indicates that the entitlements subsidy has been split between refiners and consumers, with each group benefiting substantially. If the analysis of chapter 4 is accurate, both β_2 and β_4 should be significantly negative. If petroleum policy voting is solely a function of the altruistic, general public-interest concerns of policymakers, only the coefficient of PROADA should be significantly different from zero. The model of senatorial behavior discussed above, however, implies that *both* distributional and altruistic ideological factors affect voting. Thus taking into account the available information on the refiner/consumer division of the entitlements subsidy indicates that the expectation is that $\beta_1 > 0$ and $\beta_2, \beta_3, \beta_4, \beta_5 < 0$.

6.4.2 The Determinants of Ideology Ratings

The central task at hand is to examine the factors that explain PROCRUDE. Nevertheless, if it is found that PROADA is significantly related to PROCRUDE, it will also be instructive to inquire into the sources of senatorial ideology. If, as argued above, voters rely on senatorial ideology as a second-best method of ensuring the service of their interests in a world of lumpy voter decisions and imperfect policing of senatorial voting behavior, then senatorial ideology should be related to constituent characteristics. While it is conceivable that highly particularized interests such as those of a single industry or a single labor union could determine the election of a senator, the breadth of issues on which a senator must act during a six-year term and "universal" voter participation make it likely that cross-state differences in the types of senatorial ideologies selected by voters are related to more general population characteristics. A state with a relatively large share of voters who are members of minorities, for example, might be expected to produce senators with relatively strong ideological leanings toward active intervention in minority matters. A state with relatively strong agricultural interests might be expected to elect senators amenable to governmental subsidy programs and import controls. This focus on demographic characteristics, of course, is not new to explanations of legislative voting patterns. Fenton and Chamberlayne (1969), for example, provide a systematic analysis of the general impact of constituent demographics from the perspective of political science; and Kau and Rubin (1979) report a careful econometric study of the determinants of the 1973–1974 ADA ratings of members of the US House of Representatives. The statistical analysis of ideology here parallels that of Kau and Rubin.

The variables included here as demographic discriminants are the percentage of nonwhite state population (NONWHITE); the years of average voter education (EDUC); the percentage of urban state population (URBAN); the percentage of the state manufacturing labor force that is unionized (UNION); per capita state personal income (YPERCAP); the percentage of state labor and property income from agriculture (FARM); the percentage of state labor and property income from manufacturing (MFG); the percentage of the state labor force that is blue collar (BLUCOL); the percentage of the state labor force that is white collar (WITECOL); and the average voter age (AGE).

It has been noted that even a senator without any independence from constituents may behave like an ideologue if constituents have ideological interests. As a direct measure of the liberalness of each states' voters, the percentage of the state vote going to George McGovern in the 1972 presidential election (MCGOV) is included in the analysis of PROADA. Arguably, ideology was accorded more attention in the 1972 presidential election than in any election since at least 1964. Moreover, since the outcome of the election was known with unusual certainty by election day, the individual voter could not have expected his or her vote to alter either the election's outcome or his or her own economic standing. To a large extent, as many voters acknowledged, votes were cast as ideological statements. That a vote for McGovern was a liberal vote is supported by the observations that Senator McGovern invariably voted the liberal position on the ADA-selected issues of 1975–1976 and was president of the Americans for Democratic Action from June 1976 to June 1978.

As some measure of senators' own ideological preferences, political party is included as a dummy variable (PARTY) to explain liberalness ratings. Following Stigler (1972), a political party can be viewed as a firm that specializes in the production of governmental output. For most senators, the choice of which firm to patronize (that is, party affiliation) occurred many years prior to the period of interest here and presumably reflects ideological preferences. It is also suspected that there are systematic ideological differences between Southern party affiliations and party affiliations in the rest of the country. A dummy variable (DUMSOU) is included to capture this effect.[11]

Expectations concerning some of the coefficients in the liberalness equation are straightforward. MCGOV should be positive if, as argued, constituents have ideological interests. PARTY is equal to unity for senators who are Republicans and should have a negative sign, as should DUMSOU. The type of liberalness supported by the ADA stresses civil rights and active egalitarian redistribution policies. NONWHITE should consequently have a positive sign. Farm states have traditionally produced fairly conservative (usually Republican) legislators, and FARM may be expected to have a negative effect on PROADA.

Other variables in the PROADA equation have a more ambiguous impact on PROADA. Only speculations are offered here. AGE in a state may be positively related to senatorial liberalness insofar as ADA liberals advocate policies of direct benefit to the elderly. At the same time, however, it is commonly speculated that young people have become increasingly liberal over time—with conservatives bemoaning a decline in respect for free enterprise and "traditional American values." The sign of EDUC is also ambiguous since the ADA advocates federal aid to the undereducated, but liberalness is often represented as the ideology of the well educated. The redistributive objectives of the ADA are directed to a large extent toward the very lowest ranges of the income scale. This may lead both BLUCOL and WITECOL to be negatively related to PROADA since they are likely to be net losers in such redistributions. The same reasoning may apply to UNION, although the ADA has a past history of pro-labor objectives. The sign of YPERCAP is unclear since relatively low income levels may qualify a state for more ADA-supported federal aid programs, but altruistic support for income redistribution may be a normal good. The level of manufacturing activity in a state might be expected to be negatively related to PROADA as a result of the consumerist orientation of the ADA. On the other hand, an ADA liberal candidate may have more success in states in which externalities (such as pollution) associated with manufacturing are more prevalent.[12] Thus, the sign of MFG is uncertain. Once ethnic and income variables as well as farming and manufacturing activity are accounted for, a prediction of the effects of the urban/rural makeup of a state's population on the liberalness of its senators is problematical. If it is true that liberalness is an ideology of the urbane and sophisticated, while "redneckness" is associated with rural populations, URBAN should have a positive effect on PROADA.

6.5 Political Support for Petroleum Price Regulation: Empirical Results

A comparison of pro-crude and pro-liberal voting patterns supports the general finding of Mitchell and Lopreato and Smoller that ideology can explain a great deal of energy policy. Table 6.1 shows the frequency of liberal votes and one minus the frequency of pro-crude votes (that is, the frequency of votes that are pro-refiner and pro-consumer) for each senator. The strong relationship between these measures is obvious. The correlation between liberalness and the frequency of anti-crude voting is 0.80.

Although the correlation between liberalness and anti-crude voting is

Table 6.1 US Senate Votes on Selected Ideological and Energy Issues: 1973–1977

Senator	State	Percentage liberal votes	Percentage anti-crude oil producer votes	Difference
Abourezk	SD	96.4	84.3	12.1
Allen	AL	2.9	65.7	−62.8
Baker	TE	9.1	26.1	−17.0
Bartlett	OK	3.1	2.8	0.3
Bayh	IN	100.0	100.0	0.0
Beall	MD	35.3	35.3	0.0
Bellmon	OK	16.1	2.9	13.2
Bentsen	TX	33.3	33.3	0.0
Biden	DE	84.8	100.0	−15.2
Brock	TE	23.3	6.2	17.1
Brooke	MA	87.9	97.2	−9.3
Buckley	NY	3.7	6.7	−3.0
Bumpers	AR	63.9	87.5	−23.6
Burdick	ND	69.4	66.7	2.7
Byrd, H.	VA	2.9	58.8	−55.9
Byrd, R.	WV	38.9	72.2	−33.3
Cannon	NV	32.4	84.4	−52.0
Case	NJ	91.2	96.7	−5.5
Chiles	FL	38.9	75.0	−36.1
Church	ID	91.7	100.0	−8.3
Clark	IO	100.0	100.0	0.0
Cranston	CA	88.2	88.2	0.0
Culver	IO	96.9	100.0	−3.1
Curtis	NE	0.0	13.3	−13.3
Dole	KA	15.2	11.1	4.1
Domenici	NM	14.3	10.0	4.3
Eagleton	MO	75.0	94.4	−19.4
Eastland	MS	3.4	20.7	−17.3
Fannin	AZ	8.6	3.1	5.5
Fong	HA	28.6	67.9	−39.3
Ford	KY	58.1	76.0	−17.9
Garn	UT	6.9	7.7	−0.8
Glenn	OH	57.6	95.8	−38.2
Goldwater	AZ	3.4	3.8	−0.4
Gravel	AK	85.2	19.0	66.2
Griffin	MI	8.8	28.1	−19.3
Hansen	WY	2.8	6.2	−3.4
Hart, G.	CO	93.8	92.0	1.8
Hart, P.	MI	100.0	100.0	0.0

Table 6.1 (Continued)

Senator	State	Percentage liberal votes	Percentage anti-crude oil producer votes	Difference
Hartke	IN	85.7	89.3	−3.6
Haskell	CO	94.1	96.8	−2.7
Hatfield	OR	78.8	48.4	30.4
Hathaway	ME	88.9	100.0	−11.1
Helms	NC	2.9	5.6	−2.7
Hollings	SC	42.4	93.9	−51.5
Hruska	NE	6.5	6.9	−0.4
Huddleston	KY	62.5	73.5	−11.0
Humphrey	MN	94.3	96.9	−2.6
Inouye	HA	70.4	95.8	−25.4
Jackson	WA	63.3	100.0	−36.7
Javits	NY	87.9	91.7	−3.8
Johnston	LA	22.9	37.1	−14.2
Kennedy	MA	100.0	100.0	0.0
Laxalt	NV	15.6	7.7	7.9
Leahy	VT	91.4	100.0	−8.6
Long	LA	27.3	20.6	6.7
McClellan	AR	3.6	58.1	−54.5
McClure	ID	3.3	15.7	−12.4
McGee	WY	63.0	39.4	23.6
McGovern	SD	100.0	100.0	0.0
McIntyre	NH	70.0	96.9	−26.9
Magnuson	WA	68.8	100.0	−31.2
Mansfield	MT	85.3	81.8	3.5
Mathias	MD	87.1	55.2	31.9
Metcalf	MT	93.9	91.7	2.2
Mondale	MN	96.8	100.0	−3.2
Montoya	NM	50.0	29.6	20.4
Morgan	NC	24.1	66.7	−42.6
Moss	UT	65.6	92.9	−27.3
Muskie	ME	85.7	100.0	−14.3
Nelson	WI	93.9	100.0	−6.1
Nunn	GA	16.7	69.4	−52.7
Packwood	OR	58.8	34.3	24.3
Pastore	RI	67.9	100.0	−32.1
Pearson	KA	45.5	38.2	7.3
Pell	RI	94.1	100.0	−5.9
Percy	IL	64.3	41.7	22.6
Proxmire	WI	69.4	94.4	−25.0
Randolph	WV	60.0	63.9	−3.9
Ribicoff	CN	82.4	100.0	−17.6
Roth	DE	21.2	45.2	−24.0

Table 6.1 (Continued)

Senator	State	Percentage liberal votes	Percentage anti-crude oil producer votes	Difference
Schweiker	PA	85.7	88.6	−0.9
Scott, H.	PA	42.4	46.7	−4.3
Scott, W.	VA	11.4	6.7	4.7
Sparkman	AL	17.6	43.3	−25.7
Stafford	VT	67.6	88.9	−21.3
Stennis	MS	0.0	44.4	−44.4
Stevens	AK	37.9	22.9	15.0
Stevenson	IL	74.2	97.1	−22.9
Stone	FL	36.1	61.5	−25.4
Symington	MO	67.7	100.0	−32.3
Taft	OH	29.0	16.7	12.3
Talmadge	GA	17.1	91.2	−74.1
Thurmond	SC	0.0	9.1	−9.1
Tower	TX	8.3	3.2	5.1
Tunney	CA	87.0	81.8	5.2
Weicker	CN	80.0	61.3	18.7
Williams	NJ	91.2	100.0	−8.8
Young	ND	3.2	25.0	−21.8
Mean		51.9	61.7	−9.8

Sources: Americans for Democratic Action (1976, 1977) and Congressional Quarterly, Inc., *Congressional Quarterly Almanac*.

relatively high, table 6.1 indicates substantial differences in the frequencies of pro-liberal and anti-crude voting. The average difference in these frequencies indicates that voting on petroleum policy has been more liberal than general Senate voting. This is contrary to Lopreato and Smoller's (1978) finding that House voting on energy policy has been more conservative than general House voting. Mitchell (1979) argues that a lack of knowledge concerning energy issues has led to an increasing reliance on ideology in voting and has made energy issues more polarizing than the average issue. This would imply that conservatives tend to vote more conservatively and liberals more liberally on energy issues. This conclusion, however, also is not supported by the data of table 6.1. As table 6.2 indicates, when the liberal/conservative dividing line is drawn at the average frequency of a pro-liberal vote, both conservatives and liberals tend to be more liberal on petroleum policy than on the average nonenergy issue. Of 46 conservatives, 30

Table 6.2 Liberalness on Petroleum Policy Relative to General Liberalness

	Liberals[a]	Conservatives[a]
Total	53	46
More liberal	40[b]	30
No change	1	2
More conservative	12	14

a. A senator is defined as a liberal/conservative if his liberalness rating is greater/less than the average rating. See table 6.1.
b. This includes 5 senators with liberalness and anti-crude ratings of 100% who, by definition, cannot get more liberal.

appear to have been more liberal on petroleum policy. Of 53 liberals, 41 were at least no less liberal on petroleum policy; and of the 48 liberals who were not already at "perfect" liberalness (that is, with ratings of 100%), 35 were more liberal on petroleum policy.

6.5.1 Econometric Analysis: First Pass

Notwithstanding the relatively high correlation between the frequencies of liberalness and anti-crude voting, table 6.1 exhibits numerous cases of substantial differences in the frequencies of pro-liberal and anti-crude voting. These differences suggest that ideology is not the whole story. A cursory examination of these differences indicates, for example, that those conservative senators who showed the greatest moves toward liberalness in their petroleum policy votes (Talmadge of Georgia, Allen of Alabama, and Cannon of Nevada) tend to come from states with no crude oil production, while liberal Gravel from crude-rich Alaska shows the greatest move toward a conservative, pro-crude position. These apparent responses to the wealth interests of constituents argue for a careful look at the determinants of PROCRUDE voting.

Weighted single-stage estimation of the logit forms of the oil price voting and liberalness models yields the following results (with t-statistics in parentheses):

$$\begin{aligned} \text{PROCRUDE} = {} & \underset{(0.39)}{0.085} + \underset{(5.91)}{3.596}\text{CRUDE} - \underset{(-0.83)}{0.786}\text{REFINE} \\ & - \underset{(-0.04)}{0.076}\text{SMALLREF} - \underset{(-3.35)}{9.416}\text{ENERGYUSE} \\ & - \underset{(-19.07)}{0.516}\text{PROADA}, \end{aligned} \tag{6.7}$$

$R^2 = 0.52$; F-statistic $= 20.56$; and

$$\begin{aligned}\text{PROADA} = {} & \underset{(3.11)}{0.020\text{NONWHITE}} - \underset{(-0.61)}{0.075\text{EDUC}} - \underset{(-4.60)}{0.012\text{URBAN}} \\ & + \underset{(0.04)}{0.038\text{YPERCAP}} - \underset{(-4.09)}{0.100\text{FARM}} + \underset{(5.98)}{0.044\text{MFG}} \\ & - \underset{(-7.16)}{0.135\text{BLUCOL}} - \underset{(-0.94)}{0.022\text{WITECOL}} + \underset{(4.50)}{0.125\text{AGE}} \\ & + \underset{(9.41)}{0.117\text{MCGOV}} - \underset{(-13.47)}{1.451\text{PARTY}} \\ & - \underset{(-4.34)}{0.908\text{DUMSOU}}, \end{aligned} \tag{6.8}$$

$R^2 = 0.60$; F-statistic $= 9.95$.

All of the coefficients in PROCRUDE have the expected sign, but REFINE and SMALLREF are not significantly related to PROCRUDE. Notably, the PROADA measure of senatorial ideology has a significantly negative effect on the odds that a senator will oppose current petroleum policy. CRUDE-producing interests have a significantly positive effect on PROCRUDE, while consuming interests have a significantly negative effect. This latter result is consistent with the expectation that current policies have reduced domestic product prices, although it remains a possibility that consumers are merely confused.

The liberalness of a senator, as measured by PROADA, does indeed appear to be a function of constituent characteristics.[13] NONWHITE, MFG, AGE, and MCGOV all have significantly positive effects on the ideologies manifested by senators. The significance of MCGOV supports the contention that constituents have altruistic ideological interests of their own that are independent of their more egocentric wealth interests. URBAN, UNION, FARM, and BLUCOL have a significantly negative effect on senatorial ideology. In addition to the types of redistributions typically favored by the ADA liberal, the general opposition of the ADA to defense expenditure increases may not be amenable to typical UNION, FARM, and BLUCOL interests.[14] The negative effect of URBAN on PROADA is somewhat surprising. As speculation, it may be that, holding farming interests constant, differences in the ruralness of state populations reflect differences in the nonfarm rural population. This group is noted for its extreme poverty and may be particularly interested in the redistribution policies of ADA liberals. Finally, PARTY (that is, Republican affiliation) and DUMSOU have significantly negative effects on senatorial liberalness, as expected.

The variables that do not appear to be related to senatorial ideology are YPERCAP, WITECOL, and EDUC. It is not hard to believe that white collar workers are fairly evenly divided on each side of the liberal/conservative spectrum, and cross-state differences in educational levels are apparently too small to permit statistical inference.[15] Although the positive sign of YPERCAP might be consistent with the view that altruistic redistribution is a normal good, there is a lack of statistical significance. It may be that liberalness is more a function of the areas in the tails of the income distribution than the mean.

PROADA appears as an explanatory variable in the equation explaining PROCRUDE, but the reverse, of course, is not true. Consequently, the PROCRUDE and PROADA equations have the structure of a recursive system. If there is a systematic relationship between PROCRUDE and the stochastic element in PROADA, a two-stage technique involving simultaneous equations would be warranted. In fact, it is not implausible that the two equations are related in this way. For example, unmeasured influences such as "trouble back home," aspiration to higher office, or desire for "elder statesman" status might affect senatorial voting on both energy and nonenergy issues. Two-stage estimation procedures yield the following results for PROCRUDE (with *t*-statistics in parentheses):

$$\begin{aligned}\text{PROCRUDE} = {} & \underset{(1.41)}{0.307} + \underset{(5.21)}{3.168}\text{CRUDE} - \underset{(-0.93)}{0.879}\text{REFINE} \\ & - \underset{(-0.05)}{0.084}\text{SMALLREF} - \underset{(-3.73)}{10.482}\text{ENERGYUSE} \\ & - \underset{(-16.96)}{0.845}\text{PROADA}. \qquad (6.9)\end{aligned}$$

These results do not differ markedly from the results reported in equation (6.7). All signs are as expected, but the coefficients of REFINE and SMALLREF remain insignificant. Ideology, as measured by PROADA, still has a significant effect on senatorial voting on petroleum policy, as do the interests of consumers and crude oil producers.

Strong versions of egocentric capture models of politics would predict that ideology has played no role in petroleum policy formation; and these models might try to explain away the apparent importance of PROADA in the PROCRUDE equations by arguing that PROADA is a proxy for constituents' oil-related economic interests. Estimation of the PROADA equation with CRUDE, REFINE, SMALLREF, and ENERGYUSE included does not alter any of the inferences drawn with respect to the explanatory factors examined in (6.8) (see Kalt, 1980). Of the oil-related variables, only CRUDE and SMALLREF appear to be significantly related to PROADA. Both of these variables have small, but negative, effects on senatorial ideology; and crude oil producers and small refiners may believe (not implausibly) that conservative, pro-business senators generally serve their interests. When it comes specifically to post-embargo petroleum policy, however, small refiners have benefited substantially under the policies supported by senators with high liberalness ratings. The impact of post-embargo oil policy on small refiners may reflect particularly effective lobbying efforts and may result in spite of the important role played by ideologically liberal senators. Alternatively, ADA liberalism may embody a preference for smallness per se.

Lest these, or any, speculations and conclusions regarding the causal impact of oil-related variables on the ideologies exhibited by successful Senate candidates be exaggerated, it must be noted that these variables count for very little in the dissection of ADA ratings. Despite the fact that each and every constituent can be classified as either a crude oil producer, large refiner, small refiner, or energy consumer, the variables representing these classifications have virtually no explanatory power when added to the PROADA equation. As reported in Kalt (1980), the inclusion of CRUDE, REFINE, SMALLREF, and ENERGYUSE in (6.8) only raises the fraction of the variation in PROADA that can be explained econometrically from 0.60 to 0.62. The inclusion of these variables, in fact, substantially lowers the confidence that can be placed in the attempted explanation of PROADA. [That is, the F-statistic on the expanded model falls to 7.95 from the 9.95 associated with (6.8).[16]]

6.5.2 Collinearity and Constrained Estimation: SEARCHing for More Information

The statistical results of (6.7) and (6.9) both indicate that Senate voting on oil price policy has been the consequence of ideology and the distributional interests of, at least, crude oil producers and energy consumers. These conclusions are not surprising. The insignificance of *both* large and small refining interests, however, is unexpected. If only REFINE were insignificant, it could be argued that large refiners have had no effect on the petroleum policies under investigation. Such a result would be roughly consistent with a complete passing through of the entitlements subsidy to consumers and the consequent failure of refiners to capture any of the wealth transferred from domestic crude oil producers by post-embargo regulation. Even if the marginal entitlements subsidy had been completely reflected in lower refined product prices, however, small refiners would retain the inframarginal grants made to them under the special Small Refiner Bias. Thus, SMALLREF could be expected to be significantly negative. It is a possibility, of course, that the small refiner benefits under federal regulation have been entirely due to the ideological components of policymakers' voting, rather than the political power of small refiners motivated by their own distributional concerns. This explanation, however, is hard to accept in light of *a priori* expectations about the likely political influence of a small, well-organized group with large prospective gains and losses.

The explanation for the anomalous results concerning the effects of refiner interests on the odds of a pro-crude oil producer vote lies in the high degree of collinearity between REFINE and SMALLREF and between REFINE, SMALLREF, and CRUDE. This collinearity arises because states with numerous large refiners also tend to have sizeable small refining sectors. Moreover, states with relatively large crude oil industries also tend to have large refining sectors. The presence of collinearity makes it difficult to isolate the effects of individual variables. One solution to the problem is to constrain the analysis so that it yields sharper inferences. Rather than engage in *ad hoc* searches for constraints and specifications, a more explicit Bayesian approach to the incorporation of prior knowledge into the data analysis is adopted here.

The rationale behind the introduction of prior knowledge into empirical analyses is based on the observation that the scientific investigator brings information of both a theoretical and experiential nature to the examination of any data. This knowledge is valuable and productive. It should not be (and, in fact, seldom is) completely eschewed. In the present context, for example, it seems plausible to assume that, holding ideological variables constant, a senator responds in the same way to a unit of political pressure or enticement whether from producers or refiners. The senator "sells" more output (that is, votes) to that group offering the greatest number of units of pressure or enticement. The units that senators respond to, however, are not self-evident. The analysis above has assumed that the political influence of producers and refiners is proportional to their revenues. If producing and refining interests in post-embargo policy are in opposition and if a senator responds in the same way to a dollar of revenue brought into the state whether by producers or refiners, a plausible set of constraints is $\beta_1 + \beta_2 = 0$ and $\beta_1 + \beta_3 = 0$, where β_1, β_2, and β_3 are the coefficients of CRUDE, REFINE, and SMALLREF, respectively. These constraints are referred to here as the revenue constraints.

In an era in which the political system seems to operate on a jobs standard of value, a senator may be more sensitive to the number of jobs that production and refining create in the state than to the revenues they bring in. Alternatively, a senator may be more sensitive to the net contribution (as measured, for example, by value added) of production and refining firms to the state. On a national basis, the petroleum refining industry employs approximately half (0.5231) as much labor per dollar of shipments as the crude oil production industry. Thus another plausible set of constraints on PROCRUDE is $0.5231\beta_1 + \beta_2 = 0$ and $0.5231\beta_1 + \beta_3 = 0$. These constraints are referred to here as the employment constraints. If it is value added to which senators are sensitive (rather than revenues or employment), valued-added constraints are applicable. Specifically, the national refining industry typically produces approximately one fifth (0.1928) as much value added per dollar of shipments as the crude oil production industry. The suggested value-added constraints are $0.1928\beta_1 + \beta_2 = 0$ and $0.1928\beta_1 + \beta_3 = 0$.[17]

Weighted single-stage estimations of PROCRUDE in the logit form of (6.6) and subject (alternately) to the revenue, employment, and value-added constraints are shown in table 6.3. The two-stage counterparts to the estimations reported in table 6.3 are summarized in Kalt (1980) and do not differ in any substantial way from the single-stage results. None of the imposed constraints restrict the sign of any individual coefficient or prevent individual coefficients from being zero. Rather, they imply only that β_2 and β_3 have signs opposite the sign of β_1 (if β_1, β_2, and β_3 are not zero). Nevertheless, in each case, coefficients of CRUDE, REFINE, and SMALLREF have the expected signs and are statistically significant. ENERGYUSE and, particularly, PRO-ADA also have a significant effect on the odds of a pro-crude vote; and these results are fairly insensitive to constraint specifications. The con-

Table 6.3 Constrained Estimates of PROCRUDE Coefficients
(t-Statistics in Parentheses)

Constraint/variable:	PROADA	CRUDE	REFINE	SMALLREF	ENERGYUSE	CONSTANT	T-statistics on constraints
Revenue	−0.527	4.059	−4.059	−4.059	−6.108	0.029	$\beta_1 + \beta_2 = 0$: 3.58
	(−18.18)	(7.00)	(−7.00)	(−7.00)	(−2.24)	(0.13)	$\beta_1 + \beta_3 = 0$: 2.24
Employment	−0.521	4.238	−2.217	−2.217	−8.78	0.104	$0.5231\beta_1 + \beta_2 = 0$: 1.35
	(−17.97)	(8.54)	(−8.54)	(−8.54)	(−3.16)	(0.48)	$0.5231\beta_1 + \beta_3 = 0$: 1.07
Value added	−0.517	3.681	−.710	−.710	−9.472	0.096	$0.1928\beta_1 + \beta_2 = 0$: 0.10
	(−17.82)	(8.74)	(−8.74)	(−8.74)	(−3.38)	(0.44)	$0.1928\beta_1 + \beta_3 = 0$: 0.35
Unconstrained	−0.516	3.596	−7.86	−0.076	−9.416	0.085	
	(−19.07)	(5.91)	(−0.83)	(−0.04)	(−3.35)	(0.39)	

straints, themselves, do not all perform identically. The t-statistics for each constraint are shown in table 6.3 and indicate that the value-added constraints are objected to least by the unconstrained data.

Significance tests on the constraints introduced into the analysis of PROCRUDE provide some indication of the agreement between data and prior information. The inferences to be drawn from these tests, however, are not solely data dependent. They depend on our confidence in both the data and prior information. The estimates of unconstrained coefficients represent the maximum of a data-dependent likelihood function and are the point estimates most "preferred" by the data. The imposition of prior knowledge (for example, in the form of coefficient constraints) relocates and, in general, reshapes this likelihood function as it relates to the unconstrained coefficients. Unwillingness to give up either data or prior information completely may tempt the investigator to choose some intermediate location between the data and prior points as being most nearly true. Such a decision rule would lead to the conclusion that at least all of the coefficients in PROCRUDE have the expected signs. Such a decision rule, however, leaves the selection of an intermediate location unexplained and, more important, formal methods for pooling prior information and sample information generally do not imply this simple intermediate-point conclusion.

The location and shape of the data likelihood function are described by the unconstrained least-squares point b and a family of concentric isolikelihood contours. These contours define the numerical boundaries on the possible alternatives to the least-squares coefficient values that can be selected with a given level of statistical confidence (that is, with isolikelihood). Similarly, given a set of prior constraints, the location of resulting parameter estimates can be denoted by b^* and the family of isolikelihood contours based on prior information describes a relocated likelihood function.[18]

Typical sets of data and prior isolikelihood contours for a two-variable case are shown in figure 6.1. It makes intuitive sense that an investigator willing to give up a quantity of data likelihood in exchange for an increase in prior likelihood (that is, a move from b toward b^*) should seek the maximum possible increase in prior likelihood. Graphically, this is equivalent to the selection of a point of tangency between data and prior information contours. An investigator choosing point A as a possible "true" location can have a point of equal data likelihood and higher prior likelihood by moving to point B. The locus

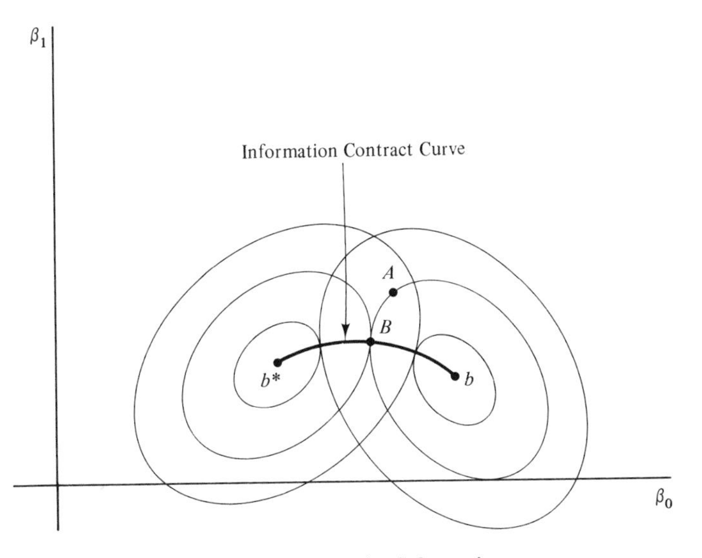

Figure 6.1 Combining sample and prior information.

of such tangencies traces out what Leamer (1978a) has called the "information contract curve" shown in figure 6.1.[19] The nonlinear nature of the contract curve implies that some possible values of β may change sign as the investigator moves between b and b^*.

Leamer (1978b) and Leonard (1977) have developed a program called SEARCH that provides explicit analysis of both the information contract curve and the set of all possible contract curves that can exist between b and b^* given different possible levels of confidence in prior information. Table 6.4 traces the contract curve between the prior point corresponding to the value-added constraint and the data point found under unconstrained estimation. The trace shows alternative points on the contract curve corresponding to alternative degrees of certainty as to the length of the confidence intervals implicit in the likelihood function based on prior information centered at b^*. These alternative degrees of certainty are expressed in terms of selected values of the standard error of the constrained regression (denoted as SIGMA). At SIGMA = 1.0, confidence intervals are taken to be as long as implied by the imposed prior (and the β estimate is the corresponding posterior point on the information contract curve). In this case, the prior standard error is taken to be unity and explanatory variables are represented as free of collinearity.[20] At SIGMA = 2.0, confidence intervals are lengthened (that is, priors are diffused) to twice their original

Table 6.4 The Information Contract Curve: Value-Added Constraint

STEP	SIGMA	DATLIKE	PRILIKE	DATDIST	PRIDIST	DATREJCT	PRIREJCT
Prior		0.9372	1.0	1.0	0.0	0.0627	0.0
1	1/32	0.9373	1.0000	0.9996	0.0003	0.0627	0.0000
2	1/8	0.9379	1.0000	0.9940	0.0055	0.0620	0.0000
3	1/2	0.9471	0.9988	0.9159	0.0794	0.0529	0.0012
4	1.0	0.9649	0.9880	0.7423	0.2489	0.0351	0.0120
5	2.0	0.9881	0.9400	0.4291	0.5632	0.0119	0.0600
6	8.0	0.9999	0.8375	0.0458	0.9532	0.0001	0.1625
7	32.0	1.0000	0.8266	0.0119	0.9879	0.0000	0.1734
Data	∞	1.0	0.8227	0.0	1.0	0.0	0.1773

Coefficient Estimates

STEP	CRUDE	REFINE	SMALLREF	ENERGYUSE	PROADA	CONSTANT
Prior	3.681	−0.710	−0.710	−9.472	−0.517	0.096
1	3.681	−0.710	−0.710	−9.472	−0.517	0.096
2	3.681	−0.711	−0.710	−9.471	−0.517	0.096
3	3.680	−0.731	−0.665	−9.463	−0.517	0.096
4	3.670	−0.758	−0.561	−9.450	−0.517	0.099
5	3.642	−0.778	−0.368	−9.437	−0.517	0.090
6	3.601	−0.786	−0.106	−9.418	−0.516	0.086
7	3.596	−0.786	−0.077	−9.417	−0.516	0.085
Data	3.596	−0.786	−0.076	−9.416	−0.516	0.085

lengths. The interpretation of the reported points is then that the investigator's confidence in the original prior has been cut in half; and, accordingly, points on the contract curve move closer to the data point.

The trace reported in table 6.4 is accompanied by summary statistics. DATLIKE is the value of the data likelihood function at the point in question, normalized so that the point most favored by the data (b) has DATLIKE = 1.0. PRILIKE reports the same values for the normalized likelihood function located at b^*. DATDIST is a measure of the distance of the point in question from the data point and is expressed as a fraction of the distance from b to b^*, where distance is measured in steps of data likelihood. PRIDIST reports a similar distance measure with respect to the prior likelihood function. DATREJCT reports the highest level of confidence at which classical statistical tests would permit the data to reject the point in question. PRIREJCT is a similar measure based on the prior-based likelihood function.[21]

The data do not exhibit a distinct preference for points near the unconstrained point b. The data-based confidence of rejection of the prior point is only 6.3%. At the posterior point corresponding to SIGMA = 1.0, the confidence of rejection is only 3.5%. Approximately 90% of the distance from the data point to the prior point must be covered before the data confidence of rejection exceeds 5%. On the other hand, the prior confidence of rejection exceeds 5% when only slightly over 50% of the distance from prior point to data point has been covered.[22] The coefficients of REFINE and SMALLREF are statistically insignificant at the data point; but these coefficients become significantly negative when the data confidence of rejection of the posterior point on the contract curve is between 5 and 5.7%.[23] In short, it is fairly easy to convince the data to move along the contract curve toward the estimates found by imposing the value-added constraints.

As figure 6.1 indicates, the information contract curve is generated by the data point b, the prior point b^*, the data-based isolikelihood contours, and prior-based assumptions about the shape of prior isolikelihood contours. If the investigator is able or willing to select only the prior location (for example, by imposition of *a priori* plausible constraints) but cannot confidently say anything else about the prior likelihood function, there is an entire family of contract curves that connect the prior and data locations. Each contract curve corresponds to a set of beliefs about the degree of accuracy and structure of information associated with the prior location. Leamer (1978a) has shown

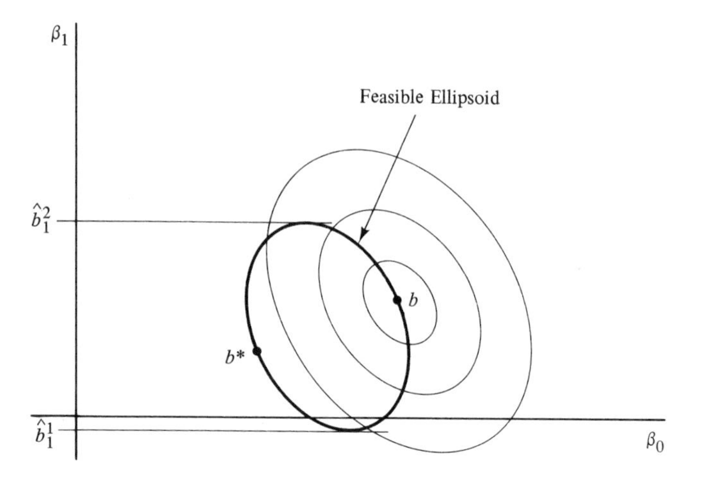

Figure 6.2 Feasible coefficients under constrained estimation.

that the family of feasible constrained coefficient estimates constitutes an ellipsoid such as shown in figure 6.2.[24] Leamer (1978a) has also shown that any possible contract curve must lie within this feasible ellipsoid and that the projection of this ellipsoid on the ith axis yields the extreme bounds (that is, feasible estimates) of the ith coefficient. For β_1 in figure 6.2, these bounds are shown as $\hat{b}_1^1$ and $\hat{b}_2^2$. The investigation of these extreme bounds is tantamount to weakening assumptions about prior precision and asking for the range of all possible estimates that might arise from combinations of prior and data locations.

For the case of the value-added constraints and weighted least-squares estimation of PROCRUDE, the extreme bounds of coefficients and constraints are reported in table 6.5. Significantly, none of the coefficients change sign, and each has the expected sign.[25] The ideology variable, consumer interest variable, and crude oil producer interest variable have particularly stable coefficients. The remarkable stability of the coefficient of PROADA indicates that this variable is virtually orthogonal to the set of other explanatory variables and that PROADA is a poor proxy for the distributional interests that constituents have in current petroleum policy. Although their effects on PROCRUDE are consistently negative, the large and small refiner interest variables seem to have somewhat unstable coefficients. This apparently reflects the collinearity problem rather than simply poor data.

Classical inference from summary statistics at the data location is of

Table 6.5 Extreme Bounds on PROCRUDE Coefficients under Value-Added Constraints
(t-Statistics in Parentheses)

Variable	Maximum	Minimum
PROADA	−.516 (−17.77)	−.517 (−17.82)
CRUDE	3.718 (7.89)	3.559 (6.24)
REFINE	−0.578 (−0.97)	−0.918 (−1.24)
SMALLREF	−0.060 (−0.03)	−0.726 (−2.44)
ENERGYUSE	−9.392 (−3.35)	−9.496 (−3.39)
CONSTANT	0.097 (0.45)	0.085 (0.39)
$0.1928\beta_1 + \beta_2 = 0$	0.113 (0.21)	−0.206 (−0.29)
$0.1928\beta_1 + \beta_3 = 0$	0.630 (0.36)	−0.013 (−0.05)

little value when the collinearity problem is present. The imposition of *a priori* reasonable information, however, can aid in the interpretation of data in such situations. In the present case, the explicit analysis of prior information has permitted systematic investigation of the evidence to be gleaned from the available data on the determinants of Senate voting on current petroleum policy. The results support the original hypotheses, namely, voting patterns on current policy are based on both distributional and ideological interests. The liberalness of senators has a strongly independent and negative effect on their support for policy measures that benefit domestic crude oil producers and harm domestic refiners and energy consumers. The economic interests of these groups, however, also have a significant independent impact on voting behavior. This is clearly true in the case of crude oil-producing and energy-consuming interests. It is somewhat less evident in the case of refining interests, but is nevertheless supported by the preponderance of evidence examined here.

6.6 Conclusion

The notion that economic regulation is designed to serve the designers' conceptions of the public interest is typically met with much skepticism on the part of economists. Numerous studies of the allocative and distributional effects of a wide variety of regulatory institutions have, of course, provided justification for a considerable amount of cynicism. This has been reinforced by the theoretical development of models of the regulatory process that leave no room for the emergence of policies designed to serve a general public interest. It has been argued here that the recognition that utility-maximizing individuals can and do behave altruistically provides the basis for a more general, but still choice-theoretic, conception of regulation and policymaking. Such a recognition, however, does not imply that regulation and policymaking are aimed exclusively at serving the public. Rather, it only implies that the altruistic, public-interest goals of individuals can play some role along with egocentric distributional objectives in the formation of governmental policy.

This proposition has been supported by the analysis of recent petroleum-related voting in the US Senate. It was argued that ideological behavior can be profitably viewed as altruistic behavior aimed at promoting the public interest; and it was found that ideology has had a major impact on policy formation. Contrary to other studies of voting on energy policy, however, ideology does not appear to be the only important explanatory variable. The interests of crude oil producers, petroleum refiners, and energy consumers are also influential in the policymaking process. The general proposition that individuals use the political process for their own ends still holds. In this case, those who receive pecuniary income in their roles as petroleum refiners and energy consumers, as well as those who receive psychic income in the form of ideological liberalism, appear to be gaining at the expense of those who receive income from the production of crude oil and in the form of ideological conservatism.

Appendix 6.A Energy Voting

The energy voting records of each senator are based on a sample of 36 oil price control-related issues on which roll call votes were registered in the United States Senate during 1973–1977. Vote tallies and summaries of issues were taken from the *Congressional Quarterly Almanac* annual listings of roll call votes. Issues included in the sample were selected on the basis of their prospective impact on producers, refiners, and energy consumers. Only those issues were included for which the prospective impact on the wealth of producers was in the opposite direction of the impact on the wealth of refiners and consumers. Issues for which the prospective impact was ambiguous within any of the three groups were excluded. In general, any measure that would tend to raise the price of crude oil was taken as favorable to the economic position of crude oil producers. Any measure that would tend to raise the marginal cost of refining, either through prospective effects on the price of crude oil or the size of the entitlements subsidy, was taken as unfavorable to the economic position of refiners and consumers.

The following is a list of the 36 votes included in the sample. The *Congressional Quarterly Almanac* identification number is shown as CQ#; it refers to the *Congressional Quarterly Almanac* for the indicated year of the vote.

1. Bartlett amendment to a Jackson substitute amendment to S1570 (Fuel Allocation) to exempt from allocation or price controls a producer of crude oil with a well producing not more than 10 barrels per day (that is, stripper oil) or with wells producing not more than a combined 1,500 barrels per day. A Yes vote favored crude oil producers; 5 June 1973. Rejected 42–51. CQ160

2. Passage of S1570 (Fuel Allocation) to direct the President to initiate a mandatory oil allocation and price control program 30 days after enactment of the bill. A No vote favored crude oil producers; 5 June 1973. Accepted 85–10. CQ162

3. Bartlett amendment to HR5777 (Fuel Allocation) to exempt from price controls and allocation programs first sale of crude oil and natural gas liquids produced from any lease whose average daily production does not exceed 10 barrels per well. A Yes vote favored crude oil producers; 31 July 1973. Rejected 43–50. CQ347

4. Bartlett amendment to a Jackson amendment to HR5777 (Fuel Allocation) to allow the President to determine whether mandatory fuel allocation and price controls serve the national interest and, if appropriate, when to begin deregulation. A Yes vote favored crude oil producers; 1 August 1973. Rejected 20–70. CQ353

5. Adoption of the conference report on S1570 (Emergency Petroleum Allocation—EPAA) to direct the President to establish a mandatory program for the allocation and pricing of petroleum. A No vote favored crude oil producers; 14 November 1973. Accepted 83–3. CQ465

6. Bartlett amendment to S2589 (National Energy Emergency Act) to authorize the President to undertake discretionary adjustment of the price of any product essential to the development, production, or delivery of fuels or other energy commodity after the Secretary of Interior had certified that there was a shortage of such commodity because of low prices. A Yes vote favored crude oil producers; 16 November 1973. Rejected 35–47. CQ473

7. Buckley amendment to an Abourezk substitute amendment (see vote

8) to S2589 (National Energy Emergency Act) to instruct conferees to substitute language to provide for administrative procedures for discretionary adjustments of oil prices to protect consumers and provide maximum incentives for domestic oil production. The latter provision would permit increases in ceiling prices. A Yes vote favored crude oil producers; 19 February 1974. Rejected 37–62. CQ31

8. Abourezk substitute motion to a Fannin motion to instruct conferees on S2589 (National Energy Emergency Act) to delete provisions that would reduce and control the price of domestic crude oil that had been exempted from controls by law or by the Cost of Living Council. A Yes vote favored crude oil producers; 19 February 1974. Rejected 37–62. CQ32

9. Adoption of the conference report on S2589 (National Energy Emergency Act) to grant the President authority to impose conservation measures and gasoline rationing and to ease clean air standards and to direct him to reduce and control the price of domestic crude oil that had been exempted from controls by law or by the Cost of Living Council. A No vote favored crude oil producers; 19 February 1974. Accepted 67–32. CQ34

10. Passage, over the President's veto, of S2589 (National Energy Emergency Act), which would authorize the President to impose conservation measures and gasoline rationing and which would rollback the price of domestic crude oil exempted from price controls. A No vote (supporting the President's veto) favored crude oil producers; 6 March 1974. Rejected 58–40 (two thirds vote of 66 required for passage). CQ53

11. Buckley amendment to HR1767 (Oil Import Fees) to limit the increase in oil import fees to crude oil only, excluding petroleum products. A Yes vote favored crude oil producers; 19 February 1975. Rejected 18–76. CQ12

12. Passage of HR1767 (Oil Import Fees) to suspend for 90 days the President's authority to impose a $3 per barrel increase in petroleum import fees. A No vote favored crude oil producers; 19 February 1975. Accepted 66–28. CQ13

13. Gravel amendment to S622 (Standby Energy Powers) to allow allocation and price control authority of the EPAA to expire 31 August 1975 and to tax excessive fuel profits. A Yes vote favored crude oil producers; 9 April 1975. Rejected 23–68. CQ122

14. Bellmon amendment to S622 (Standby Energy Powers) to phase out price controls on old oil by 1977 and to tax resulting increases in income, unless such income is reinvested in energy production and development. A Yes vote favored crude oil producers; 9 April 1975. Rejected 28–61. CQ124

15. Johnston amendment to S622 (Standby Energy Powers) to allow old oil produced by enhanced recovery methods to sell at new oil prices. A

Yes vote favored crude oil producers; 9 April 1975. Rejected 42–47. CQ125

16. Jackson motion to kill Johnston motion to reconsider the vote by which the Johnston amendment to S622 (Standby Energy Powers) was rejected (see vote 15). A No vote favored crude oil producers; 9 April 1975. Rejected 42–45. CQ126

17. Johnston motion to reconsider the vote by which the Johnston amendment to S622 (Standby Energy Powers) was rejected (see vote 15). A Yes vote favored crude oil producers; 9 April 1975. Accepted 45–41. CQ127

18. Jackson amendment to the Johnston amendment (see vote 15) to S622 (Standby Energy Powers) to limit the price of enhanced recovery old oil to $7.50 per barrel. A No vote favored crude oil producers; 9 April 1975. Accepted 46–38. CQ128

19. Bellmon amendment to S622 (Standby Energy Powers) to provide for a two-step phase-out of price controls on old oil. A Yes vote favored crude oil producers; 10 April 1975. Rejected 23–62. CQ134

20. Glenn amendment to S622 (Standby Energy Powers) to require the President to set a ceiling price for new oil at a level no higher than the price generally prevailing as of 31 January 1975. A No vote favored crude oil producers; 10 April 1975. Accepted 54–31. CQ136

21. Passage of S622 (Standby Energy Powers) to provide the President with standby emergency energy powers, to provide for congressional participation in any decision to lift price controls on domestic oil, to mandate a national energy conservation program, and to require the President to establish a ceiling price for all domestic oil. A No vote favored crude oil producers; 10 April 1975. Accepted 60–25. CQ138

22. Johnston amendment to S621 (Decontrol of Oil Prices) to exempt enhanced recovery oil from price regulation. A Yes vote favored crude oil producers; 1 May 1975. Accepted 49–37. CQ158

23. Jackson amendment to S621 (Decontrol of Oil Prices) to establish a ceiling of $7.50 per barrel on the price of enhanced recovery oil. A No vote favored crude oil producers. 1 May 1975. Rejected 38–46. CQ160

24. Passage of S621 (Decontrol of Oil Prices) to give the Congress the right to review any presidential proposal to deregulate oil prices and to require the President to set ceiling prices for all domestic crude oil not currently under price controls. A No vote favored crude oil producers; 1 May 1975. Accepted 47–36. CQ161

25. Bartlett amendment to S1849 (EPAA Extension) to extend EPAA until 31 December 1975, instead of 1 March 1976. A Yes vote favored crude oil producers; 14 July 1975. Rejected 30–53. CQ276

26. Passage of S1849 (EPAA Extension) to extend EPAA from 31 August 1975 until 1 March 1976 and to extend the coal conversion authority of the Federal Energy Administration from 30 June 1975 until

31 December 1975. A No vote favored crude oil producers; 15 July 1975. Accepted 62–29. CQ281

27. Adoption of the conference report on HR4035 (Oil Pricing Policy) to extend EPAA authority for oil price controls to 31 December 1975 and to extend for 20 days the period for congressional review and disapproval of any Administration proposal for changes in oil price controls. A No vote favored crude oil producers; 16 July 1975. Accepted 57–40. CQ287

28. Gravel motion to kill a Hollings amendment to a resolution providing funds for the Senate Select Committee on Nutrition expressing disapproval of the President's plan for gradual removal of price controls on domestic oil. A Yes vote favored crude oil producers; 30 July 1975. Accepted 50–44. CQ353

29. Jackson motion to cut off debate on a Mansfield motion to proceed to consideration of S RES 145 (Oil Price Decontrol) to disapprove the President's plan for gradual removal of price controls on domestic oil. A No vote favored crude oil producers; 30 July 1975. Rejected 54–38 (three fifths vote of 60 required to cut off debate). CQ355

30. Passage, over the President's veto, of S1849 (EPAA Extension) to extend the President's authority to control domestic oil prices from 31 August 1975 until 1 March 1976 under EPAA. A No vote (supporting the President's veto) favored crude oil producers; 10 September 1975. Rejected 61–39 (two thirds vote of 67 required for passage). CQ388

31. Muskie substitute amendment for a Mansfield amendment to HR9524 (Oil Price Controls) to extend oil price control authority from 31 August 1975 until 15 November 1975; to suspend the President's authority to submit oil price decontrol plans to Congress until 1 November 1975; and to set up procedures for expediting Senate consideration of any resolution disapproving a presidential decontrol proposal. A No vote favored crude oil producers; 26 September 1976. Accepted 72–5. CQ420

32. Passage of HR9524 (Oil Price Controls) to extend oil price control authority until 15 November 1975. A No vote favored crude oil producers; 26 September 1975. Accepted 75–5. CQ421

33. Bartlett amendment to S2872 (Federal Energy Administration) to end federal price controls on oil from stripper wells producing an average of 10 barrels or less per day. A Yes vote favored crude oil producers; 16 June 1976. Accepted 61–29. CQ288

34. Montoya amendment to S2872 (Federal Energy Administration) to end federal price controls on increased oil production achieved by applying enhanced recovery methods to existing wells after 1 February 1976; 16 June 1976. Accepted 58–35. CQ289

35. Bartlett amendment to HR5263 (Energy Tax Incentives) to include in the number of oil wells listed on an oil producing property injection wells in which water and other materials are employed to recover crude

oil, thereby increasing the amount of uncontrolled, stripper oil. A Yes vote favored crude oil producers; 31 October 1977. Accepted 62–24. CQ602

36. Long motion to kill a Bartlett amendment to HR5263 (Energy Tax Incentives) to expand the legal definition of marginally productive oil wells exempt from federal price controls. A No vote favored crude oil producers; 31 October 1977. Rejected 43–46. CQ603

Chapter 7

Conclusion: The Energy Crisis and Energy Policy in Perspective

7.1 Introduction

The US economy is nearing the completion of a full decade of struggling to adjust to the new world ushered in by OPEC in 1973. It is an open question how well this process of adjustment is proceeding overall. What is clear is that a major portion of the adaptation to the current era of rising energy prices has been and will continue to be managed, or at least significantly influenced, by the public sector. The appropriate scope of governmental involvement in energy matters is quite rightly an important topic of political and academic debate. Nowhere is this more true than in the area of energy pricing—and no other area excites such interest. The fervor that debates over energy pricing arouse should not be surprising. Hundreds of billions of dollars are at issue, and fundamental standards of social, political, and economic justice are under challenge. These substantial pecuniary and nonpecuniary stakes provide the incentives for interest groups to organize and exert their influence. As the resulting conflicts break out in the public policy arena, the well-meaning policymaker faces the task of raising discussion and understanding above the level of political catch phrases. At the same time, the scientific researcher is furnished with a surplus of opportunities for applied analysis in the economics of government regulation. This study has addressed itself to these opportunities with the hope that the results of its research will contribute to improved policy discourse.

7.2 Summary of Findings

At some stage in the late 1960s or early 1970s, federal petroleum policy underwent a major change of direction. After decades of pro-producer measures such as state pro-rationing, the depletion allowance, and import quotas, all of the important policies favoring producers were effectively abandoned. By the mid-1970s, they had been replaced by the complicated system of EPAA/EPCA price controls on crude oil production and subsidies to crude oil use. This regulatory system has survived into the 1980s. In late 1981, EPAA/EPCA price regulation will be replaced with a system of excise taxes on crude oil production and a bundle of associated redistributive expenditure policies. This Windfall Profits Tax will govern domestic oil pricing until at least the end of the decade.

The underlying theme of post-embargo federal petroleum policy has been a willingness to use price regulation to capture a substantial fraction of the increases in income that would otherwise accrue to domestic crude oil producers in the presence of rising worldwide energy prices. Since 1974, the income extracted from producers by federal price regulation has been in the range of $14–50 billion per year. Under EPAA/EPCA policies, the bulk of these sums has been redistributed to intermediate and final consumers of oil through the entitlements program. The entitlements program is the mechanism by which property rights in price-controlled domestic crude oil are established. These rights are extremely valuable because the products refined from crude oil have generally been allowed to sell domestically at uncontrolled prices. Refined product price controls of either the statutory or jawboning variety have seldom been binding. The exceptions to this —notably the periods of gasoline shortages—have been remarkable in their impact on public opinion, but they have been short-lived aberrations in the overall path of policy. Refined product price controls have apparently been political window dressing, allowing policymakers to show relevant constituents concrete measures that (appear to) prevent oil company "rip-offs" while avoiding the economic, bureaucratic, and political havoc of actual shortages. With product prices determined by marginal input costs and average input costs held considerably below marginal costs by price controls on a large fraction of crude oil inputs, access to controlled crude oil has been the source of huge amounts of economic rent. Since 1974, these rents have totalled $14–45 billion per year.

The entitlements program has been financed by the proceeds of an implicit tax on the inframarginal rents associated with access to price-controlled domestic crude oil. Large portions of the funds available to the entitlements program have been allocated to a number of special programs. These include a system of cash grants to small refiners, subsidies for the importation of certain refined products and the filling of the federal government's Strategic Petroleum Reserve, and discretionary Exceptions and Appeals grants to selected refiners. The majority of the entitlements program's funds, however, generally have been used to finance the entitlements subsidy to domestic refiners' purchases of uncontrolled (primarily imported) crude oil. On a per unit basis, this subsidy has been paying for 10–20% of the cost of imported crude oil since the entitlements program began in late 1974. To the benefit of crude oil users, the result has been an increase in crude oil imports and an expansion of domestic refined product supply.

Characterizing the entitlements program as a mechanism for distributing the rents attendant to crude oil price controls serves to emphasize the point that post-embargo federal oil policy is most appropriately viewed as an income redistribution policy. Indeed, the analysis of chapter 3 indicates that multitier crude oil price regulation has closely resembled the pricing scheme that could be expected to be adopted by a price-discriminating, surplus-extracting monopsonistic buyer of domestic crude oil. The exercise of monopsony power arises from a coalition of crude oil users and their ideological supporters. These groups have captured the federal regulatory mechanism and put it to use as their mechanism for cartel administration. Although, from an allocative view, multitier regulatory price discrimination performs better than alternative single-tier monopsonistic systems, the multitier system is not neutral in its effects on the allocation of resources on the supply side of the domestic crude oil market.

Many of the allocative impacts of crude oil price regulation are not analytically straightforward. It is clear that regulations that reduce the prices realized by producers have a negative effect on exploration and development of supply sources. For supply sources that are already developed, however, crude oil price regulations can increase or decrease oil extraction in the present period by making the present look more or less appealing than the future. The investigations of chapter 3 indicate that crude oil price controls have had a negative effect on domestic crude oil production. This is particularly clear after early 1976, when newly developed properties were brought under controls. The output of domestic crude oil that has been foregone as a result of EPAA/EPCA controls has been in the range of 0.3–1.4 million barrels per day.

The negative output effects of crude oil price controls would have been eliminated in October 1981 were it not for the Windfall Profits Tax. The WPT will effectively perpetuate multitier price regulation. Its intertemporal impact on extraction from already developed properties will depend crucially on the expected course of world oil prices. Unless this course is generally upward (in real terms) at rates in the range of 3–5% per year, the WPT will tend to shift extraction toward later periods. Production resulting from new exploration and development (except for development in the form of tertiary enhanced recovery projects) will be discouraged by the WPT's excise taxes relative to the unregulated market. Special incentives will provide significant encouragement to tertiary enhanced recovery, although the inframarginal character of these incentives will reward smallness in adopted projects and may allow development of projects that would otherwise be uneconomical. A reasonable estimate of the net output effects of the fully effective Windfall Profits Tax indicates that the WPT will reduce domestic production of crude oil by somewhere around 1 million barrels per day.

Until the Windfall Profits Tax takes full force, the entitlements program will continue to be used to redistribute the income denied to crude oil producers by price regulation. Because the entitlements subsidy lowers the marginal cost of domestic refining and thereby promotes an expansion of product output, some portion of the subsidy may ultimately accrue to the consumers of refined products in the form of lower prices. In fact, the magnitude of any subsidy-induced reduction in refined product prices has been the subject of considerable difference of opinion among both policymakers and economists. At issue is the extent to which domestic product price changes are constrained by competition from unsubsidized foreign refiners in world markets. The investigations of chapter 4 indicate that the combination of reductions in the residual US demand for imported products and the official discouragement of exporting by US refiners has caused the United States to drop out of the international markets for many products since the beginning of the entitlements program. This observation as well as consideration of the price-making ability of the United States in world product markets, analysis of actual domestic-foreign price differences, and recognition of the effects of direct import subsidies in those cases for which the United States has remained an active product importer all support the conclusion that the entitlements subsidy has substantially lowered domestic refined product prices. Most likely, about 40% of the entitlements subsidy has accrued to refined product consumers, with the remaining 60% staying with refiners.

Accounting for the incidence of both the entitlements subsidy and the special entitlements programs leads to the conclusion that, at an aggregate level, subsidized refiners have captured \$9–32 billion per year since 1975, while refined product consumers have realized gains of \$5–12 billion per year. The regulatory losses of crude oil producers have exceeded these gains of crude oil users. The difference has been an economic waste for the country as a whole. This deadweight loss has arisen primarily because the entitlements subsidy and crude oil price regulation result in larger-than-optimal levels of petroleum consumption and smaller-than-optimal levels of crude oil production, respectively. Entitlements-subsidized oil prices induce the consumers of crude oil to apply petroleum to uses that have less value to them than the wealth handed over by the country to acquire the requisite foreign crude oil. On the supply side, price regulation precludes the production of a quantity of domestic crude oil that could be had at a domestic resource cost that is less than the cost the country bears to replace the foregone output with imported oil. The combined deadweight loss of controls and entitlements most likely has been running at rates of $1–6 billion per year since 1975. Under the Windfall Profits Tax, supply-side losses alone will probably be in the range of $1–2 billion per year.

These estimated net losses from regulation are exclusive of other, less easily measured costs. Likely significant sources of these other costs include the private- and public-sector costs of regulatory administration, enforcement, and compliance; subtle costs arising from changes in the criteria, business practices, and activities that determine firms' survival and profit opportunities; and possible impairments in the domestic economy's ability to adjust to exogenous shocks in world crude oil markets. The measurable components of other costs of these types amount to somewhere around $1 billion per year. Because their distributional impacts are felt by one or more of the major groups affected by petroleum price regulation, these other costs of regulation offset the apparent gains of consumers and refiners and/or magnify the losses of producers.

In an era in which energy policy discussions are dominated by concern for the health of the nation's economy in general and the avoidance of income transfers to foreign oil suppliers in particular, post-embargo US oil price policy has reduced the national wealth and improved the already pleasant lot of crude oil-exporting countries. Although chapter 5 sought to uncover an underlying allocative rationale for these results, the outcome of the search can hardly be described as definitive. By increasing the US demand for foreign crude oil, post-embargo oil price policy might be consistent with foreign policy interests in amicable relations with certain exporting countries. More common concerns over the possible national security, environmental, and macroeconomic problems associated with reliance on imported crude oil, however, suggest that the quantitative magnitudes of regulatory deadweight losses presented in this study may be underestimated. Of course, this observation tends to bolster the conclusion that post-embargo oil price policy has been having adverse effects on the efficiency of resource allocation.

Care must be taken to recognize that the strength of the conclusion that post-embargo federal petroleum policies have been reducing the size of the national economic pie does not constitute an unqualified indictment of these policies. The clearly redistributive effects of federal oil price regulations must force the careful analyst and well-meaning policymaker to consider whether these effects are morally justified. Answers to this question ultimately flow from ethical-political value systems (that is, ideologies), rather than economics.

The political and economic roots of US petroleum policy are to be found in both the distributional interests of petroleum producers, refiners, and consumers and the ideologies of policymakers and their constituents. In particular, the analysis of chapter 6 indicates that legislative voting on US petroleum policy has been directed by the interests of a coalition of refiners, consumers, and those who find the direction of income redistribution implied by rising oil prices ideologically distasteful. The policy intervention promoted by these interests has been undertaken over the opposition of both crude oil-producing interests and those who find such intervention ideologically objectionable.

In addition to shedding light on the causes of EPAA/EPCA regulation, chapter 6 suggests some enrichment of economic theories of regulation. The sources and consequences of shirking on the part of policymakers have been examined and a policy-sector analogue to the industrial organization concept of separation of ownership and control has been outlined. Moreover, it has been possible to examine the sources and empirical importance of so-called consumption motives (in the form of altruistic ideological goals of utility-maximizing actors) in political behavior. The results strongly suggest that, at least in the case of Senate voting on post-embargo petroleum policy, the investigator who overlooks such motives and admits only the investment motive (in the form of maximization of personal wealth) will perform poorly in explaining and predicting policy actions. The extent to which this conclusion applies to other cases of economic regulation awaits further empirical research. Nevertheless, the underlying conception of political behavior, if not its general applicability, appears to be incontrovertible.

7.3 Topics for Further Research

To be sure, this study has not exhausted the fruitful areas of research into either the economics of US petroleum policy or the economics of regulation. There are numerous oil policies besides EPAA/EPCA crude oil regulations and the Windfall Profits Tax that involve potentially significant allocative and distributional consequences and merit study. Perhaps the most important general policy area yet to be extensively investigated is the regulatory treatment of distributors, marketers, and retailers. This study has focused primarily on crude oil producers and refiners. Policies such as the regulatory treatment of franchise contracts and refined product allocation regulations, however, may be having substantial impacts on downstream sectors of the domestic petroleum industry. Another area for research into the economics of petroleum policy concerns the rationales for, and effects of, mandatory oil conservation regulations such as fuel efficiency standards for automobiles and forced conversion of utility and industrial boilers to nonoil fuels. In addition, the rule makings and regulations that will accompany implementation of the Windfall Profits Tax should provide numerous opportunities for important analysis. The array of WPT-financed expenditure programs, in particular, will open up a variety of issues requiring investigation.

An important topic for research into the economics of regulation that has been neglected, for the most part, in the present study concerns the effects of economic regulation on the structure and competitiveness of industrial markets. While speculation is not lacking, there is little hard evidence on the effect of regulation on such factors as entry conditions (For example, do the costs of administration and compliance vary systematically between existing and potential firms?); the incentives for vertical integration (For example, do the deviations between allowed prices and shadow prices at various stages of the petroleum industry lead to empirically important incentives for industry participants to bypass market transactions?); and optimal firm sizes (For example, do policies such as the Small Refiner Bias that alter the size distribution of firms significantly affect technical efficiency in the industry?).

Another topic open to applied research centers around the responses of the world crude oil market and, particularly, the larger OPEC producers to federal regulation of domestic oil markets. This question has been given relatively cursory treatment in this study, but it is obvious that discussions of both the allocative/distributional effects of regulation and the specification of optimal energy policies are likely to be sensitive to its answers. Although sophisticated models of the world oil market that recognize the intertemporal aspects of producers' decisions as well as the price-making abilities of certain exporting countries are being developed, they are not yet being tied to specific elements of US policy. In a related vein, it is clear from chapter 5 that the tasks of identifying and measuring divergences between the private and social costs of imported oil are very far from complete.

Numerous aspects of the political economy of US petroleum policy remain to be examined. One of the more obvious unanswered questions centers around the causes of the noted turnaround in US petroleum policy in the late 1960s and early 1970s after the long record of pro-producer measures. Other likely topics for research include the perennial special treatment of small refiners; the singling out of the petroleum sector for regulation at the time of a general trend away from industry-specific regulation; the role that foreign policy interests have played in determining the substance of petroleum industry regulation; and the process by which the specific and ingenious combination of controls and entitlements was selected from the set of possible regulatory devices.

At a more general level, it hardly needs to be pointed out that economic theories of regulation and their empirical testing are in a stage of relative infancy. The basic analytical approach that conceives of legislators and regulators as utility maximizers possessing multiargument objective functions, rather than nonshirking captured servants of some interest group, seems promising. The major analytic challenges lie in specifying the sets of constraints faced by political actors in any given instance and, if possible, generalizing constraints so as to elevate a collection of eclectic theories to the level of a scientific paradigm.

Turning to issues in the empirical testing of theories of regulation, the approach adopted here focuses on the behavior of policymakers, rather than the attempt to explain the causes of any given instance of regulation by noting its distributional effects. While this approach has provided insight in the case at hand, it also suggests further questions of interest. As touched upon, perhaps the most significant question is whether altruistic ideological behavior is an important explanatory component in most cases of regulation, or even the general case. Subsidiary issues concern the factors that determine the importance of altruistic ideological behavior across cases (For example, is it more important at the legislative level than at the agency level?); the determinants and empirical importance of policymaker shirking (For example, does shirking increase as the policymaker's security in office—as measured, for example, by length of tenure or margin of the last electoral victory—increases?); and the importance of altruistic ideological behavior relative to other possible forms of policymaker shirking (For example, when does a policymaker take bribes, and when does a policymaker "follow the dictates of his or her own conscience"?).

7.4 US Energy Policy: Some Generalizations

The preceding discussion indicates that there is considerable room for more research. Nevertheless, the results of this study seem strong enough to support some general conclusions about US energy policy.

Most centrally, the conception of the underlying energy problem that emerges from the investigations of the economics and politics of federal petroleum policy seems to be applicable to the broader class of federal energy policies. The domestic energy "crisis" is, far more than anything else, a quarrel over income distribution—and US energy policy is the outcome of this quarrel. Indeed, it is difficult to conceive of current US energy policy as purely or primarily a response to some massive failure of domestic markets to allocate energy resources optimally in the face of external conditions acting on the US economy since the early 1970s. Such a conception is weakened by a scarcity of evidence of resource misallocation on a scale commensurate with the breadth and depth of energy market intervention, as well as the apparent absence of correspondence between the actual allocative effects of this intervention and the types of policies that might serve as remedies for the plausible suggested forms of allocative failure. A conception of US energy policy as a mechanism for redistributing wealth stands up much better.

To note that US energy policy is primarily a policy of redistribution is not to say that energy policy should be viewed as a simple case in which the groups with the most dollars at stake and the best political organization have captured the federal regulatory apparatus. Extremely important values other than pecuniary gain are at stake. The values that individuals attach to equality, the role of luck in the distribution of income, and the scope of governmental interference in the marketplace are playing causal roles in determining the course taken by energy policy. Such values are reflections of individuals' basic ideological beliefs. Undoubtedly, these ideological values do not differ in kind from the more materialistic concerns so often exhibited by members of the human race. Like these concerns, ideological values are, to some unknown extent, the result of cultural environment, and their pursuit is not invariant with respect to opportunity costs. But only a naive and simplistic determinism would argue that the behavioral manifestations of these values are, in each and every instance, the product of some underlying material interest.

The importance of ideology in the formulation of energy policy means that ideas are playing a central role in this process. But, perhaps to the disappointment of economists, who produce and peddle ideas, the intellectual debate does not turn on arguments between those who believe in the efficacy of the private sector and those who believe in the efficacy of the public sector as means of allocating resources. Rather, it is focused on the moral propriety of the distributional *outcomes* that would result under alternative approaches to the energy problem and the moral propriety of the *process* of coercively altering the distribution of income that would fall out of unregulated market exchanges.

With some notable exceptions—Alaskan natives, Appalachian coal miners, and American Indians with coal deposits on their reservations—the average domestic energy producer or owner of energy resources is higher up in the distribution of income than the average domestic consumer of energy. Thus the increasing scarcity of energy implies a redistribution toward the upper end of the income scale. The result is that the element of modern political liberalism that is such a decisive factor in policy formation is its egalitarian ethic. The element of modern political conservatism that is under challenge is its view that income acquired through noncoercive business transactions and acumen is justly acquired. The intellectual content of energy policy, then, is primarily ethical, rather than economic.

Debates over fairness, rather than efficiency, are nothing new in American politics. They are part of the ongoing debate over what kind of society the nation should have. Issues of wealth-and-poverty and freedom-and-intervention have always raised the passions and social strife that, rightly or wrongly, find expression through politics. Energy has been singled out as the latest embodiment of these issues because of the magnitude of the allocative and distributional stakes, the abruptness with which these stakes have been raised, and the fact that every voter is a direct participant in energy markets. The political conflicts that have resulted from the combination of these conditions are not susceptible to easy resolution—despite endless exhortations about "the moral equivalent of war" and the need for "a comprehensive national energy policy that serves the public interest." The crisis created by the energy developments of recent years arises because one group's ideological or economic gain from energy policy is another group's loss. The grounds for consensus on fairness in energy policy are narrow.

This rather pessimistic prognosis can be tempered somewhat. The economic analyst might supply the methods by which a policy objective selected on the grounds of fairness can be achieved most efficiently. But in a world where the (first-)best laid plans of economists often go awry on the grounds of political feasibility and administrative practicability, findings (such as in this study) that an existing policy results in net economic waste do not suffice to condemn that policy. Accordingly, any temptation to end this work with strong and detailed policy recommendations is resisted. Still, it can be pointed out that the continued pursuit of distributional goals by way of energy policy raises the possibility that all parties, even intended beneficiaries, may be made worse off by measures that reduce aggregate economic wealth through significant distortions to the allocation of resources. Improvement will come to US energy policy when distributional objectives are addressed directly through the general income tax and welfare systems and energy policy concerns itself with allocative efficiency. Price regulation and excise taxes of the type examined in this work will not contribute to an improved energy policy.

Notes

Chapter 1

1. Professor Milton Friedman (1975a,b) originally was misinformed on the entitlements program. Initial conclusions later had to be reversed.

2. The contemporary view of the Standard Oil Company received its classic statement in Tarbell (1904). The early history of both federal and state policies is reviewed in Nash (1968, pp. 11–15).

3. The history of these federal agencies is reviewed in Nash (1968). The promotion of "cooperation and mutual forebearance" is from A. C. Bedford (Chairman of the Board, Standard Oil Company of New Jersey) in a speech quoted in the *Oil and Gas Journal* (4 April 1918).

4. On the theory of common pool resources and the rule of capture, see, for example, Demsetz (1967). On the anticompetitive effects of prorationing, see, for example, Bohi and Russell (1978, pp. 248–259), Lovejoy and Homan (1967), McDonald (1971), and Mead (1976).

5. On the history of depletion allowances, see Nash (1968, pp. 34–35, 85–86). On associated economic effects see Adelman (1972, p. 233), Brannon (1974), Erickson, Spann, and Millsaps (1974), and McDonald (1963).

6. Bohi and Russell (1978, pp. 45–57).

7. Cabinet Task Force on Oil Import Control (1970, p. 260).

8. These programs are fully described in Bohi and Russell (1978).

9. Contemporary studies of the effects of MOIP included Burrows and Domencich (1970), Dam (1971), and Shaffer (1968).

10. *State Tax Guide, Statute Summaries, By Taxes, By States* (selected years).

11. Texas, for example, reached 100% allowable (that is, unrestricted production) in April 1972 and Louisiana reached 100% allowable in August 1972.

12. *Oil and Gas Journal* (13 March 1972, p. 39).

13. Johnson (1975, pp. 102–103).

14. Bohi and Russell (1978, pp. 217–220).

15. Bohi and Russell (1978, p. 220).

16. Phase IV regulations are published in full in *Economic Controls: Stabilization Guidelines* (1971–1974). EPAA regulations are published in full in *Federal Energy Guidelines* (1974–present). Also see Owen (1974).

17. Johnson (1975, pp. 110–11) and *Federal Register* (vol. 39, 30 August 1974, p. 31,623).

18. *Federal Register* (vol. 39, 30 August 1974, pp. 31,650–31,654).

19. EPCA regulations are published in full in *Federal Energy Guidelines* (1974–present).

20. Regulations governing phased decontrol are published in full in *Federal Energy Guidelines* (1974–present).

21. *Federal Register* (vol. 41, 1 April 1976, p. 13,901).

22. In August 1979, gasoline retailers were taken off the dollar-for-dollar passthrough system and placed on a fixed-cents-per-gallon (15.8¢) markup system of controls.

23. Complete descriptions of this act are found in Commerce Clearing House, Inc. (1980) and Price Waterhouse and Co. (1980).

24. Based on estimates from Democratic Study Group (1980).

Chapter 2

1. See, for example, Duchesneau (1975), Eppen (1975), Erickson (1977), Erickson and Spann (1974), Johnson et al. (1976), Teece (1976), and US Treasury Department (1973). For dissenting views see, for example, US Federal Trade Commission (1973) and Sampson (1975).

2. *Federal Register* (vol. 39, 11 November 1974, p. 39,740).

3. *Monthly Energy Review* (May 1977).

4. The implied incentive for refiners to integrate backwards into crude oil production was reduced by the buyer-supplier freeze and the buy/sell program because these regulations limited an integrated firm's ability to direct its crude oil production into its own refining operations. Moreover, any incentive for integration created by crude oil price controls that may have been present was eliminated in November 1974 when the entitlements program was introduced. As shown in section 2.4, this program implicitly taxes away the difference between refiners' marginal valuation of crude and the ceiling price. This difference, of course, is the source of incentives for vertical integration.

5. For an example of this reasoning, see Bohi and Russell (1978, p. 225).

6. In early 1979, reports of shortages and seller closings arose in both gasoline and jet fuel markets, but ended quickly in the case of jet fuel as it was decontrolled in late February 1979. Shortages in gasoline markets were apparently exacerbated by the "voluntary" price guidelines of the Council on Wage and Price Stability. For an analysis of the 1979 episode of shortages, see Verleger (1979) and Erfle and Pound (1980).

7. The banked cost regulations are described fully in *Energy Users Report* (1976, pp. 31:0306–31:0321).

8. Based on data from *Monthly Energy Review* (selected issues).

9. See US Federal Energy Administration (1976c). For comparison, only one instance of negative banks occurred in the next five months.

10. Regarding the 1973–1974 shortages, see Johnson (1975, pp. 112–114). For the 1979 episode, see Verleger (1979).

11. In 1979, spot and market-clearing world crude oil prices rose considerably above OPEC contract prices. Consequently, marginal import prices were higher than average import prices. The entitlements program equalizes refiners' average crude oil costs and, as will be shown, subsidizes crude oil imports at the margin. In 1979, the subsidized marginal cost of imports was above both the average cost of imports and the average cost of all domestically refined crude oil—and was rising rapidly.

12. Owen (1974, p. 1,313).

13. In the spring and summer of 1975, when extension of EPAA was being considered in Congress, there was considerable debate among policymakers and economists over whether, in fact, the entitlements program had the effect of an implicit subsidy. The trade press, however, recognized the implicit subsidy in the entitlements program before it was ever enacted. For a summary, see Kalt (1980, p. 143).

14. Based on data from *Monthly Energy Review* (June 1980).

15. The disequilibrium in world crude markets in 1979 and 1980 has been characterized by substantial differences in the crude oil prices charged by various countries. As a result, marginal uncontrolled imported crude prices have been considerably higher than average imported prices. The price of an entitlement in (2.14) is based on the *average* imported price; and the entitlements program inadvertently permits domestic refiners to retain rents associated with inframarginal imported crude. Average crude oil costs are not equalized across refiners and the actual subsidy is not maximized when marginal import prices exceed average import prices.

16. For a complete description of these programs see US Department of Energy, Economic Regulatory Administration (1978).

17. Beginning in June 1979, the Small Refiner Bias has been reduced for refiners with daily crude runs of less than 100 mb/d. The reduction is approximately 50% on a per barrel basis at less than 10 mb/d and declines to a reduction of one cent per barrel at 100 mb/d; *Oil and Gas Journal* (7 May 1979, p. 48).

18. Data reported here are from US Department of Energy, Economic Regulatory Administration (1978). The period preceding the entitlements program was notable for the phasing out of the Mandatory Oil Import Control Program. The MOIP had a small refiner bias of its own; hence, the pattern to the changes in the size distribution of refiners in the early 1970s is consistent with a process of market adjustment to the termination of the MOIP bias.

19. The single new grass-roots refinery that has come on stream since November 1974 was built by a new entrant, but design and site preparation were begun in 1973. Upon completion, the refinery was sold to an existing large refiner (Marathon). See Harvard University Business School (1978) and US Federal Energy Administration (1976a).

20. Frank Zarb (Administrator, Federal Energy Administration), Statement before US Senate, Committee on Interior and Insular Affairs (1976a, p. 80).

21. See note 20.

22. These special subsidies are described in *Federal Energy Guidelines* (4 June 1976, p. 13,650); *Code of Federal Regulations: Title 10* (1978, pp. 181–196); and *Oil and Gas Journal* (4 June 1979, p. 61).

23. Based on data from US Department of Energy, Economic Regulatory Administration (1978, p. 31); *Federal Register* (vols. 43 and 44, monthly entitlements notices for 1978); and *Monthly Energy Review* (selected issues).

24. US Senate, Committee on Energy and Natural Resources (1977, pp. 21–22), and *Monthly Energy Review* (August 1977, p. 69).

25. *Oil and Gas Journal* (5 December 1971, p. 73). For the description of these regulatory issues and changes, see US Federal Energy Administration (1976b); *Federal Register* (vol. 41, 3 November 1976, pp. 48,324–48,325); *Federal Register* (vol. 42, 22 March 1975, pp. 15,419–15,423); and *Energy Management* (22 December 1977, p. 3).

26. Based on data from *Federal Register* (vols. 43 and 44, monthly entitlements notices for 1978).

27. Based on *Monthly Energy Review* (September 1979) and *Code of Federal Regulations: Title 10* (1978, pp. 181–185).

28. The exhaustion of rents might be avoided through a collusion of refiners. Refiners with little or no access to controlled crude, however, would still have an incentive to expand in response to the entitlements subsidy, unless the cartel had a mechanism for directly transferring some rents to such refiners.

Chapter 3

1. Right at q_s, marginal revenue becomes negative. All output gets reclassified as old oil and total revenue falls by $(P_w - P_o)q_s$.

2. *Federal Register* (vol. 38, No. 226, 26 November 1973, p. 32,495).

3. *Twentieth Century Petroleum Statistics* (1978, p. 57).

4. The RAND Corporation (1977) presents a similar graphical analysis of the released oil provision and argues that it made the effect of EPAA regulation on domestic output ambiguous. The absence of a released oil program under EPCA, it is argued, made EPCA's effect unambiguously negative. This conclusion is disputed below in section 3.4.

5. A similar graphical analysis of the lower-upper decision is developed by Montgomery (1977).

6. The explicit value for the weighted upper-tier price in the nth month after December 1979 is $P_u^n = P_u + [0.046 \Sigma_{i=1}^{n} (1 - 0.046)^{n-1}] (P_w - P_u)$.

7. Effective after-tax prices upon decontrol for independent producers with less than 1000 b/d output are: \$25.91 for Tier One (except ANS) and \$30.49 for Tier Two.

8. EPCA lower-tier and EPAA old oil have been combined under GAP_o. This presents no problem since old oil essentially became lower tier oil in February 1976 and lower tier oil ceilings were set equal to old oil ceilings by EPCA. Figures 3.5A and 3.5B are based on year-to-year changes. A producer trying to keep abreast of the general course of the implicit tax of controls might find these changes more instructive than, say, more haphazardly varying short-term month-to-month changes in GAPs.

9. See, for example, *Petroleum Intelligence Weekly* (11 December 1978, pp. 3, 5; 25 December 1978, pp. 1–3; 15 January 1979, p. 5).

10. See, for example, *Petroleum Intelligence Weekly* (25 December 1978, p. 7; 8 January 1979, pp. 1–2; 5 February 1979, pp. 2–3, 5).

11. See, for example, *New York Times* (29 July 1979, section VI, p. 13).

12. *Daily Stock Price Record, New York Stock Exchange* (1978, 1979).

13. Data for real OPEC prices based on *Monthly Energy Review* (selected issues) and the GNP Deflator.

14. At the moment of such a policy change, $\dot{GAP}/GAP = \infty$.

15. Data on output from old properties by crude oil control category are from RAND Corporation (1977, pp. 55–57). Data on world crude oil prices are from *Monthly Energy Review* (selected issues).

16. Analogous to supply schedule S_3 in figure 3.2, a producer taking advantage of the released oil program had to have a supply price schedule, (3.4), which equaled $2P_w(t) - P_o(t)$ at an output above the BPCL. From appendix 3.A, the Lagrangian form L of π_n in (3.2) can be written as:

$$L = \int_0^\infty [2P_w(t)(q_n(t) - \text{BPCL}) + P_o(t)(2\text{BPCL} - q_n(t)) - C(q(t),t)]e^{-rt}\,dt - U\left(R_o - \int_0^\infty q(t)\,dt\right).$$

Differentiating yields the following condition for profit maximization:

$$2P_w(t) - P_o(t) = \text{MC}(q(t),t) + Ue^{rt}.$$

17. These rates are calculated by noting that GAP is the WPT itself. Thus, the task is to

solve for the expected rate of real uncontrolled price appreciation ϕ that produces a 6.3% annual rate of increase in GAP:

$$\mathrm{GAP}(0)e^{0.063t} = \tau(P_w(0)e^{\phi t} - P_b(0)e^{\theta t}),$$

where $P_w(0)$ and $P_b(0)$ are the uncontrolled price and base price incorporated in table 3.1; τ, the WPT tax rate; and θ, the annual rate of allowed real appreciation in the base price ($\theta = 0$ for Tiers One and Two and $\theta = 0.021$ for Tier Three).

18. The most cogent statement is made by Frederiksen (1978).

19. Successful monopsonistic discrimination requires the exclusion of competitive buyers. This is, in fact, accomplished through strict export controls on domestically produced oil.

20. The following is a sample of empirical estimates of long-run crude oil supply elasticities: (a) Davidson, Falk, and Lee (1974): 1.6; (b) Erickson and Spann (1973): 0.87; (c) Erickson, Spann, and Millsaps (1973): 0.90; (d) Erickson and Fisher as reported in Burrows and Domencich (1970): 0.93; (e) Erickson and Spann (1971): 0.83; (f) Standard Oil of Indiana as reported in US Senate, Committee on Interior and Insular Affairs (1976b): 0.7–1.0; (g) Independent Petroleum Association of America as reported in US Senate, Committee on Interior and Insular Affairs (1976b): 1.4; (h) Steele (1974): 1.13.

21. See, for example, US Federal Energy Administration (1976d, p. 66).

22. See Owen (1974, pp. 1,307–1,308) and US Federal Energy Administration (1976e).

23. Consider, for example, the Conference Committee statement as quoted in Owen (1974, pp. 1,307–1,308).

Chapter 4

1. *Monthly Energy Review* (January 1976).

2. *Energy Management* (19 January 1977, p. 6). Limited export quotas are granted for small amounts of products. Refiners exporting products, however, lose entitlements to that amount of crude oil that can be attributed to the exported products. Consequently, these quotas are seldom filled. The small amounts that are exported are the result of special factors (for example, long-term contracts, border trade).

3. On skepticism about price comparisons, see, for example, RAND Corporation (1977, p. 24).

4. RAND Corporation (1977, p. 24).

5. RAND Corporation (1977, p. 69).

6. Data are from *Monthly Energy Review* (June 1976) as reported by RAND Corporation (1977, p. 40). Prices are deflated by the Wholesale Price Index. The deflated data series are fully reported in Kalt (1980).

7. Data are from US President (James E. Carter) (1977a).

8. See, for example, Ruggles (1955).

9. *Monthly Energy Review* (January 1976, p. 64).

10. The quantity of imports supplied to the United States is $R_I = R_F - R_{FR}$, where R_{FR} is foreign product consumption. Hence, with P_{UR} as product price and assuming, for simplicity, no transportation costs,

$$\frac{\partial R_I}{\partial P_{UR}} = \frac{\partial R_F}{\partial P_{UR}} - \frac{\partial R_{FR}}{\partial P_{UR}}$$

and

$$\frac{\partial R_{\mathrm{I}}}{\partial P_{\mathrm{UR}}}\frac{P_{\mathrm{UR}}}{R_{\mathrm{I}}} = \frac{\partial R_{\mathrm{F}}}{\partial P_{\mathrm{UR}}}\frac{P_{\mathrm{UR}}}{R_{\mathrm{F}}}\frac{R_{\mathrm{F}}}{R_{\mathrm{I}}} - \frac{\partial R_{\mathrm{FR}}}{\partial P_{\mathrm{UR}}}\frac{P_{\mathrm{UR}}}{R_{\mathrm{FR}}}\frac{R_{\mathrm{FR}}}{R_{\mathrm{I}}}.$$

Thus, because $R_{\mathrm{FR}} = R_{\mathrm{F}} - R_{\mathrm{I}}$, this last expression is (4.1a).

11. US Federal Energy Administration (1977a, pp. 9–11).

12. The strong conclusion that the RAND study has proved the opposite of the case it asserts should not be drawn. Aggregate banked cost data are notoriously unreliable and the study's regressions have yielded little information. Of eight variables, none is significant and the probability that as a group they are significant is only 0.39.

13. See Bohi and Russell (1978, especially pp. 144–187) on special adjustments in the MOIP.

14. *Oil Import Digest* (25 August 1974, A22–A22.1).

15. *Energy Management* (1977, p. 41,103).

16. *Oil and Gas Journal* (28 May 1979, p. 48).

17. US Department of Commerce, *Export Administration Regulations* (1978, pp. 4–12 and supplements nos. 2 and 3).

18. *Code of Federal Regulations: Title 10* (1978, p. 182).

19. Import and consumption data are from *P.A.D. Districts Supply/Demand Monthly* for gasoline and middle distillates. Residual fuel data are from Bohi and Russell (1978, p. 146).

20. *Petroleum Statement Annual* (1974, pp. 3–6).

21. Data are from *P.A.D Districts Supply/Demand Monthly*. Included in the Central America-Caribbean area are Bahamas, El Salvador, Netherlands Antilles, Panama, Puerto Rico, Virgin Islands, Colombia, Trinidad-Tobago, Honduras, Venezuela, Bermuda, Costa Rica, and Guatemala.

22. Bohi and Russell (1978, pp. 125–135 and pp. 144–187).

23. Data are based on US import fees and/or customs duties, as applicable, and suggested Worldscale transportation rates for routes noted, from *Worldwide Tanker Nominal Freight Scale* (1978).

24. United States territories and possessions with refiners eligible for the entitlements subsidy on sales of products to the United States are Virgin Islands, Puerto Rico, Hawaii Foreign Trade Zone, Panama Canal Zone, and Guam. In addition, product imports from the Bahamas were granted the entitlements subsidy over most of 1975 (see *Petroleum Intelligence Weekly* (26 May 1975, p. 8)).

25. Data are from *International Petroleum Annual* (1974).

26. *Oil Buyer's Guide* (selected issues, May–November 1979).

27. Data are from *Monthly Energy Review* (June 1978).

28. Data are from *P.A.D. Districts Supply/Demand Annual* (selected issues) and *P.A.D. Districts Supply/Demand Monthly* (selected issues).

29. See, for example, Landry (1975, 1976).

30. See *Code of Federal Regulations: Title 10* (1978, p. 182) and US Department of Commerce, *Export Administration Regulations* (1978).

31. *International Petroleum Annual* (1974).

32. Using x to denote exports,

$$\frac{P_{UR} - \hat{P}_{UR}}{SUBSIDY} = \frac{\epsilon_{xr}}{\epsilon_{xr} - \eta_{xr}} \frac{\epsilon_{ur}}{\epsilon_{ur} - \eta_{ur}}.$$

33. See RAND Corporation (1977, pp. 66–67). The appropriate adjustment to n_{ir} is also made here.

34. Other econometric studies of the elasticities of product demands include Verleger and Sheehan (1976), Alt (1976), Charles River Associates (1976), and Taylor (1977).

35. Quantity data are from *International Petroleum Annual* (1977). Data for the Sino-Soviet bloc are from United Nations, *World Energy Supplies, 1950–74* (1975). Data on several minor products is not available for the Sino-Soviet bloc. These include special naphthas, lubricating oils, wax, asphalt, LPG, and petrochemicals. Throughout the analysis here, asphalt includes road oil; and naphtha and kerosene jet fuel are aggregated into a single category of jet fuel.

36. *U.S. Imports for Consumption and General Imports: TSUSA Commodity by Country of Origin* (1974).

37. For products that are exported by the United States (for example, lubricating oils), the form of (4.11) described in note 32 is applicable.

38. *Petroleum Intelligence Weekly* (17 April 1978, p. 2).

39. *Petroleum Refineries in the United States and Puerto Rico* (1978).

40. *Financial and Statistical Supplement to the 1977 Annual Report* (1978a, pp. 14–15) and *Annual Report* (1978b, pp. 17–18).

41. Very large foreign-US price differences during late 1973–early 1974 are characteristic of the data described in section 4.5.2.

42. *Petroleum Press Service* (December 1973, p. 446).

43. Rotterdam prices are converted to cents-per-gallon from dollar-per-metric ton by use of the conversion factors reported in *International Petroleum Annual* (1976). Prices are weighted averages of high and low quotations. Transportation costs are based on weighted average *Platt's Oilgram News Service* daily spot tanker rates. For the entire 1968–1976 period, these rates are available only for a UK-Caribbean route. Available rates are adjusted by the ratio of the Rotterdam-NY distance to the UK-Puerto Cordon distance. Tests in Kalt (1980) indicate this adjustment introduces no significant inaccuracies. Tariff costs are taken from sources as reported in table 4.8.

44. Owen (1974, pp. 1,283–1,284).

45. See, for example, Mancke (1975, pp. 3–16).

46. Mancke (1975, p. 14).

47. *National Journal Reports* (16 March 1974, pp. 398–399; 25 May 1974, pp. 777–782).

48. Owen (1974, p. 1,237).

49. Owen (1974, pp. 1,238–1,250).

50. The period of binding price controls is combined into one variable rather than two (as in the gasoline case) to facilitate convergence in the iterative estimation technique.

Chapter 5

1. Based on data from the *Monthly Energy Review* (June 1980), with the cost of marginal crude oil imports given by the price of uncontrolled stripper oil adjusted for acquisition costs.

2. Burrows and Domencich (1970) present evidence of −0.5 for the price elasticity of crude oil demand; and Kennedy's (1974) evidence suggests a value somewhat greater than −0.5. The demand for crude oil can be expected to be more price elastic than the general demand for energy; and Berndt and Wood (1975) estimate the elasticity of energy demand at −0.47.

3. Explicitly,

$$\text{DWL} = P_w(C' - C^*) - (1/\alpha)\,(\hat{h})^{-1/\eta}\,[C'^{\alpha} - C^{*\alpha}]$$

and

$$\text{USG} = P_w C^* + (1/\alpha)\,(\hat{h})^{-1/\eta}\,[C'^{\alpha} - C^{*\alpha}] - P_s C',$$

where $\alpha = (1 + \eta)/\eta$.

4. The incremental reduction in deadweight loss that arises from allowing lower-tier properties (under EPCA) to qualify for upper-tier prices is area $BCHG = BGD - CDH$. The incremental increase in deadweight loss from the provision which allows old or lower-tier properties to qualify for stripper prices is $FGBA = ADF - BGD$.

5. Specifically,

$$\text{DWL}_u = P_w(Q_u^* - Q_u) - (1/\delta)\,(\hat{g})^{-1/\epsilon_u}\,[Q_u^{*\delta} - Q_u^{\delta}],$$

where $\delta = (1 + \epsilon_u)/\epsilon_u$ and $\hat{g}$ is the estimated value of g (analogous to $\hat{h}$).

6. Any distortions from the stripper provision are taken to be inconsequential (see section 3.2.1). Data limitations prevent more detailed estimations. Nevertheless, it can be pointed out that by far the greatest bulk of the deadweight loss of controls has typically arisen from the effects of controls on new development. The contribution of the stripper provision, as well as old, lower- and old, upper-tier prices on existing properties, is relatively much less important.

7. Data provided by the Independent Producers Association of America indicate that producers likely to take advantage of lower tax rates for independents account for roughly 5% of lower-tier oil, 14% of upper-tier oil, and 44% of stripper oil. Independents' share of stripper production is assumed to apply to NPR and marginal oil. Producers qualifying for the independents' tax break are assumed to face associated after-tax prices on their incremental production.

8. *The Budget of the United States Government* (1980).

9. Arrow and Kalt (1979, pp. 28–31).

10. For a brief summary of national security arguments, see Bohi and Russell (1978, pp. 10–14). See, also, Cabinet Task Force on Oil Import Control (1970).

11. The Department of State's role in stabilizing and supporting the OPEC cartel is discussed in Adelman (1972–1973), *Forbes* (15 April 1976), and US Senate, Committee on Foreign Relations (1975). On general foreign policy aspects of US import policy, see, for example, Adelman (1974), Hurewitz (1976), Szyliowicz and O'Neill (1976), and Klinghoffer (1977).

12. The Department of State official who has been most public with these views has been former Director of the Department of State Office of Fuels and Energy, James E. Akins (1972–1973). See also Ball (1974), Marshall (1975), Osborne (1975).

13. *Wall Street Journal* (1 July 1977).

14. *Wall Street Journal* (27 May 1977).

Chapter 6

1. US President (James E. Carter) (1977b).

2. Mitchell (1977, p. 5).

3. Lopreato and Smoller (1978, p. 32).

4. From a narrow self-interest point of view, voting is paradoxical because the expected benefit to any voter is insignificantly different from zero, while the time and inconvenience constitute positive costs. Riker and Ordeshook (1968) have discussed altruistic motives in explaining voter participation and Stigler (1972) has noted the possibility of such motives while questioning their empirical importance.

5. Quoted in Fenno (1977).

6. With diagonal elements of an appropriately dimensioned matrix P being these weights and the elements of Σ given by (6.3), the generalized least squares variance-covariance matrix is $(X'P\Sigma^{-1}P'X)^{-1} = (X'IX)^{-1}$. For the treatment of this in estimation see Jennrich (1975).

7. Vote tallies and summaries of issues are taken from Americans for Democratic Action (1976, 1977). The two energy-related votes in this sample have been excluded. The frequency of liberal voting is measured by the ratio of liberal votes to total votes cast. This differs from the ADA's "Liberal Quotient," which is the ratio of liberal votes cast to total possible votes and thus counts the absence of a vote as equivalent to a conservative vote.

8. CRUDE is based on data from US Department of the Interior, *Minerals Yearbook* (1975); US Department of Commerce, *Survey of Current Business* (August 1977); US Department of Energy, *Monthly Energy Review* (1975 issues).

9. The value of shipments from the refining sector is from US Department of Commerce, *Annual Survey of Manufactures* (1975).

10. ENERGYUSE is based on data from US Department of the Interior, *Fuels and Energy Data: US by States and Census Divisions, 1974* (1977) and US Department of Commerce, *Fuels and Electric Energy Consumed* (1975).

11. EDUC, URBAN, YPERCAP, FARM, MFG, BLUCOL, AGE, WITECOL, MCGOV, and PARTY are from *The Almanac of American Politics* (1977). UNION and NONWHITE are from US Department of Commerce, *Statistical Abstract of the United States* (1977).

12. The ADA supports creation of an Agency of Consumer Advocacy, as well as stringent amendments to the Clean Air Act. See Americans for Democratic Action (1976, 1977).

13. The constant term in PROADA is suppressed. The PROADA variable is centered on zero by ADA design. Thus, when senators show no ideological preference, its value should be zero when none of the other explanatory variables are influential. With regard to PROCRUDE, on the other hand, there is a case for including a constant term. Lopreato and Smoller (1978) have argued that there is something about energy issues that makes them qualitatively different from the average issue and that makes representatives vote more conservatively than distributional or ideological variables would imply.

14. Of the 36 votes available for calculation of liberalness ratings, 9 involve defense spending. The ADA recommended against such spending on all 9 issues. See Americans for Democratic Action (1976, 1977).

15. The mean of EDUC is 11.85 years and the standard deviation is 0.68.

16. As reported in Kalt (1980), the inclusion of the energy-related variables in the PRO-ADA model has no substantial impact on the coefficient estimates shown in (6.8).

17. Data for constraints are from US Department of Commerce, *Census of Manufactures* (1972).

18. The data-based isolikelihood contours are defined by $(\beta - b)'N(\beta - b) = c$, where b is the least-squares location of the regression coefficients; β, the vector of true coefficients; N, the $X'X$ data precision matrix; and c, an arbitrary constant corresponding to any selected level of likelihood. With notation appropriately redefined and constrained coefficients located at b^*, the prior-based family of isolikelihood contours is described by $(\beta - b^*)'N^*(\beta - b^*) = c^*$.

19. The information contract curve $\hat{\beta}$ is the locus of tangencies between data-based and prior-based isolikelihood contours. This locus can be found by maximizing $(\beta - b)'N (\beta - b)$ subject to $(\beta - b^*)'N^*(\beta - b^*)$. Setting the Lagrangian derivatives to zero and solving for β yields

$$\hat{\beta} = (N + \lambda N^*)^{-1}(Nb + \lambda N^*b^*),$$

where λ is the Lagrangian multiplier. Graphically, the information contract curve is traced out as λ is varied.

20. See note 6 above.

21. The statistical significance of DATREJCT is addressed with the classical F-test. PRIREJCT is subject to a similar chi-squared test that uses the prior-based likelihood function. See Leamer (1978a).

22. From points not shown in table 6.4 but available from SEARCH, it is possible to be fairly precise on the values of DATREJCT and PRIREJCT. When 87.6% of the distance from the data point to the prior point has been covered, DATREJCT = 4.9%. When 89.9% of this distance has been covered, DATREJCT = 5.1%. When PRIDIST = 51.2%, PRIREJCT = 5.0%.

23. SEARCH provides coefficient standard error estimates at each point on the contract curve.

24. The family of K feasible constrained coefficient estimates computed subject to a set of g constraints $R(\beta - b^*) = 0$ (where R is $g \times k$) lies on the ellipsoid

$$\left[\beta - b^* - \frac{(b - b^*)}{2}\right]' X'X \left[\beta - b^* - \frac{(b - b^*)}{2}\right] = \frac{(b - b^*)'X'X(b - b^*)}{4}.$$

25. Although the results are not reported here, this is also true for the cases involving the revenue and employment constraints.

References

Periodicals, Reference Works, and Primary Data Sources

Barone, Michael, Grant Ujifusa, and Douglas Matthews, 1977. *The Almanac of American Politics 1978*. New York: E. P. Dutton.

Bureau of National Affairs, Inc., *Energy Users Report*. Washington, DC: Bureau of National Affairs, Inc., Serial.

Code of Federal Regulations: Title 10. Washington, DC: Government Printing Office, Annual.

Commerce Clearing House, Inc., *Economic Controls: Stabilization Guidelines*. Chicago: Commerce Clearing House, Inc., Serial.

Commerce Clearing House, Inc., *Energy Management*. Chicago: Commerce Clearing House, Inc., Serial.

Commerce Clearing House, Inc., *Federal Energy Guidelines*. Chicago: Commerce Clearing House, Inc., Serial.

Commerce Clearing House, Inc., *State Tax Guide, Statute Summaries, By Taxes, By States*. Chicago: Commerce Clearing House, Inc., Serial.

Congressional Quarterly, Inc., *Congressional Quarterly Almanac*. Washington, DC: Congressional Quarterly, Inc., Annual.

Crump, Lulie H., 1977. *Fuels and Energy Data: U.S. by States and Census Divisions, 1974*. Washington, DC, US Department of the Interior. Bureau of Mines, Information Circular Number 8739.

DeGolyer and MacNaughton Co., *Twentieth Century Petroleum Statistics*. Dallas: DeGolyer and MacNaughton Co., Annual.

Federal Register. Washington, DC: Government Printing Office, Daily.

Government Research Corporation, *National Journal Reports*. Washington, DC: Government Research Corporation, Weekly.

Herold, John S., Inc., *Petroleum Outlook*. Greenwich, CT: John S. Herold, Inc., Monthly.

International Tanker Nominal Freight Scale Association Limited and The Association of Ship Brokers and Agents (Worldscale), Inc., *Worldwide Tanker Nominal Freight Scale*. London and New York: International Tanker Nominal Freight Scale Association Limited and The Association of Ship Brokers and Agents, Inc., Annual.

McGraw-Hill, Inc., *Platt's Oilgram News Service*. New York: McGraw-Hill, Inc., Daily.

McGraw-Hill, Inc., *Platt's Oil Price Handbook and Oilmanac*. New York: McGraw-Hill, Inc., Annual.

National Petroleum Refiners Association, *Oil Import Digest*. Tulsa, OK: National Petroleum Refiners Association, Serial.

Petroleum and Energy Intelligence Weekly, Inc., *Petroleum Intelligence Weekly*. New York: Petroleum and Energy Intelligence Weekly, Inc., Weekly.

Petroleum Press Bureau, Ltd., *Petroleum Press Service*. London: Petroleum Press Bureau, Ltd., Monthly.

Petroleum Publications, Inc., *Oil Buyer's Guide*. Lakewood, NJ: Petroleum Publications, Inc., Weekly.

Petroleum Publishing Co., *Oil and Gas Journal*. Tulsa, OK: Petroleum Publishing Co., Weekly.

Standard and Poor's, *Daily Stock Price Record, New York Stock Exchange*. New York: Standard and Poor's, Quarterly.

Standard Oil Company of Indiana, 1978a. *Financial and Statistical Supplement to the 1977 Annual Report*. Chicago: Standard Oil Company of Indiana.

Standard Oil Company of Indiana, 1978b. *Annual Report*. Chicago: Standard Oil Company of Indiana.

United Nations, 1975. *World Energy Supplies, 1950–74*. New York: United Nations.

US Central Intelligence Agency, 1976. *International Oil Developments: Statistical Survey*. Washington, DC: Central Intelligence Agency.

US Department of Commerce, *Annual Survey of Manufactures*. Washington, DC: Government Printing Office, Annual.

US Department of Commerce, *Census of Manufactures*. Washington, DC: Government Printing Office, Quintennial.

US Department of Commerce, *Fuels and Electric Energy Consumed, Annual Survey of Manufactures*. Washington, DC: Government Printing Office, Annual.

US Department of Commerce, 1978. Part 377, Short Supply Controls, *Export Administration Regulations*. Washington, DC: Department of Commerce, 1 June.

US Department of Commerce, *Statistical Abstract of the United States*. Washington, DC: Government Printing Office, Annual.

US Department of Commerce, *Survey of Current Business*. Washington, DC: Office of Business Economics, Monthly.

US Department of Commerce, *U.S. Imports for Consumption and General Imports: TSUSA Commodity by Country of Origin*. Washington, DC: Government Printing Office, Annual.

US Department of Energy, *Energy Data Reports*. Washington, DC: Government Printing Office, Monthly.

US Department of Energy, *Monthly Energy Review*. Washington, DC: Government Printing Office, Monthly. [Formerly US Federal Energy Administration]

US Department of the Interior, Bureau of Mines, *International Petroleum Annual*. Washington, DC: Bureau of Mines, Annual.

US Department of the Interior, Bureau of Mines, *Minerals Yearbook*. Washington, DC: Government Printing Office, Annual.

US Department of the Interior, Bureau of Mines, *P.A.D. Districts Supply/Demand Annual*. Washington, DC: Bureau of Mines, Annual.

US Department of the Interior, Bureau of Mines, *P.A.D. Districts Supply/Demand Monthly*. Washington, DC: Bureau of Mines, Monthly.

US Department of the Interior, Bureau of Mines, *Petroleum Refineries in the United States and Puerto Rico*. Washington, DC: Bureau of Mines, Annual.

US Department of the Interior, Bureau of Mines, *Petroleum Statement Annual*. Washington, DC: Bureau of Mines, Annual.

US International Trade Commission, 1975. *Tariff Schedules of the United States Annotated*. Washington, DC: Government Printing Office.

US President, *The Budget of the United States Government*. Washington, DC: Office of Management and Budget, Annual.

Articles, Books, and Reports

Abrams, Burton A., and Russell F. Settle, 1978. The Economic Theory of Regulation and Public Financing of Presidential Elections, *Journal of Political Economy* April:245–257.

Adelman, Morris A., 1972. *The World Petroleum Market*. Baltimore: Johns Hopkins University Press.

Adelman, Morris A., 1972–1973. Is the Oil Shortage Real?: Oil Companies as OPEC Tax-Collectors, *Foreign Policy* Winter:69–107.

Adelman, Morris A., 1974. Politics, Economics, and World Oil, *American Economic Review* May:58–67.

Akins, James, 1972–1973. This Time the Wolf Is Here, *Foreign Affairs* Winter:69–107.

Alchian, Armen, and Harold Demsetz, 1972. Production, Information Costs, and Economic Organization, *American Economic Review* December:777–795.

Alt, Christopher, 1976. *National Petroleum Product Supply and Demand, 1976–1978*. Washington, DC: Federal Energy Administration, May.

Alt, Christopher, Anthony Bopp, and George Lady, 1976. Short Term Forecasts of Energy Supply and Demand in A. Bradley Askin, and John Kraft, eds., *Econometric Dimensions of Energy Demand and Supply*. Lexington, MA: Lexington Books, pp. 81–90.

Americans for Democratic Action, 1976 and 1977. *ADA Legislative Newsletter*. Washington, DC: Americans for Democratic Action.

Arrow, Kenneth J., and Joseph P. Kalt, 1979. *Petroleum Price Regulation: Should We Decontrol?* Washington, DC: American Enterprise Institute.

Average Price of OPEC Oil to Rise Soon, But Details Are Unclear, Saudi Aide Says, *Wall Street Journal* 27 May 1977.

Ball, George W., 1974. Our Bankrupt Oil Policy, *Newsweek* 21 October:25.

Becker, Gary S., 1958. Competition and Democracy, *Journal of Law and Economics* October:105–109.

Bedford, A. C., 1918. Quoted in *Oil and Gas Journal* 4 April.

Berndt, Ernst R., and D. O. Wood, 1975. Technology, Prices, and the Derived Demand for Energy, *Review of Economics and Statistics* August:259–268.

Bohi, Douglas R., and Milton Russell, 1978. *Limiting Oil Imports*. Baltimore: Johns Hopkins University Press.

Box, George E. P., and Gwilym M. Jenkins, 1976. *Time Series Analysis: Forecasting and Control*. San Francisco: Holden-Day, Inc.

Brannon, Gerard M., 1974. *Energy Taxes and Subsidies; A Report to the Energy Policy Project of the Ford Foundation*. Cambridge, MA: Ballinger Publishing Co.

Breton, Albert, 1974. *The Economic Theory of Representative Government*. Chicago: Aldine Publishing Company.

Bronfenbrenner, Martin, 1971. *Income Distribution Theory*. Chicago: Aldine-Atherton Press.

Buchanan, James M., 1954. Individual Choice in Voting and the Market, *Journal of Political Economy*:334–343.

Buchanan, James M., and Gordon Tullock, 1965. *The Calculus of Consent*. Ann Arbor: University of Michigan Press.

Burness, H. Stuart, 1976. On the Taxation on Nonreplenishable Natural Resources, *Journal of Environmental Economics and Management* December: 289–311.

Burrows, James C., and Thomas A. Domencich, 1970. *An Analysis of the United States Oil Import Quota*. Lexington, MA: Heath Lexington Books.

Cabinet Task Force on Oil Import Control, 1970. *The Oil Import Question: A Report on the Relationship of Oil Imports to the National Security*. Washington, DC: Government Printing Office, February.

Carlton, Dennis W., 1978. "Valuing Market Benefits and Costs in Related Output and Input Markets" (unpublished). Chicago: University of Chicago.

Carter Raises Hope OPEC Might Forgo Further Price Rise but Venezuela Differs, *Wall Street Journal* 1 July 1977.

Chapel, Steven W., 1976. *The Oil Entitlements Program and Its Effects on the Domestic Refining Industry*. Santa Monica, CA: RAND Corporation, September.

Charles River Associates, 1976. *Price Elasticities of Demand for Transportation Fuels*. Cambridge, MA: Charles River Associates, Inc., May.

Commerce Clearing House, Inc., 1980. *Crude Oil Windfall Profit Tax Act of 1980, Law and Explanation*. Chicago: Commerce Clearing House, Inc.

Cox, James, and Arthur W. Wright, 1978. The Effects of Crude Oil Price Controls, Entitlements and Taxes on Refined Product Prices and Energy Independence, *Land Economics* February:1–15.

Dam, Kenneth W., 1971. Implementation of Import Quotas: The Case of Oil, *Journal of Law and Economics* April:1–60.

Davidson, Paul, Laurence H. Falk, and Hoesung Lee, 1974. Oil: Its Time Allocation and Project Independence, *Brookings Papers on Economic Activity*:411–418.

Democratic Study Group, 1980. *Windfall Profit Tax Conference Report Fact Sheet*. Washington, DC: Democratic Study Group.

Demsetz, Harold, 1967. Toward a Theory of Property Rights, *American Economic Review, Papers and Proceedings* May:347–359.

Demsetz, Harold, 1968. Why Regulate Utilities, *Journal of Law and Economics* April:55–65.

Don't Blame the Oil Companies: Blame the State Department, *Forbes* 15 April 1976.

Downs, Anthony, 1957. An Economic Theory of Political Action in a Democracy, *Journal of Political Economy* April:135–150.

Duchesneau, Thomas D., 1975. *Competition in the U.S. Energy Industry*. Cambridge, MA: Ford Foundation.

Eppen, Gary, 1975. *Energy: The Policy Issues*. Chicago: University of Chicago Press.

Erfle, Stephen, and John Pound, 1980. "U.S. Oil Markets in Transition: A Political Pressure Theory of the Market" (unpublished). Cambridge, MA: Energy and Environmental Policy Center Working Paper, Harvard University, July.

Erickson, Edward W., 1971. Supply Response in a Regulated Industry: The Case of Natural Gas, *Bell Journal of Economics and Management Science* Spring:94–121.

Erickson, Edward W., and Robert M. Spann, 1973. Joint Costs and Separability in Oil and Gas Exploration in Milton F. Searl, ed., *Energy Modeling*. Washington, DC: Resources for the Future, Inc., pp. 209–250.

Erickson, Edward W., 1974. The U.S. Petroleum Industry in Edward W. Erickson and Leonard Waverman, eds., *The Energy Question, An International Failure of Policy*. Toronto: University of Toronto Press, vol. 2, pp. 5–24.

Erickson, Edward W., 1977. Charges of Domestic Energy Monopoly: The Dog in the Manger of U.S. Energy Policy in Frank N. Traeger, ed., *Oil, Divestiture and National Security*. New York: National Strategy Information Center, pp. 41–65.

Erickson, Edward W., and Stephen W. Millsaps, 1973. Percentage Depletion and the Price and Output of Domestic Crude Oil in US House, Ways and Means Committee, *General Tax Reform*. Washington, DC: Government Printing Office, pp. 1,318–1,320.

Erickson, Edward W., and Stephen W. Millsaps, 1974. Oil Supply and Tax Incentives, *Brookings Papers on Economic Activity*:449–493.

European Economic Community, 1978. "Platts Prices and the Rotterdam Spot Market" (unpublished). Brussels: Prepared for the Directorate-General of Energy, Brussels.

Fenno, Richard F., 1978. U.S. House Members and Their Constituencies: An Exploration, *American Political Science Review* September:883–917.

Fenton, John H., and Donald W. Chamberlayne, 1969. The Literature Dealing with the Relationships Between Political Processes, Socioeconomic Conditions and Public Policies in the American States: A Bibliographical Essay, *Polity* Spring:388–404.

Frederiksen, Mark J., 1978. "The Theory of Multi-Tiered Price Regulation and Its Application to Domestic Crude Oil Production," delivered at Western Economic Association Meetings, Anaheim, California, June.

Frederiksen, Mark J., 1979. "The Effects of Multi-Tiered Price Controls on Domestic Crude Oil Production" (unpublished). Washington, DC: George Washington University, 31 July.

Friedman, Milton, 1975a. Two Economic Fallacies, *Newsweek* 12 May:83.

Friedman, Milton, 1975b. Subsidizing OPEC Oil, *Newsweek* 23 June:75.

Gart, John J., and James R. Zweifel, 1967. On the Bias of Various Estimators of the Logit and its Variance with Application to Quantal Bioassay, *Biometrika*:181–187.

Hall, Robert E., and Robert S. Pindyck, 1977. The Conflicting Goals of National Energy Policy, *Public Interest* Spring:3–15.

Harvard University Business School, 1978. "Energy Corporation of Louisiana (A): Synopsis and Analysis," Case No. 9-678-054. Boston: Harvard University Business School.

Henderson, David R., 1978. "The Effects of Price Controls on Allocation Over Time," Working Paper Series No. GPB 78-3. Rochester: Graduate School of Management, University of Rochester, June.

Herfindahl, Orrus C., and Allen V. Kneese, 1974. *Economic Theory of Natural Resources*. Columbus, OH: Merrill Publishing Co.

Hirshleifer, Jack, 1976. Comment on Peltzman (1976), *Journal of Law and Economics* August:241–244.

Hotelling, Harold, 1931. The Economics of Exhaustible Resources, *Journal of Political Economy* April:137–175.

Houthakker, Hendrik S., and Michael Kennedy, 1978. "A Long-Run Model of World Energy Demands, Supplies and Prices," Discussion Paper No. 673. Cambridge, MA: Harvard Institute of Economic Research, Harvard University, December.

Hurewitz, Jacob, ed., 1976. *Oil, the Arab-Israel Dispute and the Industrial World*. Boulder, CO: Westview Press.

ICF, Inc., 1979. *Imperfect Competition in the International Energy Market: A Computerized Nash-Cournot Model*. Washington, DC: ICF, Inc. [Submitted to the US Department of Energy, May 1979.]

Jacobs, E. Allen, 1980. "Petroleum Markets Under Price and Allocation Controls" (unpublished). Cambridge, MA: Massachusetts Institute of Technology, December.

Jennrich, R. I., 1975. Maximum Likelihood Estimation by Means of P3R, *BMD Communications* (UCLA Health Sciences Computing Facility) 3 July.

Johnson, William A., 1975. The Impact of Price Controls on the Oil Industry: How to Worsen an Energy Crisis in Gary D. Eppen, ed., *Energy: The Policy Issues*. Chicago: University of Chicago Press, pp. 99–121.

Johnson, William A., Richard E. Messick, Samuel V. Vactor, and Frank Wyant, 1976. *Competition in the Oil Industry*. Washington, DC: George Washington University Energy Policy Research Project.

Jordan, William A., 1972. Producer Protection, Prior Market Structure and The Effects of Government Regulation, *Journal of Law and Economics* April:151–176.

Kalt, Joseph P., 1980. "Federal Regulation of Petroleum Prices: A Case Study in the Theory of Regulation" (unpublished doctoral thesis). Los Angeles: University of California, Los Angeles.

Kau, James B., and Paul H. Rubin, 1979. Self-Interest, Ideology, and Logrolling in Congressional Voting, *Journal of Law and Economics* October:365–384.

Kennedy, Michael, 1974. An Economic Model of the World Oil Market, *Bell Journal of Economics and Management Science* Autumn:540–579.

Kissinger, Henry A., 1979. *White House Years*. Boston: Little, Brown, and Company.

Klinghoffer, Arthur J., 1977. *The Soviet Union and International Oil Politics*. New York: Columbia University Press.

Landry, James E., General Counsel, Air Transport Association of America, accompanied by Bert W. Rein, counsel, and John E. Gillick, counsel, 1975. Statement in US House, Subcommittee on Energy and Power, *Energy Conservation and Oil Policy*. Washington, DC: Government Printing Office, pp. 2,013–2,021.

Landry, James E., General Counsel, Air Transport Association of America, 1976. Statement in US Senate, Committee on Interior and Insular Affairs, *Energy Actions No. 3 and 4, Middle Distillate Control*. Washington, DC: Government Printing Office, pp. 51–56.

Leamer, Edward E., 1978a. *Specification Searches: Ad Hoc Inference with Non-Experimental Data*. New York: John Wiley.

Leamer, Edward E., 1978b. Search, *Perspective* (UCLA Academic Computing Service) April.

Lee, Dwight R., 1978. Price Controls, Binding Constraints, and Intertemporal Economic Decision Making, *Journal of Political Economy* April:293–301.

Leijonhufvud, Axel, 1971. *Keynes and the Classics*. London: The Institute of Economic Affairs.

Leonard, Herman, 1977. "SEARCH: Manual for Bayesian Inference" (unpublished). Cambridge, MA: Harvard University, June.

Lichtblau, John H., 1977. Pricing U.S. Oil Products, *Wall Street Journal*, 11 November, editorial page.

Lopreato, Sally C., and Fred Smoller, 1978. "Explaining Energy Votes in the Ninety-Fourth Congress" (unpublished). Austin: Center for Energy Studies, University of Texas at Austin, June.

Lovejoy, Wallace F., and Paul T. Homan, 1967. *Oil Conservation Regulation*. Baltimore: Johns Hopkins University Press.

MacAvoy, Paul W., 1971. The Regulation-Induced Shortage of Natural Gas, *Journal of Law and Economics* April:167–199.

MacAvoy, Paul W., ed., 1977. *Federal Energy Administration Regulation: Report of the Presidential Task Force*. Washington, DC: American Enterprise Institute.

"Major Reorganization Going on in U.S. Intelligence Community," *New York Times* 29 July 1979.

Mancke, Richard, 1975. *Performance of the Federal Energy Office*. Washington, DC: American Enterprise Institute.

Marshall, Eliot, 1975. Oil Double-Talk, *The New Republic* 5 and 12 July:10–13.

McDonald, Stephen L., 1963. *Federal Tax Treatment of Income from Oil and Gas*. Washington, DC: Brookings Institution.

McDonald, Stephen L., 1971. *Petroleum Conservation in the United States: An Economic Analysis*. Baltimore: Johns Hopkins University Press.

Mead, Walter J., 1976. Petroleum: An Unregulated Industry? in Robert J. Kalter and William A. Vogely, eds., *Energy Supply and Government Policy*. Ithaca, NY: Cornell University Press, pp. 130–160.

Mitchell, Edward J., 1974. *U.S. Energy Policy: A Primer*. Washington, DC: American Enterprise Institute.

Mitchell, Edward J., 1977. *Energy and Ideology*. Washington, DC: American Enterprise Institute, October.

Mitchell, Edward J., 1979. The Basis of Congressional Energy Policy, *Texas Law Review* March:591–613.

Montgomery, W. David, 1977. "The Transition to Uncontrolled Crude Oil Prices" (unpublished), Social Science Working Paper Number 186. Pasadena: California Institute of Technology, October.

Nash, Gerald R., 1968. *United States Oil Policy: 1890–1964*. Pittsburgh: University of Pittsburgh Press.

Nelson, Charles R., 1973. *Applied Time Series Analysis for Managerial Forecasting*. San Francisco: Holden-Day, Inc.

Nozick, Robert, 1974. *Anarchy, State, and Utopia*. New York: Basic Books.

Olson, Mancur, 1971. *The Logic of Collective Action*. New York: Shocken Books.

Osborne, John, 1975. Bowing to OPEC, *The New Republic* 19 July:12–14.

Owen, Charles R., 1974. The History of Petroleum Price Controls in Office of Economic

Stabilization, US Treasury Department, *Historical Working Papers; 8/15/71 to 4/30/74*. Washington, DC: US Treasury Department, pp. 1,223–1,340.

Peltzman, Sam, 1976. Toward a More General Theory of Regulation, *Journal of Law and Economics* August:211–248.

Pindyck, Robert S., 1978. Gains to Producers from the Cartelization of Exhaustible Resources, *Review of Economics and Statistics* May:238–251.

Pindyck, Robert S., 1979. Some Long-Term Problems in OPEC Oil Pricing, *Journal of Energy and Development* Spring:259–272.

Pindyck, Robert S., and Esteban Hnyilicza, 1976. Pricing Policies for a Two-Part Exhaustible Resource Cartel: The Case of OPEC, *European Economic Review* August:139–154.

Posner, Richard A., 1974. Theories of Economic Regulation, *Bell Journal of Economics and Management Science* Autumn:335–358.

Posner, Richard A., 1975. The Social Cost of Monopoly and Regulation, *Journal of Political Economy* August:807–827.

Price Waterhouse and Co., 1980. *The Crude Oil Windfall Profit Tax*. New York: Price Waterhouse and Co.

RAND Corporation, 1977. *Petroleum Regulation: The False Dilemma of Decontrol*. Santa Monica, CA: RAND Corporation, January.

Riker, William H., and Peter C. Ordeshook, 1968. A Theory of the Calculus of Voting, *American Political Science Review* March:25–42.

Ruggles, Richard, 1955. The Nature of Price Flexibility and the Determinants of Relative Price Changes in the Economy in National Bureau of Economic Research, *Business Concentration and Public Policy*. Princeton, NJ: Princeton University Press, pp. 441–495.

Salant, Stephen W., 1976. Exhaustible Resources and Industrial Structure: A Nash-Cournot Approach to the World Oil Market, *Journal of Political Economy* October:1,079–1,093.

Sampson, Anthony, 1975. *The Seven Sisters, the Great Oil Companies and the World They Shaped*. New York: Viking Press.

Scherer, Frederic M., Alan Beckenstein, Erich Kaufer, and R. Dennis Murphy, 1975. *The Economics of Multiplant Operations*. Cambridge, MA: Harvard University Press.

Shaffer, Edward, 1968. *The Oil Import Program of the United States*. New York: Praeger.

Sheerin, John C., 1977. "The Economics of Resource Exhaustion Under Price Controls" (unpublished doctoral thesis). Boulder: University of Colorado.

Steele, Henry, 1969. Statement before US Senate, Judiciary Committee, *Hearings on Government Intervention in the Market Mechanism: The Petroleum Industry*. Washington, DC: Government Printing Office, pp. 208–222.

Steele, Henry, 1974. Cost Trends and the Supply of Crude Oil in the U.S. in Michael Macrakis, ed., *Energy: Demand, Conservation, and Institutional Problems*. Cambridge, MA: MIT Press, pp. 303–317.

Stigler, George J., 1971. The Theory of Economic Regulation, *Bell Journal of Economics and Management Science* Spring:3–21.

Stigler, George J., 1972. Economic Competition and Political Competition, *Public Choice* Fall:91–106.

Stigler, George J., and James K. Kindahl, 1970. *The Behavior of Industrial Prices*. New York: National Bureau of Economic Research.

Sweeney, James L., 1977. Economics of Depletable Resources: Market Forces and Intertemporal Bias, *Review of Economic Studies* February:125–141.

Szyliowicz, Joseph S., and Bard E. O'Neill, 1976. *Petropolitics and the Atlantic Alliance*. Washington, DC: National Security Affairs Monograph, November.

Tarbell, Ida M., 1904. *History of the Standard Oil Company*. New York: McClure, Phillips and Co.

Taylor, Lester D., 1977. The Demand for Energy: A Survey of Price and Income Elasticities in William D. Nordhaus, ed., *International Studies of the Demand for Energy*. Amsterdam: North-Holland, pp. 1–26.

Teece, David J., 1976. "Vertical Integration and Vertical Divestiture in the U.S. Petroleum Industry," Research Paper Number 300. Stanford: Graduate School of Business, Stanford University, March.

Tiebout, Charles M., 1956. A Pure Theory of Local Expenditures, *Journal of Political Economy* October:416–424.

Tolley, George S., and John D. Wilman, 1977. The Foreign Dependence Question, *Journal of Political Economy* April:323–347.

US Congress, Congressional Budget Office, 1979. *The Windfall Profits Tax: A Comparative Analysis of Two Bills*. Washington, DC: Congressional Budget Office, November.

US Department of Energy, Economic Regulatory Administration, 1978. *Small Refiner Bias Analysis: Final Report*. Washington, DC: Department of Energy, January.

US Federal Energy Administration, 1976a. *Trends in Refinery Capacity and Utilization*. Washington, DC: Federal Energy Administration, June.

US Federal Energy Administration, 1976b. Hearing on The Domestic Crude Oil Advantage Under the Entitlements Program, Washington, DC, 8 September.

US Federal Energy Administration, 1976c. *Critique of the Library of Congress' Analysis of FEA's Findings on Middle Distillate Control*. Washington, DC: Federal Energy Administration, 19 June.

US Federal Energy Administration, 1976d. *National Energy Outlook*. Washington, DC: Government Printing Office.

US Federal Energy Administration, 1976e. "Stripper Well Crude Oil—First Sale Price Exemption and Modified Treatment Under the Domestic Crude Oil Allocation Program: Notice of Proposed Rule Making and Public Hearing" (mimeographed), 7 September.

US Federal Energy Administration, 1977a. *Banked Costs and Market Structure and Behavior*. Washington, DC: Federal Energy Administration, 30 June.

US Federal Energy Administration, 1977b. *Preliminary Findings and Views Concerning the Exemption of Motor Gasoline from the Mandatory Petroleum Allocation and Price Regulations*. Washington, DC: Federal Energy Administration.

US Federal Trade Commission, 1973 *Preliminary FTC Staff Report on Its Investigation of the Petroleum Industry*. Washington, DC: US Senate Committee on Interior and Insular Affairs.

US President (James E. Carter), 1977a. *Economic Report of the President*. Washington, DC: Government Printing Office.

US President (James E. Carter), 1977b. *The National Energy Plan*. Washington, DC: Government Printing Office, April.

US Senate, Committee on Energy and Natural Resources, 1977. *Regulation of Domestic Crude Oil Prices*. Washington, DC: Government Printing Office, March.

US Senate, Committee on Foreign Relations, 1975. *Multinational Oil Corporations and U.S. Foreign Policy*. Washington, DC: Government Printing Office, 2 January.

US Senate, Committee on Interior and Insular Affairs, 1976a. *Energy Action No. 2: Small Refiners' Entitlements*. Washington, DC: Government Printing Office.

US Senate, Committee on Interior and Insular Affairs, 1976b. *Estimates of the Economic Cost of Producing Crude Oil*. Washington, DC: Government Printing Office.

US Treasury Department, 1973. *Staff Analysis of the Preliminary FTC Staff Report on Its Investigation of the Petroleum Industry*. Washington, DC: Government Printing Office.

Verleger, Philip K., Jr., 1979. "A Review of the U.S. Petroleum Crisis of 1979" (unpublished), delivered at the Brookings Panel Meeting, Brookings Institution, Washington, DC, 4–5 October.

Verleger, Philip K., Jr., 1980. "An Assessment of the Economic Incentive for Enhanced Oil Recovery" (unpublished). New Haven: School of Organization and Management, Yale University, August.

Verleger, Philip K., Jr., and Dennis P. Sheehan, 1976. A Study of the Demand for Gasoline in Dale W. Jorgenson, ed., *Econometric Studies of U.S. Energy Policy*. New York: North-Holland, pp. 177–241.

Weidenbaum, Murray L., 1979. "The Trend of Government Regulation of Business" (unpublished). St. Louis: Center for the Study of American Business, Washington University, July.

Wright, Arthur W., 1978. The Case of the United States: Energy as a Political Good, *Journal of Comparative Economics*:144–176.

Zellner, Arnold, and Tong H. Lee, 1965. Joint Estimation of Relationships Involving Discrete Random Variables, *Econometrica* April:382–394.

Index